APOPTOSIS
AND CANCER CHEMOTHERAPY

Apoptosis and Cancer Chemotherapy

Edited by

John A. Hickman

and

Caroline Dive

University of Manchester, UK

Humana Press
Totowa, New Jersey

Library of Congress Cataloging-in-Publication Data

Apoptosis and cancer chemotherapy/edited by John A. Hickman and Caroline Dive.
    p.   cm.—(Cancer drug discovery and development; 5)
  Includes bibliographical references and index.
  ISBN 0-89603-743-6 (alk. paper)
  1. Cancer—Chemotherapy. 2. Apoptosis. 3. Carcinogenesis. I. Hickman, John A.
II. Dive, Caroline. III. Series.
  [DNLM: 1. Neoplasms—drug therapy. 2. Neoplasms—physiopathology. 3.
Apoptosis—drug effects. 4. Apoptosis—physiology. 5. Drug resistance, neoplasm.
QZ 267 A6435 1999]
RC271.C5A69 1999
616.99'4061—dc21
DNLM/DLC
for Library of Congress   99-12556
                          CIP

FOREWORD

The past few years have witnessed an astonishing international effort that established the role of some 20 new molecules in apoptosis and added activation or suppression of apoptosis to the accepted biological functions of a great many others already familiar in cancer biology. Some of these molecules are receptors, transducing cytokine-mediated signals; others appear to intensify or diminish the risk that a compromised cell will fire its apoptosis effector mechanism. All are of interest as potential targets for tumor therapy, and some may prove to be control points influenced in the pathogenesis of cancer and other diseases as diverse as viral infection, neurodegenerative disorders, and stroke. Sometimes, in the midst of these developments, a kind of euphoria appears to have gripped the research community, with the expectation that apoptosis will afford explanations to many unsolved questions in cellular regulation. This book, in a series of thoughtful and provocative articles—some from established leaders in the field, and others from younger scientists—seeks to redress the balance.

One central issue is the role of apoptosis in defining the response of authentic tumor populations to chemotherapeutic agents. It is easy to construct experiments in which large differences appear in the initiation of apoptosis in populations of tumor cells exposed in vitro to a variety of potential therapies. But it is not always justifiable to interpret these results as indicating tumor resistance or sensitivity in the therapeutic sense. Clonogenic assays, applied to the same populations, sometimes produce different answers, presumably because they address a different end point *(1,2)*. Rather than enumerating the cells that die (a process sensitive in the experimental context to the kinetics of death) they identify those cells with the capacity to survive and replicate, even in the austere conditions of low-density culture. The clonogenic cells often represent a tiny fraction of the original population, not readily measured by the techniques used to enumerate cell kill, but potentially highly significant. Nonetheless, when such clonogenic cells are studied in detail, a high proportion bears mutations that are the fingerprints of repaired DNA damage *(3,4)*. These fingerprints clearly indicate the transient presence of DNA damage of a type known to activate apoptosis. The question therefore remains how such long-term surviving cells sustained such damage without becoming committed to death.

This question is only part of another, more general one. Although the molecular interactions of the terminal effector pathway of apoptosis have now been described in great detail, we know surprisingly little of the mechanisms coupling DNA damage to the activation of the pathway in the first place. New reagents are becoming available that indicate that critical damage-signaling molecules, such as p53, may become phosphorylated, perhaps at sites that are specific to their particular types of DNA injury. But it is also clear that the cellular context in which these changes take place greatly influences the outcome of such signaling. Thus, in hepatocytes damaged by ultraviolet light the unequivocal, immediate p53 activation is coupled to cell-cycle arrest (5,6), whereas in some other cell types the same immediate injury and p53 activation lead to apoptosis. Moreover, powerful p53-independent death mechanisms are present in some cell types but not others. And, in the tissue context, these signaling and activation pathways might be profoundly influenced by local paracrine factors (7), which in tumors may emanate from adjacent normal or neoplastic cells, including the vascular stroma and infiltrating lymphocytes.

New questions are also suggested from the profusion of new signaling, modulating, and effector molecules now implicated in apoptosis. How redundant are the dozen odd members of the caspase family? Are caspases activated in different cellular locations, and if so does this have a biological meaning? There are many beautiful studies implicating the mitochondria as a source of caspase activation, but it is already probable that other intracellular sources may in appropriate circumstances be just as significant, including the cell membrane (unequivocally a source of ceramide following irradiation) or the nucleus. Should we be searching for activation mechanisms centered on the cytoskeleton and the endoplasmic reticulum also? Some provocative but now quite long-established data indicate that apoptosis-suppressor molecules, such as bcl-2, can have radically different effects when targeted to different intracellular membranes (8).

Finally, cell biology has a trick of producing entirely new options, at times when these are least expected. The caspase cascade appears to afford a highly satisfactory explanation for the long-described cluster of structural events, involving coordinate changes in cell surface, nucleus, cytosol, and cytoskeleton, that led long ago to the identification of apoptosis on morphological grounds. But it is not clear that caspase activation is the only means whereby these changes may be affected, nor even that other styles and modes of "programmed" cell death may not exist.

It is with questions such as these that this book is concerned.

Andrew H. Wyllie

REFERENCES

1. Brunet CL, Gunby RH, Benson RSP, Hickman JA, Watson AJM, Brady G. Commitment to cell death measured by loss of clonogenicity is separable from the appearance of apoptotic markers. *Cell Death Different.* 1998; 5: 107–113.
2. Longthorne VL, Williams GT. Caspase activity is required for commitment to fas-mediated apoptosis. *EMBO J.* 1997; 16: 3805–3812.
3. Griffiths SD, Clarke AR, Healy LE, Ross G, Ford AM, Hooper ML, Wyllie AH, Greaves M. Absence of p53 promotes propagation of mutant cells following genotoxic damage. *Oncogene* 1997; 14: 523–531.
4. Corbet SW, Clarke AR, Gledhill S, Wyllie AH. p53-dependent and independent links between DNA damage, apoptosis and mutation frequency in ES cells. *Oncogene* 1998; in press.
5. Bellamy COC, Prost S, Wyllie AH, Harrison DJ. UV but not γ-radiation induces specific transcriptional activity of p53 in primary hepatocytes. *J. Path.* 1997; 183: 177–181.
6. Bellamy COC, Clarke AR, Wyllie AH, Harrison DJ. p53 deficiency in liver reduces local control of survival and proliferation but does not affect apoptosis after DNA damage. *FASEB J.* 1997; 11: 591–599.
7. Harrington FA, Bennet MR, Fanidi A, Evan GI. C-myc induced apoptosis in fibroblasts is inhibited by13, specific cytokines. *EMBO J.* 1994; 13: 3286–3295.
8. Zhu WJ, Cowie A, Wasfy GW, Penn LZ, Leber B, Andrews DW. Bcl-2 mutants with restricted subcellular location reveal spatially distinct pathways for apoptosis in different cell types. *EMBO J.* 1996; 15: 4130–4141.

PREFACE

Semplice, non semplicistico

What a pleasure to put together a book on this topic! We are delighted that so many of our good friends and colleagues have given their time not only to review their field of interest, but, as we requested, to muse on some of the conundrums that make this a field with many intriguing enigmas! There are cautionary caveats in many of the chapters and it seems that the field is moving into a well-earned reflective phase. When new concepts galvanize a community because they genuinely throw light on some long-standing problem—cynics call it the creation of a bandwagon—there is always a danger that the elegance of the concept engenders too simplistic and generalized a perception of the detail. It is something of a relief to read here that major players in the field are themselves concerned by many aspects of the models that have emerged. However, despite these concerns and calls for reflection, this is not a field in retreat. Quite the opposite. New data continue to intrigue and to move these models ahead. We learn, unsurprisingly, that we must accomodate complexity—a *sine qua non* in biology. We hope that the reader will take from these chapters a sense of excitement. It is somewhat glib and hackneyed to write that this new understanding of what makes cells so tough—defining why they won't die after chemotherapy—holds promise for novel forms of therapeutic intervention sometime in the future. We are sure it does. But, like Ted Weinert *(1)*, we also think this may take a decade or two. Why so optimistic? Because modulation of cell survival pathways is a central process of carcinogenesis—how else can tumors survive with an increasingly unstable genome? Changing survival by the astute use of new agents should impact on a central aspect of tumor pathology. However, we are all aware of just how long it takes to move new concepts into clinical practice—just look at the signal transduction inhibitors. And guess what? The problems of selectivity are the same as faced us a decade or two ago.

The book is called *Apoptosis* and *Cancer Chemotherapy*. Accordingly, we have invited a number of reviews that address some of the key mechanistic questions concerning the control of apoptosis and the molecules that will either be targets for drugs in the future or may influence the efficacy of current therapies. These provide challenging perspectives on fields of growing complexity. There are, thankfully, also dis-

senting voices who caution those who are caught in the apoptotic mael-
strom to take stock of their commitment to particular paradigms and not
to forget some well-tested concepts from an honorable past. We hope
that, taken together, this provides a balanced view of the field as it is now.
Of course, we have not managed to persuade everyone that we had
wished to write to do so. Some were too busy, some never answered
letters and E-mails, and some were in the midst of moving a laboratory.
Others felt the field was reviewed to death anyway. We don't agree with
that and hope that these unique perspectives brought to bear on this
unfolding tale will stimulate the naïve reader to join the pursuit, and the
more hardened campaigners will refresh their own perspective on a field
that sometimes appears dauntingly complex.

Finally, we are grateful to Andrew Wyllie for providing his pan-
oramic view of the field and to Joe Bertino for ending the book with a
clinician's call that we have everything to go for! And of course, we
would like to sincerely thank our authors. We hope that the reader will
appreciate the extra effort that has gone into preparing these critical and
contemplative reviews. We are very grateful to Carol Miles and Reneé
Holland who helped edit manuscripts, and to all at Humana Press and
Beverly Teicher for their patience and guidance throughout.

John A. Hickman
Caroline Dive

REFERENCE

1. Weinert T. DNA damage and checkpoint pathways: molecular anatomy
   and interactions with repair. *Cell* 1998; 94: 555–558.

CONTENTS

CONTRIBUTORS

BRUNO ANTONSSON • *Serono Pharmaceutical Research Institute, Plan-les-Ouates/Geneva, Switzerland*

DEBORAH K. ARMSTRONG • *The Johns Hopkins Oncology Center, Baltimore, MD*

RENATO BASERGA • *Kimmel Cancer Center, Thomas Jefferson University, Philadelphia, PA*

HELEN M. BEERE • *La Jolla Institute for Allergy and Immunology, San Diego, CA*

GARRETT M. BRODEUR • *Division of Oncology, The Children's Hospital of Philadelphia, PA*

J. MARTIN BROWN • *Department of Radiation Oncology, Cancer Biology Research Laboratory, Stanford University School of Medicine, Stanford, CA*

ROBERT BROWN • *CRC Department of Medical Oncology, CRC Betason Laboratories, Glasgow, Scotland*

ROBIN BROWN • *GlaxoWellcome Medicines Research Centre, Stevenage, UK*

ELIZABETH M. BRUCKHEIMER • *Department of Molecular Pathology, M. D. Anderson Cancer Center, University of Texas, Houston, TX*

SANDRA J. CAMPBELL • *Department of Cellular and Molecular Pathology, Ninewells Hospital and Medical School, University of Dundee, Scotland*

VALERIE P. CASTLE • *Division of Hematology-Oncology, University of Michigan, Ann Arbor, MI*

CHRISTINE M. CHRESTA • *CRC Molecular and Cellular Pharmacology Group, School of Biological Sciences, University of Manchester, UK*

NANCY E. DAVIDSON • *The Johns Hopkins Oncology Center, Baltimore, MD*

KLAUS-MICHAEL DEBATIN • *University Children's Hospital, Ulm, Germany*

NICHOLAS DENKO • *Department of Radiation Oncology, Mayer Cancer Biology Research Laboratory, Stanford University, Stanford, CA*

CAROLINE DIVE • *Cancer Research Campaign, Molecular and Cellular Pharmacology Group, School of Biological Sciences, University of Manchester, UK*

Wafik S. El-Deiry • *Laboratory of Molecular Oncology and Cell Cycle Regulation, Departments of Medicine, Genetics, and Cancer Center, Howard Hughes Medical Institute, University of Pennsylvania School of Medicine, Philadelphia, PA*

Amato J. Giaccia • *Department of Radiation Oncology, Mayer Cancer Biology Research Laboratory, Stanford University, Stanford, CA*

Bjorn T. Gjertsen • *Department of Molecular Pathology, M. D. Anderson Cancer Center, University of Texas, Houston, TX*

Douglas R. Green • *La Jolla Institute for Allergy and Immunology, San Diego, CA*

Hillary A. Hahm • *The Johns Hopkins Oncology Center, Baltimore, MD*

Peter A. Hall • *Department of Cellular and Molecular Pathology, Ninewells Hospital and Medical School, University of Dundee, Scotland*

John A. Hickman • *School of Biological Sciences, University of Manchester, UK*

Tsuyoshi Honda • *Department of Molecular Pathology, M. D. Anderson Cancer Center, University of Texas, Houston, TX*

Yi-Te Hsu • *Biochemistry Section, Surgical Neurology Branch, National Institute of Neurological Disorders and Stroke, National Institutes of Health, Bethesda, MD*

Alan L. Johnson • *Department of Biological Sciences, University of Notre Dame, IN*

Surender Kharbanda • *Department of Cancer Pharmacology, Dana-Farber Cancer Institute, Harvard Medical School, Boston, MA*

Constantinos Koumenis • *Department of Radiation Oncology, Mayer Cancer Biology Research Laboratory, Stanford University, Stanford, CA*

Donald Kufe • *Department of Cancer Pharmacology, Dana-Farber Cancer Institute, Harvard Medical School, Boston, MA*

Scott W. Lowe • *Cold Spring Harbor Laboratory, Cold Spring Harbor, NY*

Jean-Claude Martinou • *Serono Pharmaceutical Research Institiute, Plan-les-Ouates/Geneva, Switzerland*

Timothy J. McDonnell • *Department of Molecular Pathology, M. D. Anderson Cancer Center, University of Texas, Houston, TX*

Nora M. Navone • *Department of Genitourinary Medical Oncology, M. D. Anderson Cancer Center, University of Texas, Houston, TX*

MARY O'NEILL • *Department of Cellular and Molecular Pathology, Ninewells Hospital and Medical School, University of Dundee, Scotland*

D. MARK PRITCHARD • *School of Biological Sciences, University of Manchester, UK*

JOHN C. REED • *The Burnham Institute, La Jolla, CA*

MARIANA RESNICOFF • *Kimmel Cancer Center, Thomas Jefferson University, Philadelphia, PA*

SIAN T. TAYLOR • *Cancer Research Campaign Molecular and Cellular Pharmacology Group, School of Biological Sciences, University of Manchester, UK*

JONATHAN L. TILLY • *Vincent Center for Reproductive Biology, Massachusetts General Hospital, and Department of Obstetrics, Gynecology, and Reproductive Biology, Harvard Medical School, Boston, MA*

RALPH WEICHSELBAUM • *Department of Radiation and Cellular Oncology, University of Chicago, IL*

BRADLY G. WOUTERS • *Department of Radiation Oncology, Cancer Biology Research Laboratory, Stanford University School of Medicine, Stanford, CA*

RICHARD YOULE • *Biochemistry Section, Surgical Neurology Branch, National Institute of Neurological Disorders and Stroke, National Institutes of Health, Bethesda, MD*

ZHI-MIN YUAN • *Department of Cancer Pharmacology, Dana-Farber Cancer Institute, Harvard Medical School, Boston, MA*

1 Does Apoptosis Contribute to Tumor Cell Sensitivity to Anticancer Agents?

J. Martin Brown, PhD
and Bradly G. Wouters, PhD

CONTENTS

ABSTRACT

The widely held view that tumor cells treated with anticancer agents die from apoptosis, and that cells resistant to apoptosis are resistant to cancer treatment, is incorrect. Two principal factors have given rise to this dogma. First, cell killing has often been assessed in short-term assays that are influenced more by the rate, than the overall level, of cell death. Second, conclusions have been extrapolated from normal cells transformed with dominant oncogenes to tumor cells, without taking into account that tumor cells have invariably undergone selection to an apoptotically resistant phenotype. If clonogenic

From: *Apoptosis and Cancer Chemotherapy*
Edited by: J. A. Hickman and C. Dive © Humana Press Inc., Totowa, NJ

survival is used to assess cell killing, and real tumor cells are used, then apoptosis, and the genes controlling it, such as p53 and bcl-2, play little or no role in the sensitivity of tumor cells of nonhematological origin to anticancer drugs and radiation.

INTRODUCTION: IS THIS A SERIOUS QUESTION?

The concept that tumor cells exposed to radiation or anticancer drugs usually die from apoptosis has become such a widely held tenet of modern cancer treatment that it seems somewhat ridiculous to ask the question of whether apoptosis contributes to the outcome of cancer therapy. However, the tenet is on shaky ground. Though apoptosis is an extremely important process to study, understand, and exploit in future cancer treatment, its demonstrated role in presentday cancer therapy is mostly limited to tumors of hematological, particularly of lymphoid, origin. This chapter focuses on cells of nonhematological origin, and critically reviews the data underlying the hypothesis that these cancer cells, when treated with radiation or chemotherapeutic drugs, die of apoptosis, and that cells resistant to apoptosis are resistant to cell kill by these agents.

As is discussed throughout this book, many genes have been identified that affect the extent to which certain cell types undergo apoptosis during normal development, and following pathological stress. Together with the belief that apoptosis plays a major role in cell killing by DNA-damaging agents, these genetic studies lead naturally to the current belief that tumors with mutations in p53, high levels of Bcl-2, or high ratios of Bcl-2/Bax, should be resistant to cancer treatment. Because there is now a wealth of data from clinical studies in which outcome has been correlated with the status of these and other genes affecting apoptosis, this hypothesis would seem an easy one to test. However, a major problem with such analyses is that it is often impossible to separate treatment sensitivity from patient prognosis. For example, tumors with mutated p53 can be more anaplastic and metastatic, and can have a higher proportion of proliferating cells and, in general, a more aggressive phenotype than similar tumors with wild-type p53. This can lead to a worse prognosis for patients whose tumors have mutated p53, independent of the intrinsic sensitivity of the tumor cells to cytotoxic therapy *(1)*. Because such effects confuse the specific question of the role of apoptosis in treatment sensitivity, this review is restricted to the narrower question of whether the level of apoptosis and/or genes controlling apoptosis affect the sensitivity of cancer cells to killing by genotoxic agents. The current view is that this is the case. Weinberg *(2)* summed up this position as follows:

> *For years, it was assumed that radiation therapy and many chemo-therapeutic drugs killed malignant cells directly, by wreaking widespread havoc in their DNA. We now know that the treatments often harm DNA to a relatively minor extent. Nevertheless, the affected cells perceive that the inflicted damage cannot be repaired easily, and they actively kill themselves. This discovery implies that cancer cells able to evade apoptosis will be far less responsive to treatment.*

Widely quoted in support of this position, and in particular the role of mutated p53 in radiation and anticancer drug resistance, are pioneering studies with oncogenically transformed fibroblasts from embryos of p53 wild-type (p53$^{+/+}$) and p53 knockout mice (p53$^{-/-}$) *(3••,4••)*. Other supporting evidence includes the highly significant associations of mutated p53 with drug resistance in the National Cancer Institute (NCI) 60 cell-screening panel *(5•)*. However, despite the seemingly strong case that cells die from cancer treatment because of apoptosis chiefly controlled by wild-type p53, much current data do not appear to fit. For example, several authors have reported that large changes in apoptosis do not lead to any changes in eventual cell killing *(6•–9••)*, or that p53 mutations and/or a high Bcl-2:Bax ratio do not necessarily lead to clinical resistance *(1,10)*. The authors believe there are two principal reasons for the current confusion:

1. Many investigators have used methods for assessing the extent of cell killing to anticancer drugs and radiation, based on early functional changes (such as dye uptake), or on growth inhibition, rather than on clonogenic survival. These can lead to incorrect assessments of overall cell kill.
2. Conclusions derived from normal cells transformed with dominant oncogenes, such as *E1A,* and *myc,* have been extrapolated to tumor cells, despite the fact that apoptosis, particularly the early apoptosis characteristic of cells of lymphoid origin, often is an insignificant mode of cell death for the cells of the majority of solid tumors.

Both of these issues are explored in more depth in the following sections.

MEASURING CELL KILLING: A MAJOR PROBLEM IN MANY CURRENT STUDIES

How should cell killing (or its counterpart, cell survival) be measured after a toxic insult? This would not seem to be a particularly difficult problem: A dead cell has many morphological features, loses metabolic functions, and, in particular, fails to exclude dyes such as propidium iodide and trypan blue. If death is caused by apoptosis, then a number of well-characterized features occur, including chromatin condensation and fragmentation, formation of nucleosomal DNA ladders, and exposure of phosphotidyl serine in the outer cell membrane that can be detected with annexin V. Thus, dead cells can be readily identified and their proportion in a population quantitated. However, identification of dead cells, once they have died, is not the problem. The problem is that cells do not die immediately after treatment: They can take many hours to many days to die, and this is highly dependent on the cell type and the toxic agent being investigated *(11,12)*. In addition, proliferation of surviving cells in the population can further complicate the picture. Thus, in most circumstances with assays involving the whole cell population, there is no time at which an accurate assessment of cell-kill can be made by counting the number, or proportion, of cells alive or dead in the population.

This is illustrated in Fig. 1, in which two hypothetical cell types, treated with an agent that produces 90% cell-kill, die with different kinetics: one (A) rapidly, so that most of the cells have died within 1 d; and the other (B) more slowly. The figure shows that an assessment of the percent viable cells, or percent dead cells, made at any time would conclude that population A was more sensitive to treatment than population B. A similar inappropriate conclusion would be reached if A and B were two treatments on the same cell population, but treatment A caused more rapid cell-kill than treatment B. Despite this, many, if not most, investigators assessing cellular sensitivity to genotoxic agents measure viability at 2–4 d after treatment by total population staining (the MTT or XTT assays), or they assess the extent of cell death from the proportion of cells incorporating a dye excluded by live cells, such as trypan blue or propidium iodide.

The problem of correctly assessing the surviving fraction following a given treatment was solved for mammalian cells in the mid-1950s by Puck and Markus *(13••)*, who developed the technique of cloning individual cells in vitro. The ability of a single cell to grow into a colony (usually defined as more than 50 cells) is an assay that tests every cell in the population for its ability to undergo unlimited division. It is, for mammalian cells, the exact counterpart of assays measuring bacterial or yeast survival after treatment with cytotoxic agents. The assay has become widely used for assessing the response to cytotoxic agents of cells in vitro, as well as cells in normal tissues and tumors in vivo. When the logarithm of the percent surviving cells determined

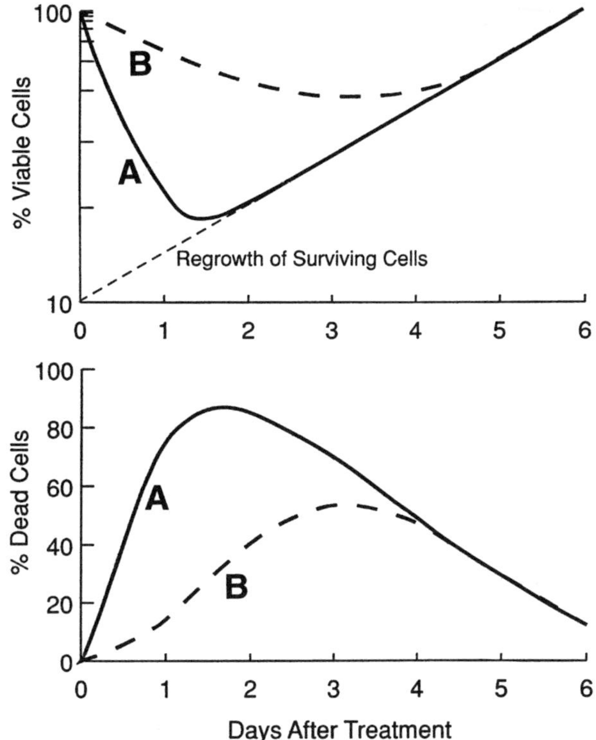

Fig. 1. Responses of two hypothetical cell types that die with different kinetics, treated with an agent that produces 90% cell-kill. Cell population A dies rapidly, so that most of the cells have died within 1 d, and the other cell population, B, dies more slowly. For both populations, viability decreases as cells are killed, and then increases again as survivors proliferate. The panels show that measurement of the percent viable cells, or percent dead cells, made at any time would conclude that population A was more sensitive to treatment than population B. A similar inappropriate conclusion would be reached if A and B were two treatments on the same cell population, but treatment A caused more rapid cell-kill than treatment B.

by clonogenic assay is plotted against the dose of agent used, a straight line (sometimes with an initial shoulder region) is usually obtained, implying an exponential relationship between dose and cell-kill. For both radiation and anticancer drugs, this log cell-kill hypothesis for cell survival has been demonstrated under many different conditions to be effective in predicting the radiation or chemotherapy dose needed to cure experimental mouse tumors or multicellular spheroids *(14–16•)*. Referring again to Fig. 1, plating the cells out following treatment to measure colony formation would lead to the correct survival estimate of 10% for cell populations A and B.

Therefore, the ability of cells to undergo unlimited proliferation as tested by their ability to form a colony has become the gold standard for assessment of cellular sensitivity. It is, however, important to be precise about the meaning of this: It is the best way to assess the proportion of cells surviving a particular treatment under the experimental conditions used. It does not mean that all responses of tumors can be reliably predicted using a clonogenic assay of the cells treated in vitro. This was recently illustrated in studies by Waldman et al. *(8••)* and the authors *(9••)*. HCT116 human colon carcinoma cells, isogenic for the *p53* target-gene *p21*[WAF1] (designated *p21*[+/+] or *p21*[−/−]), had the same response when tested by clonogenic survival in vitro (though they differed greatly in their level of apoptosis under the same conditions). However, when tumors were grown in vivo from the same cells, those from p21[−/−] cells were much more sensitive to radiation, as measured by growth delay, than their wild-type counterpart. This has recently been suggested as an inability of clonogenic survival to reflect tumor sensitivity *(8••,17,18)*. However, if clonogenic survival is assessed from the irradiated tumors, then there is close agreement between this assay and the in vivo growth-delay assay for the two cell lines *(9••)*. The difference, therefore, between the in vitro and in vivo data is the result of different environmental conditions of the cells at the time of treatment.

When used with care (for example, it is vital that single cells and not clumps are plated, and that sufficient time is given for all colonies to appear), there is little doubt that the clonogenic assay is the most reliable assay for assessing cell killing after genotoxic agents *(19••)*. With this in mind, now the question may be asked about whether apoptosis is a reliable determinant of cell killing to radiation and anticancer drugs when assayed by clonogenic survival. Though there are many examples in the literature, particularly for cells of hematologic origin, in which apoptosis and cell killing assayed by clonogenic survival are well correlated, there are also many examples in which this is not the case. Figure 2, for example, shows the authors' results, in which the response of HCT116 cells that are either p21[+/+] or p21[−/−] have been assessed both for apoptosis and for clonogenic survival under identical treatment conditions to radiation, to etoposide, and to the bioreductive drug, tirapazamine. It is clear from these data, and from other examples in the literature *(6,7,20•,21•)*, that the extent of apoptosis is not a reliable indicator of cellular sensitivity to anticancer agents. The message from Fig. 2 is that cells can die from apoptosis at different rates, and that the dominant mode of cell death may or may not be apoptosis.

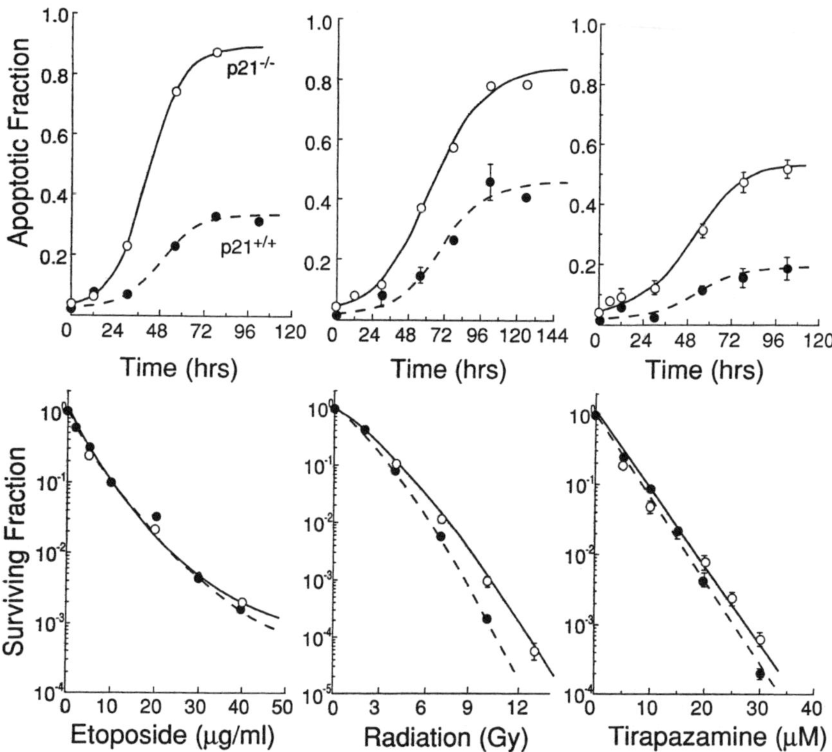

Fig. 2. The response of HCT116 cell lines isogenic for the CDK inhibitor *p21*[WAF1] are shown following three different genotoxic treatments: etoposide, tirapazamine, and radiation. The *upper frames* show that the loss of p21 *(solid lines)* sensitizes cells to death by apoptosis for each of these treatments in comparison to the wild-type cell line *(dashed lines)*. However, the overall sensitivity, as assessed by clonogenic assay (lower frames), is unchanged.

In the next section, one particular aspect of genotype and sensitivity, namely the impact of wild-type or mutant p53 on cellular sensitivity to genotoxic damage, is examined.

DOES P53 STATUS AFFECT SENSITIVITY OF CELLS OF NONLYMPHOID ORIGIN TO GENOTOXIC DAMAGE?

Although there are many studies of the influence of p53 status on the sensitivity of cells to genotoxic agents, after the above discussion on the unsuitability of many of the assays used to measure cell killing, the authors have considered only those that have used clonogenic survival, and have also focused only on those studies using ionizing radiation. This is because, first, different anticancer drugs have different mechanisms of action that could involve p53 directly, independent of apoptosis (e.g., involving nucleotide excision repair), and, second, because investigators use a wide variety of exposure conditions for drugs, but use much more defined radiation exposure conditions. There are also more studies with radiation than with all anticancer drugs combined.

In reviewing the literature on the question of the influence of p53 on the radiation response of cells, the authors applied the following criteria: Only cells or tissues of nonhematologic origin have been considered; and only investigations in which clonogenic survival was used to assess cell killing were considered. Also excluded from the analysis are studies in which wild-type p53 was massively overexpressed in cells prior to radiation using viral vectors. Typically, this results in radiation-induced apoptosis and radiation sensitization *(22,23)*.

In the literature, the data conforming to these criteria that are relevant to this question fall into two principal categories. First are those papers in which a group of nongenetically matched cell lines have been assessed for statistical differences in radiation sensitivity between those that have wild-type p53 and those that have mutant p53. Of the 27 publications in the total pool, 10 fall into this category, of which three find p53 mutated cells more radioresistant *(24–26)*, three find p53-mutations make no difference to radiation sensitivity *(27–29)*, and four find p53-mutated cells more radiation-sensitive *(30–33)*. Illustrative data from the two largest published series, each assessing radiation sensitivity and p53 status in cell lines or primary cultures from human cancers, are shown in Fig. 3.

The second category includes publications in which investigators have used genetically matched cell lines in vitro or tissues in vivo from $p53^{+/+}$ vs $p53^{-/-}$ mice, or have used cells with or without expression of the human papillomavirus (HPV) protein E6, or have used transfection of mutant p53 or conditionally expressed p53. In this group, five investigations found p53-null/mutated cells more radioresistant *(34–38)*, nine found no difference in radiation sensitivity between p53 wild-type or null/mutant cells *(37,39–46)*, and three found p53-null or mutant cells more sensitive that wild-type cells *(41•,47,48)*. Clearly, these data do not justify the conclusion that p53-mutated or -null cells are more resistant to radiation-induced cell-kill than are p53 wild-type cells.

For anticancer drugs, there have been few studies using clonogenic survival to determine the influence of p53 on the sensitivity of cells of nonhematological origin. However, even here, the available data

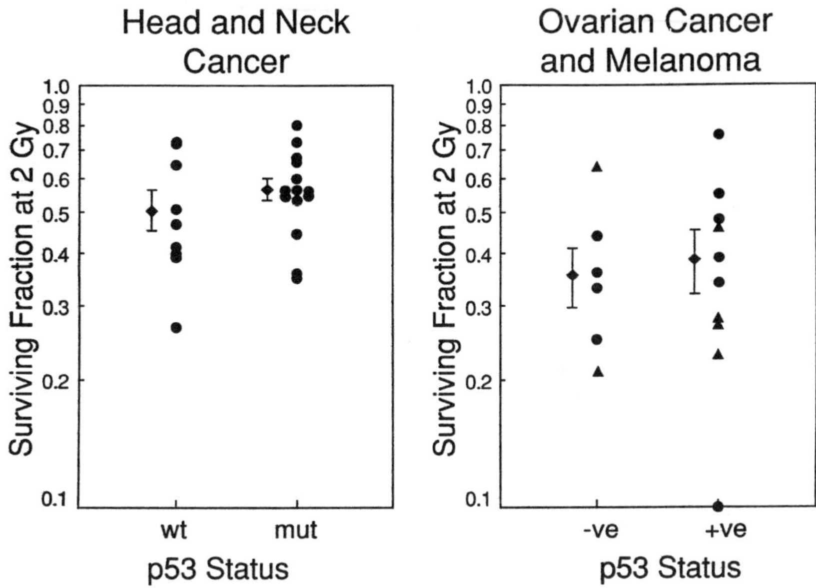

Fig. 3. Results from the two largest studies with human cancer cells on *p53* status and cellular sensitivity to ionizing radiation. Cell sensitivity (the surviving fraction after 2 Gy) was assessed by colony formation. (**A**) Results shown are from primary tumor cell lines derived from head and neck cancers *(28)*. In these studies, the *p53* gene was sequenced to determine those tumors expressing wild-type or mutant *p53*. (**B**) Results show sensitivity of p53 expressing vs p53 nonexpressing primary cultures from ovarian cancers *(circles)* and melanoma *(triangles) (27)*. In these studies, high levels of p53 expression were used as a surrogate for p53 mutation. The filled diamond shows the mean +/−1 standard error of the mean for each population.

also support the position that mutated p53 does not lead to drug resistance when assayed by clonogenic survival *(43•,49)*. In fact, mutations in p53 have been shown to confer sensitivity to drugs whose toxicity is modulated by nucleotide excision repair, such as nitrogen mustard and cisplatin *(49)*.

Given the rather strong evidence that p53 status does not affect the sensitivity of cells to radiation and anticancer drugs, it is pertinent to ask how such a belief arose. Several factors have probably contributed.

1. Lymphocytes, and lymphoma cells, are highly sensitive to anticancer therapy, and invariably die a rapid apoptotic death. The same is true of certain other tissues, such as the salivary gland, which also has a high sensitivity to radiation *(50)*.

2. In early studies with genetically matched, E1A-transformed, fibroblasts from wild-type and p53 knockout mice, cell killing was assessed by dye-exclusion 1–2 d after treatment *(3,4)*. These results indicated a huge influence of p53 loss in resistance to apoptosis, but did not measure clonogenic survival.

3. As mentioned earlier, conclusions derived from normal cells transformed with dominant oncogenes have been extrapolated to tumor cells. The following section points out that these transformed normal cells are hypersensitive both to apoptosis and to killing as measured by clonogenic survival. This is not the case for tumor cells in general.

4. The correlation of mutated p53 with resistance to anticancer drugs found in the NCI screen of 60,000 compounds tested against 60 different cell lines is actually a correlation, not of cell-kill, but of 2-d growth inhibition, with p53 status. Since wild-type p53 is required for growth arrest following DNA damage *(51••)*, such a correlation of sensitivity to growth inhibition with wild-type p53 is expected. In fact, because there is little or no correlation of growth inhibition with cytotoxicity *(52•)*, these data do not provide evidence for the involvement of p53 in drug sensitivity to cell killing.

Although wild-type p53 can influence the decision about whether or not cells undergo apoptosis following genotoxic insult, and this leads to differences in the rate at which cells die and leave the population, the available data do not support a role for p53 in determining the overall level of cell-kill in tumor cells. Assays for cell killing that can be affected by the rate at which cells die have led to erroneous conclusions about the influence of p53 on overall cell killing. Having said this, there may be subtle influences of p53 abrogation on radiosensitivity in some cells types. For example, Yount et al. *(37)* showed that, although abrogating p53 function did not change the overall radiosensitivity of human glioblastoma cells, it did increase the resistance of synchronized cells when irradiated in early G1.

ARE THERE SITUATIONS IN WHICH APOPTOSIS CAN CONTRIBUTE TO SENSITIVITY TO CYTOTOXIC AGENTS?

Whether apoptosis can contribute to the overall cellular cytotoxicity of present anticancer therapies and whether apoptosis does contribute to such therapies in solid human tumors are clearly different questions. There are several examples in the literature in which apoptosis does contribute to the overall sensitivity of cells to treatment with radiation or chemotherapeutic agents, as assessed either by in vivo treatments

(3,53•) or by clonogenic assays in vitro *(7,38,53–55;* Wouters and Brown, unpublished data). However, a major difference between these studies and others, which have failed to find a link between apoptosis and overall sensitivity, has been the choice of the cells used in the study. For cells of nonhematopoetic origin, the cases in which apoptosis does contribute to overall sensitivity have mostly been those that used cells derived from normal tissues genetically engineered to express dominant oncogenes. Introduction of oncogenes, such as *E1A (56), myc (53,57)* or HPV *E7 (58),* renders them dramatically more sensitive to apoptosis in response to various genotoxic or nongenotoxic stresses, and can also make these cells hypersensitive to overall killing by radiation *(3,7,54)* and by various chemotherapeutic agents (unpublished data). Apoptosis in these cases is characteristic of lymphoid cells: Death is usually rapid after treatment, is dependent on the status of p53 *(4,59),* and can be inhibited by overexpression of Bcl-2 *(53,60–62••).* For example, in mouse embryo fibroblasts (MEFs) transformed with *E1A* and Ha-*ras,* cells with wild-type *p53* undergo rapid apoptosis in response to genotoxic and nongenotoxic stress, but similarly transformed cells from *p53* knockout mice do not *(4,62).* This dramatic difference in apoptosis also translates into an increased sensitivity of the transformed p53$^{+/+}$ MEFs in terms of the overall sensitivity as determined either by clonogenic survival (Wouters and Brown, unpublished data) or by tumor response in vivo *(3••).*

It is important to realize that the loss of p53 in the transformed MEFs results not so much in resistance to treatment as much as it does in an elimination of the unusually hypersensitive response (in this case, presumably caused by apoptosis) that is found following expression of E1A. The p53-null-transformed MEFs revert to an overall sensitivity that is closer to the untransformed parental cells. In other words, in normal mouse embryo fibroblasts, *p53* modulates the hypersensitive phenotype induced by oncogenic stimuli.

An illustration of the role that apoptosis can play in minimally transformed cells was recently demonstrated using Rat1 cells engineered to conditionally express *myc (53•).* In this study, induced expression of *myc* increased radiation sensitivity to both apoptosis and clonogenic survival, and this increased sensitivity was inhibited by expression of *bcl*-2. The authors of this study also showed a direct role for apoptosis during treatment of these cells in vivo. When grown as tumors in mice, expression of *myc* enhanced the sensitivity of the tumors to fractionated doses of radiation, and this was abrogated by expression of *bcl*-2, indicating that apoptosis was directly influencing tumor response to radiation. This experiment thus provides direct evidence that expression of oncogenes in cells originally isolated from normal tissues can induce an apoptotically sensitive phenotype that is hypersensitive to treatment with anticancer therapies.

Studies such as these, using engineered minimally transformed cells, have especially shaped the current opinion that apoptosis is an important determinant of the response of human tumor cells to cancer therapy. As discussed earlier, this has led to the notion that a tumor cell with mutant p53 will be more resistant than the same cell with wild-type p53, or that increased expression of Bcl-2 in cancer cells will be protective. Acceptance of this hypothesis assumes that cells derived from human tumors will behave in ways similar to the oncogenically transformed cells described above. However, a clear distinction between tumor cells and genetically engineered cells from normal tissues is the requirement of the tumor cells, during their evolution, to overcome any apoptotically sensitive state that may have been induced by initial oncogenic transformation. The selection against this apoptotically sensitive state is driven by the selective forces produced by microenvironmental stresses, such as hypoxia *(62••)*, reduced growth factor, and nutrient supply *(57)*, and the requirement for anchorage-independent survival *(63,64; see* Fig. 4).

It is therefore likely that the majority of tumors have evolved beyond the point at which they may have been apoptotically hypersensitive to genotoxic and nongenotoxic stress. This phenotypic evolution can occur with or without alteration of such genes as *p53* or those in the *bcl-2* family, but is associated with a loss of the rapid induction of apoptosis following genotoxic treatment. This selection does not necessarily eliminate the cell's ability to carry out apoptosis; in some cases, the majority of tumor cells may still die by this process, although usually in a more delayed manner, occurring several days after treatment, and often after cell division *(9,65•)*. Selection against the apoptotically sensitive phenotype is probably the explanation for the findings in this chapter that apoptosis in human tumor cells has minimal impact on the overall cellular sensitivity to present anticancer therapy. Several of these studies showed that, even in cases in which the level of apoptosis can be modified in human tumor cell lines by *p53 (42,44)*, *p21 (8••,9••)*, or *bcl-2 (20•,21•,66)*, the overall sensitivity to various genotoxic agents was unrelated to the level of apoptosis. This finding indicates that, in these cases, apoptosis following genotoxic treatment occurs only in cells that have already lost their capacity to continue growth, often as a result of unrepaired or misrepaired DNA lesions *(see* Fig. 5).

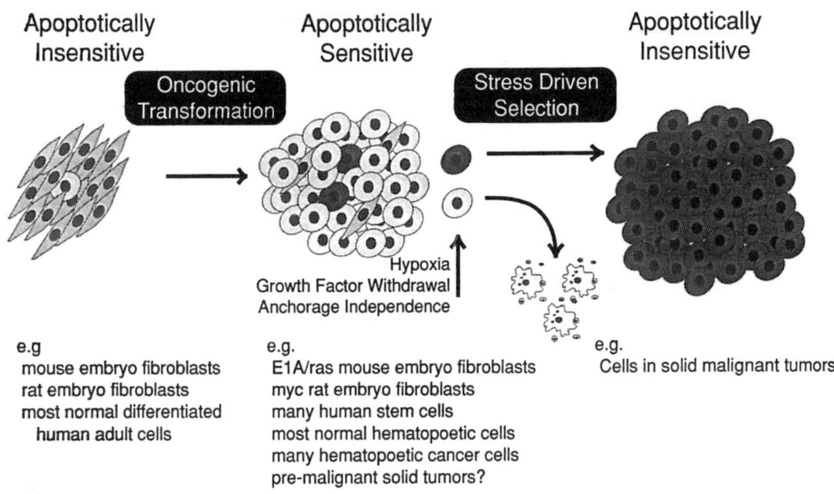

Fig. 4. Malignant evolution and apoptotic sensitivity. Most differentiated adult cells, as well as normal mouse and human fibroblast cell lines, are resistant to the induction of apoptosis by genotoxic agents **(left).** These cells typically undergo permanent arrest or senescence following treatment. Initial oncogenic transformation increases the proliferative potential and can dramatically sensitize these cells to apoptosis **(middle).** In general, these cells are more sensitive to apoptosis induced by both genotoxic and nongenotoxic stress (e.g., hypoxia, growth factor withdrawal, or anchorage independence). Cells in this state have also been shown to be hypersensitive to overall cell killing, as assessed by clonogenic assay. Examples of cells that can be classified within this apoptotically sensitive state include the E1A and *ras*-transformed MEFs, the myc-expressing Rat1 cells, many human stem cells, and hematopoetic tumor cells. Most solid human tumors, and tumor-derived cell lines, have evolved beyond this apoptotically hypersensitive state because of the selective pressure arising from various forms of nongenotoxic stress found within the tumor microenvironment **(right).** In response to genotoxic stress, the predominant mode of cell death for cells in this state may or may not be apoptosis. Manipulations in the levels of apoptosis by means of genetic changes has only been shown to effect cells that are apoptotically sensitive (middle). Dramatic changes in the level of apoptosis in cells from human tumors often have no effect on overall survival.

The studies with oncogenically transformed normal cells, however, do suggest the exciting possibility of improving cancer therapy by exploiting apoptosis. Development of therapeutic agents that are able to revert human tumors to the apoptotically sensitive phenotype associated with initial transformation would allow use of current cancer therapies with improved therapeutic potential. The key to the development of such therapies is an increased understanding of the genetic changes

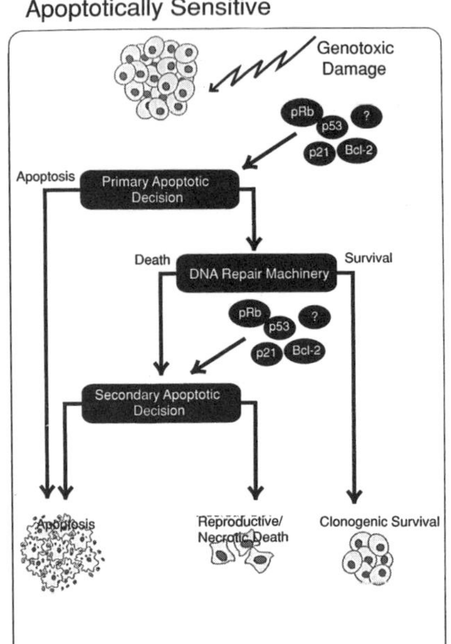

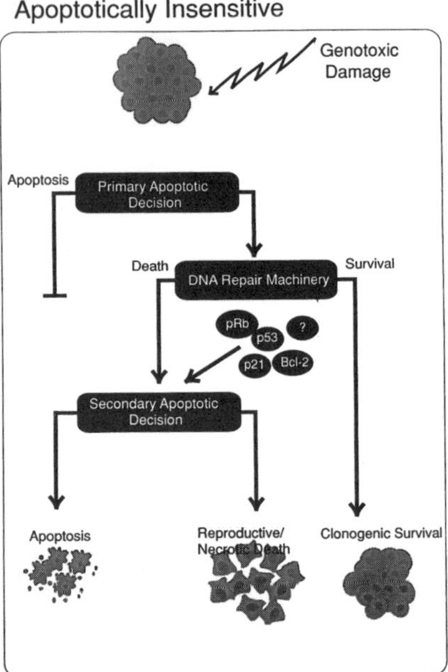

Fig. 5. Two models are shown that illustrate the pathways to cell death in apoptotically sensitive and apoptotically insensitive cells (as defined in Fig. 4). For apoptotically sensitive cells, genotoxic damage can signal an immediate apoptotic response. This signal is dependent on many genes, including *p53, p21, bcl-2,* and *pRb.* Cells that avoid death at this point undergo DNA repair, resulting in cells that have clonogenic potential, and those that do not. In those cells destined to die at this point, a secondary apoptotic decision is made. Cells now may either undergo apoptosis, or die by other means, such as reproductive or necrotic cell death. For apoptotically insensitive cells, which comprise the majority of cells from solid tumors, the primary apoptotic decision point is disabled. Following repair, cells that are destined to die still undergo a secondary apoptotic decision point, and thus the predominant mode of cell death may or may not be apoptosis. The genes controlling apoptosis at both decision points are similar, and thus it is possible to modulate the levels of apoptosis in cells within the apoptotically insensitive state by genetic means (e.g., introduction of Bcl-2). However, modulation of apoptosis at this secondary point effects only the mode of cell death, and will thus have no significant effect on the overall fraction of surviving cells.

that occur both during initial oncogenic transformation, and during the selection process that follows. Equally important will be the use of assays that correctly identify therapies that effect the overall level of cell-kill, as opposed to those that simply measure changes in the kinetics or mode of cell death.

ACKNOWLEDGMENTS

The authors thank Amato Giaccia for helpful comments. This work was supported by U.S. National Cancer Institute Grant CA 15201 (J.M.B.), and by a National Cancer Institute of Canada Research Fellowship (B.G.W.).

REFERENCES

Papers of particular interest have been highlighted as:
* of special interest
•• of outstanding interest

1. Elledge RM, Green S, Howes L, Clark GM, Berardo M, Allred DC, et al. bcl-2, p53, and response to tamoxifen in estrogen receptor-positive metastatic breast cancer: a Southwest Oncology Group study. *J Clin Oncol* 1997; 15: 1916–1922.
2. Weinberg RA. How cancer arises. *Sci Am* 1996; 275: 62–70.
••3. Lowe SW, Bodis S, McClatchey A, Remington L, Ruley HE, Fisher DE, Housman DE, Jacks T. p53 status and the efficacy of cancer therapy in vivo. *Science* 1994; 266: 807–810.
   Mouse embryo fibroblasts transformed with the adenovirus E1A *gene and activated* ras *form tumors when injected into immunodeficient mice. Tumors derived from* p53 *wild-type mouse fibroblasts show greater levels of apoptosis and respond more dramatically and rapidly to ionizing radiation and Adriamycin than tumors derived from fibroblasts from* p53 *knockout mice. This and the following paper were important in establishing the paradigm that inactivation of p53 may lead to treatment resistance.*
••4. Lowe SW, Ruley HE, Jacks T, Housman DE. p53-dependent apoptosis modulates the cytotoxicity of anticancer agents. *Cell* 1993; 74: 957–967.
   This seminal paper shows that the adenovirus E1A *gene sensitizes mouse embryo fibroblasts to apoptosis induced by a variety of anticancer agents including ionizing radiation. It also shows that p53 is required for this activity.*
•5. Weinstein JN, Myers TG, O'Connor PM, Friend SH, Fornace AJ, Jr., Kohn KW, et al. An information-intensive approach to the molecular pharmacology of cancer. *Science* 1997; 275: 343–349.
   Shows that there is a strong correlation between wild-type p53 *and the sensitivity of cells in the NCI drug screen to anticancer drugs. The assay used to assess sensitivity, however, is growth inhibition after 2 d, not clonogenic survival.*
•6. Aldridge DR, Arends MJ, Radford IR. Increasing the susceptibility of therat 208F fibroblast cell line to radiation-induced apoptosis does not alter its clonogenic survival dose-response. *Brit J Cancer* 1995; 71: 571–577.
   One of the first papers to show that an increased susceptibility to radiation-induced apoptosis does not necessarily lead to increased cell kill determined by clonogenic survival.
7. Han JW, Dionne CA, Kedersha NL, Goldmacher VS. p53 status affects the rate of the onset but not the overall extent of doxorubicin-induced cell death in rat-1 fibroblasts constitutively expressing c-Myc. *Cancer Res* 1997; 57: 176–182.

••8. Waldman T, Zhang Y, Dillehay L, Yu J, Kinzler K, Vogelstein B, Williams J. Cell-cycle arrest versus cell death in cancer therapy. *Nature Med* 1997; 3: 1034–1036.
Shows that loss of the p21 gene sensitizes cell to apoptosis, but not to clonogenic cell killing after treatment with DNA-damaging agents in vitro. However, its loss did sensitize tumors derived from these cells to radiation.

••9. Wouters BG, Giaccia AJ, Denko NC, Brown JM. Loss of p21Waf1/Cip1 sensitizes tumors to radiation by an apoptosis- independent. *Cancer Res* 1997; 57: 4703–4706.
HCT116 human colon carcinoma cells lacking p21 are sensitive to apoptosis induced by DNA damaging agents, and are radiosensitive as tumors in vivo, though not as cells in vitro. However, the in vivo sensitivity is not a result of the increased level of apoptosis.

10. Baxendine-Jones J, Campbell I, Ellison D. p53 status has no prognostic significance in glioblastomas treated with radiotherapy. *Clin Neuropathol* 1997; 16: 332–336.

11. Thompson LH, Suit HD. Proliferation kinetics of x-irradiated mouse L cells studied with time-lapse photography. II. *Int J Radiat Biol Related Stud Phys Chem Med* 1969; 15: 347–362.

12. Schneiderman MH, Hofer KG, Schneiderman GS. An in vitro 125IUdR-release assay for measuring the kinetics of cell death. *Int J Radiat Biol* 1991; 59: 397–408.

••13. Puck TT, Markus PI. Action of x-rays on mammalian cells. *J Exper Med* 1956; 103: 653–666.
Seminal paper showing that single mammalian cells could be cultured like bacterial cells in vitro and that colony formation could be used to assess sensitivity to ionizing radiation.

14. Durand RE. Cure, regression and cell survival: a comparison of common radiobiological endpoints using an in vitro tumour model. *Br J Radiol* 1975; 48: 556–571.

15. Reinhold HS, De Bree C. Tumour cure rate and cell survival of a transplantable rat rhabdomyosarcoma following x-irradiation. *Eur J Cancer* 1968; 4: 367–374.

•16. Skipper HE, Schabel FM, Wilcox WS. Experimental evaluation of potential anticancer agents, XIII. On the criteria and kinetics associated with "curability" of experimental leukemias. *Cancer Chemother Rep* 1964; 35: 1–111.

17. Lamb JR, Friend SH. Which guesstimate is the best guesstimate? Predicting chemotherapeutic outcomes. *Nature Med* 1997; 3: 962,963.

18. Gura T. Cancer models: systems for identifying new drugs are often faulty. *Science* 1997; 278: 1041,1042.

••19. Tannock IF. Biological properties of anticancer drugs. In: *Basic Science of Oncology*, Tannock IF, Hill RP, eds. 2nd ed., McGraw-Hill: New York. 1992; pp. 302–316.
Excellent discussion of the different ways in which killing by anticancer drugs has been assessed by different investigators. The discussion concludes that colony formation (the clonogenic assay) is the best method for measuring overall cell-kill.

•20. Lock RB, Stribinskiene L. Dual modes of death induced by etoposide in human epithelial tumor cells allow Bcl-2 to inhibit apoptosis without affecting clonogenic survival. *Cancer Res* 1996; 56: 4006–4012.

•21. Yin DX, Schimke RT. BCL-2 expression delays drug-induced apoptosis but does not increase clonogenic survival after drug treatment in HeLa cells. *Cancer Res* 1995; 55: 4922–4928.

22. Gallardo D, Drazan KE, McBride WH. Adenovirus-based transfer of wild-type p53 gene increases ovarian tumor radiosensitivity. *Cancer Res* 1996; 56: 4891–4893.
23. Gjerset RA, Sobel RE. Treatment resistance, apoptosis, and p53 tumor suppressor gene therapy. In: *Encyclopedia of Cancer,* Bertino JR, ed. Academic Press: San Diego. 1997; pp. 1785–1791.
24. Peacock J, Chung S, Benchimol S, Hill RP. Mutant p53 increases radioresistance in rat embryo fibroblasts simultaneously transfected with HPV16-E7 and/or activated H-ras. *Oncogene* 1994; 9: 1527–1536.
25. McIlwrath AJ, Vasey PA, Ross GM, Brown R. Cell cycle arrests and radiosensitivity of human tumor cell lines: dependence on wild-type p53 for radiosensitivity. *Cancer Res* 1994; 54: 3718–3722.
26. Siles E, Villalobos M, Valenzuela MT, Nunez MI, Gordon A, McMillan TJ, Pedraza V, Ruiz de Almodovar JM. Relationship between p53 status and radiosensitivity in human tumour cell lines. *Br J Cancer* 1996; 73: 581–588.
27. Zaffaroni N, Benini E, Gornati D, Bearzatto A, Silvestrini R. Lack of a correlation between p53 protein expression and radiation response in human tumor primary cultures. *Stem Cells* 1995; 13: 77–85.
•28. Brachman DG, Beckett M, Graves D, Haraf D, Vokes E, Weichselbaum RR. p53 mutation does not correlate with radiosensitivity in 24 head and neck cancer cell lines. *Cancer Res* 1993; 53: 3667–3669.
Experiments show that the radiosensitivity of cell lines derived from head and neck cancers does not correlate with p53 status.
29. Jung M, Notario V, Dritschilo A. Mutations in the p53 gene in radiation-sensitive and -resistant human squamous carcinoma cells. *Cancer Res* 1992; 52: 6390–3693.
30. Ribeiro JC, Barnetson AR, Fisher RJ, Mameghan H, Russell PJ. Relationship between radiation response and p53 status in human bladder cancer cells. *Int J Radiat Biol* 1997; 72: 11–20.
31. Zolzer F, Hillebrandt S, Streffer C. Radiation induced G1-block and p53 status in six human cell lines. *Radiother Oncol* 1995; 37: 20–28.
32. Biard DS, Martin M, Rhun YL, Duthu A, Lefaix JL, May E, May P. Concomitant p53 gene mutation and increased radiosensitivity in rat lung embryo epithelial cells during neoplastic development. *Cancer Res* 1994; 54: 3361–3364.
33. Servomaa K, Kiuru A, Grenman R, Pekkola-Heino K, Pulkkinen JO, Rytomaa T. p53 mutations associated with increased sensitivity to ionizing radiation in human head and neck cancer cell lines. *Cell Prolif* 1996; 29: 219–230.
34. Pardo FS, Su M, Borek C, Preffer F, Dombkowski D, Gerweck L, Schmidt EV. Transfection of rat embryo cells with mutant p53 increases the intrinsic radiation resistance. *Radiat Res* 1994; 140: 180–185.
35. Gupta N, Vij R, Haas-Kogan DA, Israel MA, Deen DF, Morgan WF. Cytogenetic damage and the radiation-induced G1-phase checkpoint. *Radiat Res* 1996; 145: 289–298.
36. Bristow RG, Jang A, Peacock J, Chung S, Benchimol S, Hill RP. Mutant p53 increases radioresistance in rat embryo fibroblasts simultaneously transfected with HPV16-E7 and/or activated H-ras. *Oncogene* 1994; 9: 1527–1536.
37. Yount GL, Haas-Kogan DA, Vidair CA, Haas M, Dewey WC, Israel MA. Cell cycle synchrony unmasks the influence of p53 function on radiosensitivity of human glioblastoma cells. *Cancer Res* 1996; 56: 500–506.

38. Tsang NM, Nagasawa H, Li C, Little JB. Abrogation of p53 function by transfection of HPV16 E6 gene enhances the resistance of human diploid fibroblasts to ionizing radiation. *Oncogene* 1995; 10: 2403–2408.

39. DeWeese TL, Walsh JC, Dillehay LE, Kessis TD, Hedrick L, Cho KR, Nelson WG. Human papillomavirus E6 and E7 oncoproteins alter cell cycle progression but not radiosensitivity of carcinoma cells treated with low-dose-rate radiation. *Int J Radiat Oncol Biol Phys* 1997; 37: 145–154.

40. Hendry JH, Adeeko A, Potten CS, Morris ID. p53 deficiency produces fewer regenerating spermatogenic tubules after irradiation. *Int J Radiat Biol* 1996; 70: 677–682.

•41. Hendry JH, Cai WB, Roberts SA, Potten CS. p53 deficiency sensitizes clonogenic cells to irradiation in the large but not the small intestine. *Radiat Res* 1997; 148: 254–259.

42. Huang H, Li CY, Little JB. Abrogation of P53 function by transfection of HPV16 E6 gene does not enhance resistance of human tumour cells to ionizing radiation. *Int J Radiat Biol* 1996; 70: 151–160.

•43. Slichenmyer WJ, Nelson WG, Slebos RJ, Kastan MB. Loss of a p53-associated G1 checkpoint does not decrease cell survival following DNA damage. *Cancer Res* 1993; 53: 4164–4168.
    Shows that sensitivity to ionizing radiation is not affected by loss of the G_1 checkpoint. Despite these and other data, the notion persisted for a long time that cells that do not arrest in G_1 are more sensitive to killing by DNA-damaging agents.

44. Westphal CH, Rowan S, Schmaltz C, Elson A, Fisher DE, Leder P. atm and p53 cooperate in apoptosis and suppression of tumorigenesis, but not in resistance to acute radiation toxicity. *Nat Genet* 1997; 16: 397–401.

45. Zellars RC, Naida JD, Davis MA, Lawrence TS. Effect of p53 overexpression on radiation sensitivity of human colon cancer cells. *Radiat Oncol Investig* 1997; 5: 43–49.

46. Merritt AJ, Allen TD, Potten CS, Hickman JA. Apoptosis in small intestinal epithelial from p53-null mice: evidence for a delayed, p53-independent G2/M-associated cell death after gamma-irradiation. *Oncogene* 1997; 14: 2759–2766.

47. Kawashima K, Mihara K, Usuki H, Shimizu N, Namba M. Transfected mutant p53 gene increases X-ray-induced cell killing and mutation in human fibroblasts immortalized with 4-nitroquinoline 1-oxide but does not induce neoplastic transformation of the cells. *Int J Cancer* 1995; 61: 76–79.

48. Pellegata NS, Antoniono RJ, Redpath JL, Stanbridge EJ. DNA damage and p53-mediated cell cycle arrest: a reevaluation. *Proc Natl Acad Sci USA* 1996; 93: 15209–15214.

49. Fan S, Chang JK, Smith ML, Duba D, Fornace AJ, Jr., O'Connor PM. Cells lacking CIP1/WAF1 genes exhibit preferential sensitivity to cisplatin and nitrogen mustard. *Oncogene* 1997; 14: 2127–2136.

50. Stephens LC, Schultheiss TE, Price RE, Ang KK, Peters LJ. Radiation apoptosis of serous acinar cells of salivary and lacrimal glands. *Cancer* 1991; 67: 1539–1543.

••51. Kastan MB, Onyekwere O, Sidransky D, Vogelstein B, Craig RW. Participation of p53 protein in the cellular response to DNA damage. *Cancer Res* 1991; 51: 6304–6311.
    Seminal paper established the paradigm that the G1 arrest following ionizing radiation is mediated by wild-type p53.

•52. Brown JM. NCI's anticancer drug screening program may not be selecting for clinically active compounds. *Oncol Res* 1997; 9: 213–215.
Points out that the 60-cell-line screen for anticancer drugs used by the NCI measures growth delay rather than cell killing, and therefore may not be selecting for agents that would be effective anticancer drugs.

•53. Rupnow BA, Murtha AD, Alarcon RM, Giaccia AJ, Knox SJ. Direct evidence that apoptosis enhances tumor responses to fractionated radiotherapy. *Cancer Res* 1998; 58: 1779–1784.
Shows that apoptosis can affect the sensitivity of minimally transformed cells growing as tumors to fractionated irradiation.

54. Guo M, Chen C, Vidair C, Marino S, Dewey WC, Ling CC. Characterization of radiation-induced apoptosis in rodent cell lines. *Radiat Res* 1997; 147: 295–303.

55. Ling CC, Guo M, Chen CH, Deloherey T. Radiation-induced apoptosis: effects of cell age and dose fractionation. *Cancer Res* 1995; 55: 5207–5212.

56. Lowe SW, Jacks T, Housman DE, Ruley HE. Abrogation of oncogene-associated apoptosis allows transformation of p53-deficient cells. *Proc Natl Acad Sci USA* 1994; 91: 2026–2030.

57. Evan GI, Wyllie AH, Gilbert CS, Littlewood TD, Land H, Brooks M, et al. Induction of apoptosis in fibroblasts by c-myc protein. *Cell* 1992; 69: 119–128.

58. Pan H, Griep AE. Altered cell cycle regulation in the lens of HPV-16 E6 or E7 transgenic mice: implications for tumor suppressor gene function in development. *Genes Devel* 1994; 8: 1285–1299.

••59. Hermeking H, Eick D. Mediation of c-myc-induced apoptosis by p53. *Nature* 1994; 265: 2091–2093.
Seminal paper that demonstrated the ability of p53 to induce apoptosis in cells transformed with the oncogene c-myc.

60. Bissonnette RP, Echeverri F, Mahboubi A, Green DR. Apoptotic cell death induced by c-myc is inhibited by bcl-2. *Nature* 1992; 359: 552–554.

61. Fanidi A, Harrington EA, Evan GI. Cooperative interaction between c-myc and bcl-2 proto-oncogenes. *Nature* 1992; 359: 554–556.

••62. Graeber TG, Osmanian C, Jacks T, Housman DE, Koch CJ, Lowe SW, Giaccia AJ. Hypoxia-mediated selection of cells with diminished apoptotic potential in solid tumours. *Nature* 1996; 379: 88–91.
Established the paradigm that hypoxia in solid tumors can select for cells with mutant p53 by selective killing by apoptosis of cells with wild-type p53.

63. McGill G, Shimamura A, Bates RC, Savage RE, Fisher DE. Loss of matrix adhesion triggers rapid transformation-selective apoptosis in fibroblasts. *J Cell Biol* 1997; 138: 901–911.

64. Nikiforov MA, Hagen K, Ossovskaya VS, Connor TM, Lowe SW, Deichman GI, Gudkov AV. p53 modulation of anchorage independent growth and experimental metastasis. *Oncogene* 1996; 13: 1709–1719.

•65. Waldman T, Lengauer C, Kinzler KW, Vogelstein B. Uncoupling of S phase and mitosis induced by anticancer agents in cells lacking p21. *Nature* 1996; 381: 713–716.

66. Kyprianou N, King ED, Bradbury D, Rhee JG. bcl-2 over-expression delays radiation-induced apoptosis without affecting the clonogenic survival of human prostate cancer cells. *Int J Cancer* 1997; 70: 341–348.

2 p53, Apoptosis, and Chemosensitivity

Scott W. Lowe, PhD

CONTENTS

ABSTRACT

Basic cancer research provides a theoretical foundation for rational approaches to improve cancer diagnosis, prognosis, and therapy. For example, the p53 tumor suppressor can facilitate apoptosis induced by anticancer agents, implying that p53 mutations should contribute to radiation and drug resistance in human tumors. Indeed, p53 mutations do correlate with drug resistance in some tumor types, and reintroduction of p53 into p53 mutant tumors enhances chemosensitivity in vitro and in vivo. However, in many settings, the

From: *Apoptosis and Cancer Chemotherapy*
Edited by: J. A. Hickman and C. Dive © Humana Press Inc., Totowa, NJ

relationship between p53, apoptosis, and chemosensitivity remains ambiguous. Although methodological shortcomings clearly contribute to this confusion, the situation is further confounded by the fact that p53's apoptotic activity depends on context. Hence, the chemotherapeutic agent, tumor type, and genetic background may each impact on the relationship between p53 and chemosensitivity. An improved understanding of this complexity will be necessary to guide future efforts in exploiting p53 or other cancer genes for improved cancer treatment.

INTRODUCTION

Resistance to the cytotoxic agents used to treat cancer remains a significant problem in oncology. Because most anticancer agents were discovered through empirical screens, efforts to overcome resistance are hindered by limited understanding of why these agents are ever effective. The role of apoptosis in radiation and drug toxicity provides a compelling explanation for drug sensitivity and resistance. This view suggests that responsive tumors readily undergo apoptosis in response to cytotoxic agents, and that resistant tumors may have acquired mutations that suppress apoptotic programs. Many oncogenes and tumor-suppressor genes modulate apoptosis; hence, the role of apoptosis in radio- or chemosensitivity provides a mechanistic link between cancer genetics and cancer therapy. One of the most provocative examples of this relationship involves the p53 tumor suppressor.

p53 AND CELLULAR RESPONSE TO DNA DAMAGE

The p53 tumor suppressor has a remarkable number of biological activities, including central roles in cell-cycle checkpoints, apoptosis, senescence, maintenance of genomic integrity, and control of angiogenesis. In turn, p53 can be activated by a large number of cellular stresses, including ribonucleotide depletion, hypoxia, oxidative stress, and certain mitogenic oncogenes *(1–3)*. Perhaps the best-studied activator of p53 is DNA damage, which can promote p53-dependent arrest or apoptosis *(4•,5•,6•)*. Since radiation and many chemotherapeutic agents either directly or indirectly damage DNA, one can easily imagine that *p53* status might affect the outcome of cancer therapy. Still, the precise nature of this impact is not intuitively obvious.

p53 is a fundamental component of a DNA-damage-inducible G1 checkpoint *(6•)*, implying that p53 promotes cell-cycle arrest to facilitate DNA repair. If, as is often speculated, anticancer agents preferentially kill inappropriately proliferating cells, the inability to arrest following

DNA damage might contribute to the tumor selectivity of radiation therapy. Although indirect evidence in certain tumor cells is consistent with this possibility *(7•)*, this hypothesis could not be confirmed in isogenic wild-type and p53-null murine fibroblasts *(8)*. Apoptosis is another outcome of p53 activation following DNA damage. For example, radiation induces apoptosis in normal, but not p53-deficient, thymocytes, both in vitro and in vivo *(4,5)*. These studies raise the possibility that cells lacking p53 might more readily survive certain forms of cancer therapy.

p53 AND CHEMOSENSITIVITY

Studies using oncogenically transformed murine fibroblasts from normal and p53-deficient mice provide compelling evidence that the p53-dependent apoptotic program can affect the outcome of cancer therapy. In this setting, disruption of p53 function reduces apoptosis, thereby increasing cell survival and producing a multidrug-resistant phenotype *(9•)*. Similar effects are observed in vivo, in which fibrosarcomas derived from oncogenically transformed fibroblasts expressing p53 display massive apoptosis and regress, but tumors derived from *p53*-deficient cells show little apoptosis and continue to grow *(10••)*. Moreover, acquired *p53* mutations are associated with resistance and relapse of tumors derived from *p53*-expressing cells. Although highly artificial, these experiments demonstrate that genetic control of cell death can have a dramatic impact on the efficacy of cancer therapy and make the clear prediction that *p53* mutations—by reducing apoptosis—should contribute to drug resistance in tumors.

Consistent with this model, *p53* mutations correlate with reduced drug toxicity in panels of cell lines from certain tumor types, including Burkitt's lymphoma *(11•)*, astrocytomas *(12)*, gliomas *(13)*, melanomas *(14)*, and testicular carcinomas *(15)*. Moreover, dysfunctional p53 correlates with decreased sensitivity to anticancer agents in broad analysis of the 60 National Cancer Institute (NCI) cell lines derived from various tumor types, although the extent of this effect was agent-dependent *(16••)*. Because comparisons between human tumor lines are necessarily correlative, other studies directly induce or suppress p53 function, to examine its effects in cells of the same genetic background. In some settings, inactivation of p53 with exogenous genes, such as dominant-negative p53 mutants, Mdm-2, or human papillomavirus (HPV) E6 protein, reduces the efficacy of cytotoxic agents *(17–19)*, although p53 inactivation can also have the opposite effect *(see below)*. Perhaps the most striking responses, however, are seen following reintroduction of p53 into p53-mutant tumor lines. Here, a large number of studies document marked increases in radiation- and drug-induced apoptosis following transfer of *p53* genes, both in vitro and in xenographs *(20••,21–24•)*. This is exactly what would be predicted if p53 enhances chemosensitivity.

p53 MUTATIONS AND DRUG RESISTANCE IN HUMAN TUMORS

Although cell culture studies are important and are experimentally tractable, the only relevant measure of drug efficacy is tumor regression, ultimately, in human patients. Consistent with the observations from some model systems, *p53* status is strongly linked to drug resistance in several tumor types. The most striking examples occur in lymphoid malignancies, including non-Hodgkin's lymphoma, acute myeloid leukemia, myelodysplastic syndrome, and chronic lymphocytic leukemia. When these patients are classified by *p53* status, tumor response, and survival, those patients harboring tumors with *p53* mutations are much less likely to enter remission, and are more likely to die, than those with tumors harboring wild-type *p53* genes *(25••,26•, 27)*. Furthermore, in acute lymphoblastic leukemia, *p53* mutations in primary tumors are exceedingly rare, and most patients typically respond to therapy. However, a subfraction of patients relapse, and approx 30% of relapsed tumors harbor mutant *p53*. Of those who relapse, patients with *p53* mutant tumors are less likely to enter a second remission, compared to patients with tumors harboring wild-type *p53 (28•)*. Enrichment for cells with *p53* mutations is exactly what would be predicted if *p53* mutations conferred a survival advantage to cells during chemotherapy.

p53 mutations are linked with drug resistance in carcinomas, as well. In breast cancer, *p53* mutations are a strong predictor of relapse and death *(29••)*, and are associated with resistance to several therapeutic regimens *(30•,31••,32)*. In Wilms' tumor, *p53* mutations are associated with a more aggressive (anaplastic) tumor type that is refractory to therapy and displays reduced apoptosis *(33)*. *p53* mutations are associated with treatment failure or drug resistance in other solid tumors, including metastatic colon *(34•)*, esophogeal *(35)*, testicular *(15)*, and ovarian carcinomas *(36••)*. Bax is a proapoptotic member of the Bcl-2 family that can act as a p53 effector in apoptosis. Low Bax protein expression correlates with drug resistance and radiosensitivity in breast cancer *(37)*, which is consistent with the impact of p53 on anticancer agent sensitivity in this tumor type.

In addition to apoptosis, at least two other aspects of p53 biology may contribute to drug resistance. First, mutant p53 can upregulate expression of P-glycoprotein (PGP), a membrane pump implicated in multidrug resistance because of its ability to prevent intracellular accumulation of certain drugs *(38)*. Consequently, *p53* mutations may indirectly promote drug resistance by enhancing PGP levels. However, *p53* mutations do not correlate with elevated PGP expression in human tumors *(25••,39)* and both radiation and non-PGP substrate drugs can display p53-dependent toxicity *(40)*. Second, because *p53* mutations relax genomic integrity, it is conceivable that secondary mutations actually produce drug resistance. However, this seems unlikely, because reintroduction of wild-type *p53* has been shown to enhance chemosensitivity *(see above)*. Nevertheless, genomic instability can only compound the problem.

MANY STUDIES FAIL TO CORRELATE p53 MUTATIONS WITH REDUCED ANTICANCER AGENT CYTOTOXICITY

Although the correlations described above are striking, many studies on human tumor lines or patients find no correlation between p53 status and radio- or chemosensitivity *(see* Chapter 1). In fact, others see an opposite effect *(41•,42•)*. For example, bladder cancer tumors that are positive for p53 by immunohistochemistry (a surrogate marker for p53 mutation) are more responsive to a variety of treatments than are tumors with undetectable p53 *(43••)*. Similarly, in bladder cancer cell lines, cells expressing mutant p53 are more sensitive to radiation than their wild type counterparts *(44)*. These observations are consistent with the notion that disruption of p53-dependent cell-cycle checkpoints can enhance chemosensitivity *(7•)*.

What can explain this apparent paradox? Much has been made of the variability between short-term and long-term clonogenic survival assays *(see* Chapter 1). The issue is further confounded by the fact that clonogenic survival in vitro may not accurately reflect cell-kill in vivo *(45••); see also* Chapter 1 for an opposing view), perhaps because autocrine or paracrine survival factors, which have a major impact on apoptosis, are, in turn, influenced by cell density or microenvironment. Still, p53 can enhance drug-induced toxicity in clonogenic survival assays *(21•,46•,47)* or, more importantly, in vivo *(10••,20••,48••)*. A more likely explanation for this paradox is methodological, because current technologies are unable to properly classify tumors based on p53 functional status. Finally, the diversity of p53 activities and subsequent consequences of *p53* loss may vary between settings. Because defects in damage-induced checkpoints may enhance chemosensitivity, but defects in apoptosis promote drug resistance, the clinical impact of *p53* mutation may be determined by which effect predominates. This, in turn, may be influenced by tumor type, chemotherapeutic agent, or genetic background.

p53 DETECTION AND THE IMPACT OF THE p53 PATHWAY

In correlative studies using human cell lines or tumors, the ability to make statistically relevant comparisons requires that samples be accurately classified based on *p53* status. Most clinical studies rely on p53 immunohistochemistry as a surrogate marker for *p53* mutations, but this method is subjective, and necessarily misclassifies a subset of tumors *(49••)*. Direct sequencing of *p53* is more accurate and can identify associations between *p53* mutations and clinical parameters that immunohistochemistry cannot *(49••)*, but is tedious and subject to normal cell contamination. In studies with limited material, misclassification of even a few tumors can affect the ability to uncover statistically significant differences.

Even with accurate diagnosis of *p53* status, two mechanistic issues can dramatically affect the interpretation of the results. First, tumors harboring different *p53* mutations may not behave identically. Mutations in the *p53* gene can occur in over 100 codons, and may or may not be accompanied by loss of the wild-type allele *(1,2)*. p53's apoptotic activity is highly sensitive to *p53* dosage *(4•,5•)*, and individual *p53* mutations can have distinct functional consequences *(50,51)*. Indeed, recent studies suggest that certain *p53* mutations result in tumors with a particularly aggressive clinical course *(52,53)*. Second, p53 function can be compromised by various mechanisms, including gene mutation, deletion, extragenic mutations in the p53 pathway, or heteromeric protein interactions *(1,2)*. For example, following DNA damage, p53 is activated by kinases that produce changes in p53 phosphorylation and conformation *(54,55)*. Upon activation, p53 regulates a series of target genes, including the p21[CIP1/WAF1] cyclin-dependent kinase inhibitor and Bax. p53 also associates with a series of cellular or viral proteins that modulate its activity, including Mdm2, HPV E6, p33[ING,] and p19[ARF] *(1,2,56•,57•)*. Extragenic mutations that affect p53 function occur in human cancer; for example, HPV E6 in cervical carcinomas; Mdm-2 in a variety of tumors, and Bax in heredity nonpolyposis colorectal cancer-type colon cancer *(1,2,58••)*. In these settings, *p53* status would be misclassified by all current technologies. Given these caveats, it is almost impossible to interpret negative results in clinical studies.

CHEMOTHERAPEUTIC AGENT AND DOSE

Chemotherapeutic agents can show a variable dependence on p53 for toxicity *(16••)*. For example, microtubule-targeting agents do not damage DNA, and can effectively kill p53 mutant cells *(12,19,47•,59)*, although this effect is not universal *(60)*. Another issue relates to dosage: In all settings, cytotoxic agents induce apoptosis in the absence of p53, if provided at sufficiently high doses. This implies that p53 is not an essential component of the apoptotic machinery, but apparently determines the threshold of damage at which this machinery is engaged. Even HL60 cells, a p53-deficient tumor line that is often considered drug sensitive, becomes much more drug-sensitive upon re-expression of p53 *(23•)*. Hence, the impact of p53 on chemosensitivity may be influenced by whether or not a drug can be administered at sufficient levels to achieve a p53-independent effect. Finally, few studies have examined the effects of p53 on the cytotoxicity of combination therapy—the typical strategy used in patients. A better understanding of these issues may facilitate the design of more effective treatment protocols.

Still, discrepancies relating p53 to chemosensitivity are observed between studies using the same agent. For example, inactivation of p53 enhances cisplatin toxicity in normal fibroblasts *(41•);* by contrast, E1A-expressing fibroblasts lacking p53 are more resistant than their p53 normal counterparts *(61)*. Hence, in fibroblasts, a signal gene can change the impact of p53 on chemosensitivity. In the MCF7 breast carcinoma and U2OS osteosarcoma lines, inactivation of wild-type p53 with HPV E6 or a dominant-negative p53 mutant enhances cisplatin toxicity *(42•)*. By contrast, inactivation of p53 reduces cisplatin toxicity in murine thymocytes *(47•)*. Finally, enforced p53 expression enhances cisplatin toxicity in many settings, including in xenographs derived from lung carcinoma cells *(20••,24•)*. Thus, the role of p53 in chemosensitivity may depend on context.

p53'S APOPTOTIC ACTIVITY
IS CONTEXT-DEPENDENT

The prediction that p53 should enhance chemosensitivity is based on its apoptotic function; however, other functions of p53 may not promote chemosensitivity. Perhaps positive correlations between p53 and chemosensitivity will only occur in circumstances in which p53's apoptotic function predominates. In this regard, a major determinant of p53's apoptotic activity is cell type. In most normal tissues, p53 does not produce apoptosis at sublethal levels of ionizing radiation, but rather, may promote cell-cycle arrest and/or facilitate DNA repair. However, some tissues, including cells of hematopoietic origin and intestinal stem cells, are intrinsically wired for damage-induced apoptosis, and, in these settings, apoptosis depends particularly on p53 *(4•,5•,48••,62–65•)*. Hence, in tumors derived from these tissues, p53 may be in an apoptotic mode and may be a strong factor in determining cellular chemosensitivity. This appears to be the case in hematologic malignancies *(see above)*.

Tissue-independent changes also can affect p53's apoptotic activity, including p53 levels or genetic background. In an osteosarcoma line, low levels of p53 promote cell-cycle arrest, but higher levels promote apoptosis *(66)*. Alternatively, in a series of colon carcinoma lines, ectopic p53 expression produces either growth arrest or apoptosis, depending on intrinsic genetic factors *(67••)*. Presumably, the impact of genetic background reflects the fact that certain oncogenic changes can reveal p53's apoptotic activity in cell types in which p53 normally facilitates growth arrest *(9•)*. For example, both the *E1A* and *myc* oncogenes induce p53 protein and promote p53-dependent apoptosis in fibroblasts *(68•,69•)*, and these changes also promote radio- and chemosensitivity *(9•,70,71••)*. E1A's proapoptotic activity depends on its ability to inactivate the Rb tumor suppressor, and perhaps the E2Fs; indeed, loss of Rb and overexpression of E2F-1 can promote p53-dependent apoptosis *(72–74••)*. Still, Rb deficiency can also promote p53-independent cell death in certain contexts *(73)*. Recent studies demonstrate that the ability of these oncogenic changes to activate p53 depends chiefly on p19[ARF] *(74••,75••)*, an alternative reading-frame product of the *INK4a* tumor-suppressor locus *(76)*. Although p19[ARF] acts in a DNA-damage-independent signaling pathway to p53 *(75••,77••)*, it can synergize with radiation and chemotherapeutic drugs to facilitate p53-dependent apoptosis *(75••)*. In contrast, the Bcl-2 oncogene can suppress p53's apoptotic function while allowing growth arrest *(78)*. Hence, the outcome of p53 activation in tumors does not simply reflect tissue of origin, but also may be determined by the cumulative impact of the oncogenic mutations in each tumor cell. These issues, in turn, may have a substantial impact on p53's relationship to chemosensitivity.

THERAPEUTIC STRATEGIES

An improved understanding of the relationship between p53 and chemosensitivity may lay the groundwork for new cancer therapies. In tumors that retain p53 function, it may be possible to superactivate p53 to promote apoptosis in combination with standard chemotherapy. For example, strategies to disrupt the p53–Mdm-2 negative feedback loop *(79•,80•)*, or to mimic p19^{ARF} *(75••)*, have been presented. In tumors with p53 mutations, reintroduction of p53 should synergize with standard agents to induce apoptosis and promote tumor regression. Indeed, gene therapy strategies using this approach are currently in clinical trials *(22)*. Although gene therapy suffers from gene delivery limitations, strategies that combine this with standard therapies are particularly attractive as patients continue to receive systemic therapy. Subsequent research may identify suitable targets for small-molecule inhibition, allowing reactivation of the p53 apoptotic program and chemosensitivity in the absence of p53.

On the other hand, normal hematologic cells and certain stem cells of the small intestine are remarkably sensitive to apoptosis following radiation or chemotherapy *(see above)*. It is interesting that damage to these tissues accounts for many of the most debilitating side effects of cancer therapy. Ironically, the presence of p53 in these normal tissues may be deleterious for cancer patients, because in these settings, p53 activity contributes to the deletion of normal cells *(48••)*. Perhaps agents that suppress p53 function in these tissues would make effective chemoprotective agents.

CONCLUSIONS

Despite extraordinary advances in the understanding of cancer, basic cancer research has yet to make a substantial impact on the treatment of human malignancy. Most cancer patients continue to receive highly toxic drugs derived from empirical screens, and the best therapy remains complete surgical resection of the tumor. Still, underlying the massive effort to identify the molecular defects in cancer cells is the premise that this information will eventually produce better diagnostic and prognostic tools, and will ultimately suggest rational therapeutic strategies. The role of p53 in apoptosis and chemosensitivity provides a provocative link between factors that influence tumor development and those involved in drug toxicity. Current clinical studies suggest that p53 status may be a useful predictive tool in some settings, although a better understanding the complexity of p53 regulation and activity will be necessary for this information to be broadly useful in diagnosis or prognosis. Nevertheless, reintroduction of p53 into p53-defective tumors enhances apoptosis and the cytotoxicity following treatment with a broad range of anticancer agents. Hence, strategies to enhance p53 activity in tumor cells, by gene therapy or by the appropriate manipulation of downstream targets, are likely to have therapeutic benefit.

ACKNOWLEDGMENTS

The author apologizes to those investigators whose work, because of space constraints, was not cited. The author also thanks M. E. McCurrach and R. R. Wallace-Brodeur for writing assistance, and acknowledges support from a Sidney Kimmel Scholar Award and grant CA13106 from the National Cancer Institute.

REFERENCES AND FURTHER READINGS

• of special interest
•• of outstanding interest

1. Ko LJ, Prives C. p53: puzzle and paradigm. *Genes Dev* 1996; 10: 1054–1072.
2. Levine, AJ. p53, the cellular gatekeeper for growth and division. *Cell* 1997; 88: 323–331.
3. Weinberg RA. Cat and mouse games that genes, viruses, and cells play. *Cell* 1997; 88: 573–575.
•4. Lowe SW, Schmitt EM, Smith SW, Osborne BA, Jacks T. p53 is required for radiation-induced apoptosis in mouse thymocytes. *Nature* 1993; 362: 847–849.
•5. Clarke AR, Purdie CA, Harrison DJ, Morris RG, Bird CC, Hooper ML, Wyllie AH. Thymocyte apoptosis induced by p53-dependent and independent pathways. *Nature* 1993; 362: 849–852.
6. Kastan MB, Zhan Q el DW, Carrier F, Jacks T, Walsh WV, Plunkett BS, Vogelstein B, Fornace AJ. Mammalian cell cycle checkpoint pathway utilizing p53 and GADD45 is defective in ataxia-telangiectasia. *Cell* 1992; 71: 587–597.
•7. Waldman T, Lengauer C, Kinzler KW, Vogelstein B. Uncoupling of S phase and mitosis induced by anticancer agents in cells lacking p21. *Nature* 1996; 381: 713–716.
8. Slichenmyer WJ, Nelson WG, Slebos RJ, Kastan MB. Loss of a p53-associated G1 checkpoint does not decrease cell survival following DNA damage. *Cancer Res* 1993; 53: 4164–4168.
•9. Lowe SW, Ruley HE, Jacks T, Housman DE. p53-dependent apoptosis modulates the cytotoxicity of anticancer agents. *Cell* 1993; 74: 957–967.
••10. Lowe SW, Bodis S, McClatchey A, Remington L, Ruley HE, Fisher DE, Housman DE, Jacks T. p53 status and the efficacy of cancer therapy in vivo. *Science* 1994; 266: 807–810.
    Using genetically controlled experimental tumors, this study provides direct in vivo evidence that p53 status can influence the outcome of cancer therapy.
•11. Fan S, El-Deiry WS, Bae I, Freeman J, Jondle D, Bhatia K, et al. p53 gene mutations are associated with decreased sensitivity of human lymphoma cells to DNA damaging agents. *Cancer Res* 1994; 54: 5824–5830.

12. Iwadate Y, Tagawa M, Fujimoto S, Hirose M, Namba H, Sueyoshi K, Sakiyama S, Yamaura A. Mutation of the p53 gene in human astrocytic tumours correlates with increased resistance to DNA-damaging agents but not to anti-microtubule anticancer agents. *Br J Cancer* 1998; 77: 547–551.

13. Iwadate Y, Fujimoto S, Tagawa M, Namba H, Sueyoshi K, Hirose M, Sakiyama S. Association of p53 gene mutation with decreased chemosensitivity in human malignant gliomas. *Int J Cancer* 1996; 69: 236–240.

14. Li G, Tang L, Zhou X, Tron V, Ho V. Chemotherapy-induced apoptosis in melanoma cells is p53 dependent. *Melanoma Res* 1998; 8: 17–23.

15. Houldsworth J, Xiao H, Murty VV, Chen W, Ray B, Reuter VE, Bosl GJ, Chaganti RS. Human male germ cell tumor resistance to cisplatin is linked to TP53 gene mutation. *Oncogene* 1998; 16: 2345–2349.

•• 16. Weinstein JN, Myers TG, O'Connor PM, Friend SH, Fornace AJ, Jr., Kohn KW, et al. An information-intensive approach to the molecular pharmacology of cancer. *Science* 1997; 275: 343–349.
Comprehensive study examining the impact of p53 and other factors on drug sensitivity and resistance in the NCI 60 cell lines.

17. Mcilwrath AJ, Vasey PA, Ross GM, Brown R. Cell cycle arrests and radiosensitivity of human tumor cell lines: Dependence on wild-type p53 for radiosensitivity. *Cancer Res* 1994; 54: 3718–3722.

18. Kondo S, Barnett GH, Hara H, Morimura T, Takeuchi J. MDM2 protein confers the resistance of a human glioblastoma cell line to cisplatin-induced apoptosis. *Oncogene* 1995, 10: 2001–2006.

19. Fan S, Cherney B, Reinhold W, Rucker K, O'Connor PM. Disruption of p53 function in immortalized human cells does not affect survival or apoptosis after taxol or vincristine treatment. *Clin Cancer Res* 1998; 4: 1047–1054.

•• 20. Fujiwara T, Grimm EA, Mukhopadhyay T, Zhang WW, Owen-Schaub LB, Roth JA. Induction of chemosensitivity in human lung cancer cells in vivo by adenovirus-mediated transfer of the wild-type p53 gene. *Cancer Res* 1994; 54: 2287–2291.
Documents the efficacy of p53 *gene therapy in combination with cisplatin in a xenograph lung carcinoma model, and provides proof-of-concept data for combining p53 and chemotherapy in clinical trials.*

• 21. Blagosklonny MV, El-Deiry WS. Acute overexpression of wt p53 facilitates anticancer drug-induced death of cancer and normal cells. *Int J Cancer* 1998; 75: 933–940.

22. Roth JA. Gene replacement strategies for lung cancer. *Curr Opin Oncol* 1998; 10: 127–312.

• 23. Ju JF, Banerjee D, Lenz HJ, Danenberg KD, Schmittgen TC, Spears CP, et al. Restoration of wild-type p53 activity in p53-null HL-60 cells confers multidrug sensitivity. *Clin Cancer Res* 1998; 4: 1315–1322.

• 24. Gjerset RA, Turla ST, Sobol RE, Scalise JJ, Mercola D, Collins H, Hopkins PJ. Use of wild-type p53 to achieve complete treatment sensitization of tumor cells expressing endogenous mutant p53. *Mol Carcinog* 1995; 14: 275–285.

•• 25. El Rouby S, Thomas A, Costin D, Rosenberg CR, Potmesil M, Silber R, Newcomb EW. p53 gene mutation in B-cell chronic lymphocytic leukemia is associated with drug resistance and is independent of MDR1/MDR3 gene expression. *Blood* 1993; 82: 3452–3459.
First study demonstrating a strong association between p53 *mutations and drug resistance in human tumors.*

••26. Wattel E, Preudhomme C, Hecquet B, Vanrumbeke M, Quesnel B, Dervite I, Morel P, Fenaux P. p53 mutations are associated with resistance to chemotherapy and short survival in hematologic malignancies. *Blood* 1994; 84: 3148–3157. *p53 mutations correlate drug resistance in diverse hematologic malignancies. Indeed, the consequences of p53 mutations are ominous: No patient with a p53-mutant tumor survived the examination period.*

27. Wilson WH, Teruya-Feldstein J, Fest T, Harris C, Steinberg SM, Jaffe ES, Raffeld M. Relationship of p53, bcl-2, and tumor proliferation to clinical drug resistance in non-Hodgkin's lymphomas. *Blood* 1997; 89: 601–609.

•28. Diccianni MB, Yu J, Hsiao M, Mukherjee S, Shao LE, Yu AL. Clinical significance of p53 mutations in relapsed T-cell acute lymphoblastic leukemia. *Blood* 1994; 84: 3105–3112.

••29. Kovach JS, Hartmann A, Blaszyk H, Cunningham J, Schaid D, Sommer SS. Mutation detection by highly sensitive methods indicates that p53 gene mutations in breast cancer can have important prognostic value. *Proc Natl Acad Sci USA* 1996; 93: 1093–1096
Using direct sequencing of p53 cDNAs from tumors, this group documents a clear association between p53 mutations and poor prognosis in breast cancer. Previous studies using p53 immunohistochemistry were ambiguous.

•30. Bergh J, Norberg T, Sjogren S, Lindgren A, Holmberg L. Complete sequencing of the p53 gene provides prognostic information in breast cancer patients, particularly in relation to adjuvant systemic therapy and radiotherapy. *Nat Med* 1995; 1: 1029–1034.

••31. Aas T, Borresen AL, Geisler S, Smith-Sorensen B, Johnsen H, Varhaug JE, Akslen LA, Lonning PE. Specific P53 mutations are associated with de novo resistance to doxorubicin in breast cancer patients. *Nat Med* 1996; 2: 811–814. *This study uses p53-sequencing methodology to correlate p53 mutations with resistance to doxorubicin in breast cancer patients. Although the sample size is rather small, this study eliminates one complication prevalent in other clinical studies: All patients were treated with the same regimen.*

32. Clahsen PC, Duval C, Pallud C, Mandard AM, Delobelle DA, van den Broek L, Sahmoud TM, van de Vijver MJ. p53 protein accumulation and response to adjuvant chemotherapy in premenopausal women with node-negative early breast cancer. *J Clin Oncol* 1998; 16: 470–479.

33. Bardeesy N, Beckwith JB, Pelletier J. Clonal expansion and attenuated apoptosis in Wilms' tumors are associated with p53 gene mutations. *Cancer Res* 1995; 55: 215–219.

•34. Benhattar J, Cerottini JP, Saraga E, Metthez G, Givel JC. p53 mutations as a possible predictor of response to chemotherapy in metastatic colorectal carcinomas. *Int J Cancer* 1996; 69: 190–192.

35. Moreira LF, Naomoto Y, Hamada M, Kamikawa Y, Orita K. Assessment of apoptosis in oesophageal carcinoma preoperatively treated by chemotherapy and radiotherapy. *Anticancer Res* 1995; 15: 639–644.

••36. Buttitta F, Marchetti A, Gadducci A, Pellegrini S, Morganti M, Carnicelli V, et al. p53 alterations are predictive of chemoresistance and aggressiveness in ovarian carcinomas: a molecular and immunohistochemical study. *Br J Cancer* 1997; 75: 230–235.
Correlates p53 mutations with resistance to cisplatin-based therapy in aggressive ovarian carcinoma. In this setting, loss of p53 function does not appear to

enhance cisplatin-based toxicity, although in some other settings it does (see *refs*. 41,42).

37. Krajewski S, Blomqvist C, Franssila K, Krajewska M, Wasenius VM, Niskanen E, Nordling S, Reed JC. Reduced expression of proapoptotic gene BAX is associated with poor response rates to combination chemotherapy and shorter survival in women with metastatic breast adenocarcinoma. *Cancer Res* 1995; 55: 4471–4478.

38. Chin KV, Ueda K, Pastan I, Gottesman MM. Modulation of activity of the promoter of the human MDR1 gene by Ras and p53. *Science* 1992; 255: 459–462.

39. Preudhomme C, Lepelley P, Vachee A, Soenen V, Quesnel B, Cosson A, Fenaux P. Relationship between p53 gene mutations and multidrug resistance (mdr1) gene expression in myelodysplastic syndromes. *Leukemia* 1993; 7: 1888–1890.

40. O'Connor PM, Jackman J, Jondle D, Bhatia K, Magrath I, Kohn KW. Role of the p53 tumor suppressor gene in cell cycle arrest and radiosensitivity of Burkitt's lymphoma cell lines. *Cancer Res* 1993; 53: 4776–4780.

•41. Hawkins DS, Demers GW, Galloway DA. Inactivation of p53 enhances sensitivity to multiple chemotherapeutic agents. *Cancer Res* 1996; 56: 892–898.

•42. Fan S, Smith ML, Rivet DJ, Duba D, Zhan Q, Kohn KW, Fornace AJ, Jr., O'Connor PM. Disruption of p53 function sensitizes breast cancer MCF-7 cells to cisplatin and pentoxifylline. *Cancer Res* 1995; 55: 1649-1654.

••43. Cote RJ, Esrig D, Groshen S, Jones PA, Skinner DG. p53 and treatment of bladder cancer. *Nature* 1997; 385: 123–125.
Using increased p53 immunohistochemical staining as a surrogate marker for p53 mutations, the authors describe an association between p53 aberrations and improved response to chemotherapy in bladder carcinoma. Although the sample size is small, this paper provides a clear example of the complex relationship between p53 and chemosensitivity. For a more detailed discussion, see the reply by Lowe and Jacks (Nature *1997; 385: 125–126).*

44. Ribeiro JC, Barnetson AR, Fisher RJ, Mameghan H, Russell PJ. Relationship between radiation response and p53 status in human bladder cancer cells. *Int J Radiat Biol* 1997; 72: 11–20.

••45. Waldman T, Zhang Y, Dillehay L, Yu J, Kinzler K, Vogelstein B, Williams J. Cell-cycle arrest versus cell death in cancer therapy. *Nat Med* 1997; 3: 1034–1036.
By documenting clear discrepancies between tumor response in vivo and clonogenic survival assays, the authors call into question the utility of these assays in predicting drug cytotoxicity.

•46. Griffiths SD, Clarke AR, Healy LE, Ross G, Ford AM, Hooper ML, Wyllie AH, Greaves M. Absence of p53 permits propagation of mutant cells following genotoxic damage. *Oncogene* 1997; 14: 523–531.

•47. Vasey PA, Jones NA, Jenkins S, Dive C, Brown R. Cisplatin, camptothecin, and taxol sensitivities of cells with p53-associated multidrug resistance. *Mol Pharmacol* 1996; 50: 1536–1540.

••48. Pritchard DH, Potter CS, Hickman JA. The relationships between p53-dependent apoptosis, inhibition of proliferation, and 5-fluorouracil-induced histopathology in murine intestinal epithelia. *Cancer Res* 1998; 58: 5453–5465.
Demonstrates that p53 loss increases long-term cell survival in colonic epithelium in mice treated with 5-fluorouracil. The effect is caused, in part, by the reduced capacity of p53-deficient cells to undergo apoptosis. Also, the fact that p53 loss leads to less damage of normal tissue provides proof-of-concept data that inhibitors of p53 function might act as chemoprotective agents.

••49. Sjogren S, Inganas M, Norberg T, Lindgren A, Nordgren H, Holmberg L, Bergh J. p53 gene in breast cancer: prognostic value of complementary DNA sequencing versus immunohistochemistry. *J Natl Cancer Inst* 1996; 88: 173–182. *Provides a clear example of how the use of different methodological approaches can alter the interpretation of clinical studies.*

50. Rowan S, Ludwig RL, Haupt Y, Bates S, Lu X, Oren M, Vousden KH. Specific loss of apoptotic but not cell-cycle arrest function in a human tumor derived p53 mutant. *EMBO J* 1996; 15: 827–838.

51. Friedlander P, Haupt Y, Prives C, Oren M. Mutant p53 that discriminates between p53-responsive genes cannot induce apoptosis. *Mol Cell Biol* 1996; 16: 4961–4971.

52. Taubert H, Meye A, Wurl P. Prognosis is correlated with p53 mutation type for soft tissue sarcoma patients. *Cancer Res* 1996; 56: 4134–4136.

53. Vega FJ, Iniesta P, Caldes T, Sanchez A, Lopez JA, de Juan C, et al. p53 exon 5 mutations as a prognostic indicator of shortened survival in non-small-cell lung cancer. *Br J Cancer* 1997; 76: 44–51.

54. Shieh SY, Ikeda M, Taya Y, Prives C. DNA damage-induced phosphorylation of p53 alleviates inhibition by MDM2. *Cell* 1997; 91: 325–334.

55. Siliciano JD, Canman CE, Taya Y, Sakaguchi K, Appella E, Kastan MB. DNA damage induces phosphorylation of the amino terminus of p53. *Genes Dev* 1997; 11: 3471–3481.

•56. Garkavtsev I, Grigorian IA, Ossovskaya VS, Chernov MV, Chumakov PM, Gudkov AV. Candidate tumour suppressor p33[ING1] cooperates with p53 in cell growth control. *Nature* 1998; 391: 295–298.

•57. Kamijo T, Weber JS, Zambetti G, Zindy F, Roussel MF, Sherr CJ. Interactions of the ARF tumor suppressor with p53 and Mdm2. *Proc Natl Acad Sci USA* 1998; 95: 8292–8297.

••58. Rampino N, Yamamoto H, Ionov Y, Li Y, Sawai H, Reed JC, Perucho M. Somatic frameshift mutations in the BAX gene in colon cancers of the microsatellite mutator phenotype. *Science* 1997; 275: 967–969. *This study is interesting for many reasons, and provides a clear example of how mutations in the p53 pathway can occur in the absence of* p53 *gene mutations. The fact that* bax *and* p53 *mutations appear mutually exclusive implies, that in this tumor type, these genes function in the same pathway. Clinical studies examining the relationship between p53 and chemosensitivity based on* p53 *gene status would incorrectly classify tumors with* bax *mutations as p53-normal.*

59. Zhang CC, Yang JM, White E, Murphy M, Levine A, Hait WN. The role of MAP4 expression in the sensitivity to paclitaxel and resistance to vinca alkaloids in p53 mutant cells. *Oncogene* 1998; 16: 1617–1624.

60. Lanni JS, Lowe SW, Licitra EJ, Liu JO, Jacks T. p53-independent apoptosis induced by paclitaxel through an indirect mechanism. *Proc Natl Acad Sci USA* 1997; 94: 9679–9683.

61. McCurrach ME, Connor TM, Knudson CM, Korsmeyer SJ, Lowe SW. *bax*-deficiency promotes drug resistance and oncogenic transformation by attenuating p53-dependent apoptosis. *Proc Natl Acad Sci USA* 1997; 94: 2345–2349.

62. Lotem J, Sachs L. Hematopoietic cells from mice deficient in wild-type p53 are more resistant to induction of apoptosis by some agents. *Blood* 1993; 82: 1092–1096.

63. Clarke AR, Gledhill S, Hooper ML, Bird CC, Wyllie AH. p53 dependence of early apoptotic and proliferative responses within the mouse intestinal epithelium following gamma-irradiation. *Oncogene* 1994; 9: 1767–1773.

64. Merritt AJ, Potten CS, Kemp CJ, Hickman JA, Balmain A, Lane DP, Hall PA. Role of p53 in spontaneous and radiation-induced apoptosis in the gastrointestinal tract of normal and p53-deficient mice. *Cancer Res* 1994; 54: 614–617.

•65. Merritt AJ, Allen TD, Potten CS, Hickman JA. Apoptosis in small intestinal epithelial from p53-null mice: evidence for a delayed, p53-independent G2/M-associated cell death after gamma-irradiation. *Oncogene* 1997; 14: 2759–2766.

66. Chen X, Ko LJ, Jayaraman L, Prives C. p53 levels, functional domains, and DNA damage determine the extent of the apoptotic response of tumor cells. *Genes Dev* 1996; 10: 2438–2451.

••67. Polyak K, Waldman T, He TC, Kinzler KW, Vogelstein B. Genetic determinants of p53-induced apoptosis and growth arrest. *Genes Dev* 1996; 10: 1945–1952.
Even within a single tumor type, the cellular response to p53 is highly variable. This study begins to dissect this inherent genetic variability. Adenovirus-mediated transfer of p53 induces either cell-cycle arrest or apoptosis in a series of colon carcinoma lines; cell fusion studies indicate that apoptosis is the dominant outcome. This paper supports the view that oncogenic mutations can have a diverse impact on p53's apoptotic activity.

•68. Lowe SW, Ruley HE. Stabilization of the p53 tumor suppressor is induced by adenovirus 5 E1A and accompanies apoptosis. *Genes Dev* 1993; 7: 535–545.

•69. Hermeking H, Eick D. Mediation of c-Myc-induced apoptosis by p53. *Science* 1994; 265: 2091–2093.

70. Lotem J, Sachs L. Regulation by bcl-2, c-myc, and p53 of susceptibility to induction of apoptosis by heat shock and cancer chemotherapy compounds in differentiation-competent and -defective myeloid leukemic cells. *Cell Growth Differ* 1993; 4: 41–47.

••71. Rupnow BA, Murtha AD, Alarcon RM, Giaccia AJ, Knox SJ. Direct evidence that apoptosis enhances tumor responses to fractionated radiotherapy. *Cancer Res* 1998; 58: 1779–1784.
Activation of c-Myc activity in vivo can enhance cellular radiosensitivity through an apoptotic mechanism.

72. Morgenbesser SD, Williams BO, Jacks T, DePinho RA. p53-dependent apoptosis produced by Rb-deficiency in the developing mouse lens. *Nature* 1994; 371: 72–74.

73. Macleod KF, Hu Y, Jacks T. Loss of Rb activates both p53-dependent and independent cell death pathways in the developing mouse nervous system. *EMBO J* 1996; 15: 6178–6188.

••74. Zindy F, Eischen CM, Randle DH, Kamijo T, Cleveland JL, Sherr CJ, Roussel MF. MYC signaling via the ARF tumor suppressor regulates p53-dependent apoptosis and immortalization. *Genes Dev* 1998; 12: 2424–2433.
This study, and ref. 75 show that the p19^{ARF} tumor suppressor mediates oncogene activation of p53, that ARF-null cells have an attenuated p53 response.

••75. de Stanchina E, McCurrach ME, Zindy F, Shieh S-Y, Ferbeyre G, Samuelson AV, et al. E1A signaling to p53 involves the p19ARF tumor suppressor. *Genes Dev* 1998; 12: 2434–2442.

DNA damage and oncogenes signal p53 through parallel pathways; however, simultaneous activation of p53 by both pathways synergize to promote apoptosis. Hence, p19^ARF attenuates p53 function and mutations at the INK4a locus (encoding p16 and p19^ARF) are common in human tumors. How might the presence of INK4a/ARF mutations impact the classification of tumors as p53-mutant or p53-normal in clinical studies relating p53 to chemosensitivity?

76. Haber DA. Splicing into senescence: the curious case of p16 and p19ARF. *Cell* 1997; 91: 555–558.

••77. Kamijo T, Zindy F, Roussel MF, Quelle DE, Downing JR, Ashmun RA, Grosveld G, Sherr CJ. Tumor suppression at the mouse INK4a locus mediated by the alternative reading frame product p19ARF. *Cell* 1997; 91: 649–659.
Provides compelling evidence that the alternate reading-frame product of the INK4a locus acts as a tumor suppressor. Note the similarity between the phenotype of ARF-null and p53-null mice.

78. Chiou SK, Rao L, White E. Bcl-2 blocks p53-dependent apoptosis. *Mol Cell Biol* 1994; 14: 2556–2563.

•79. Bottger A, Bottger V, Sparks A, Liu WL, Howard SF, Lane DP. Design of a synthetic Mdm2-binding mini protein that activates the p53 response in vivo. *Curr Biol* 1997; 7: 860–869.

•80. Chen L, Agrawal S, Zhou W, Zhang R, Chen J. Synergistic activation of p53 by inhibition of MDM2 expression and DNA damage. *Proc Natl Acad Sci USA* 1998; 95: 195–200.

3 The Role of p53 in Chemosensitivity

Wafik S. El-Deiry, MD, PhD

CONTENTS

ABSTRACT

The p53 tumor suppressor is a major determinant of chemo- and radiosensitivity of cancer cells. Loss of p53-dependent apoptosis, the most common deficiency in human cancer, correlates with resistance to therapy. Using bioinformatic and functional genomic approaches, the p53 pathway leading to apoptosis is being unraveled, and includes signals that ultimately activate caspases, alter cytochrome-c release from mitochondria, and generate toxic oxygen species. Recently, a link between p53 and the tumor necrosis factor (TNF)-related apoptosis-inducing ligand (TRAIL) death receptor KILLER/DR5 expression has been identified as a potential p53-dependent mechanism for signaling caspase activation after the exposure of mammalian cells to DNA-damaging drugs. The upstream pathway from DNA damage to p53 stabilization appears to involve kinases,

From: *Apoptosis and Cancer Chemotherapy*
Edited by: J. A. Hickman and C. Dive © Humana Press Inc., Totowa, NJ

such as ataxia telangiectasia-mutated gene or DNA-PK, which phosphorylate p53, thereby inhibiting Mdm-2 binding and p53 degradation, as well as acetylation events that activate latent sequence-specific DNA binding and transactivation by p53. The new information regarding specific phosphorylation and acetylation events, as well as elucidation of novel downstream mediators, offers multiple inroads into drug design, to modulate or bypass p53 function in killing cancer cells.

INTRODUCTION

There are many excellent reviews covering the basics of p53 structure and function, and the reader is encouraged to refer to them for older references *(1••,2–4•)*. Several lines of evidence have implicated p53 as a major determinant of chemo- and radiosensitivity. Studies of cells derived from the p53 knockout mice have suggested that p53 loss leads to resistance to apoptosis induced by exposure to ionizing radiation or a variety of chemotherapeutic drugs. Thymocytes from p53−/− mice are radioresistant, and fibroblasts that have been transformed by ras and *E1A* are resistant to apoptosis induced by adriamycin, cisplatin, and 5-fluorouracil (5-FU). Data from the National Cancer Institute's cell screen of over 60 human cancer and leukemia cell lines, treated with more than 60,000 compounds, including a set of 123 standard antineoplastic agents, has suggested that the vast majority of clinically useful chemotherapeutic drugs are most effective in killing wild-type p53-expressing cells *(5,6•)*. In some, but not all, cases, targeting p53 protein for degradation by human papillomavirus (HPV) E6 protein has led to altered chemosensitivity *(7–9)*.

It has become clear from chemosensitivity testing of human cancer cell lines as a function of p53 status that there are important tissue specific differences in cancer cell sensitivity. For example, certain malignant hematopoietic cells are more sensitive to DNA-damaging drugs than solid-tumor cell lines, regardless of p53 status. Also, p53 status does not always correlate with chemosensitivity. It is not too difficult to identify cell lines derived from a given tissue that express wild-type p53, and yet appear to be more resistant to DNA-damaging drugs, compared to cell lines from the same tissue type that express mutant p53 or no p53. Such observations suggest either that cancer cells evolve changes that suppress the role of wild-type p53 in chemosensitivity, or that p53 status may not always be critically important. The former possibility is favored for several reasons. First, p53 is the most commonly mutated gene in human cancer, and it is clear that loss of p53 correlates with malignant tumor progression and drug resistance in vivo in many tumor types *(3)*. Second, p53 is a potent inducer of apoptosis, and there is ample evidence that cells can evolve mechanisms to inactivate its function without mutation, for example, by Mdm-2 overexpression. Thus, it is not too difficult to imagine scenarios in which critical events upstream or downstream of wild-type p53 could inhibit its activity as a tumor suppressor. Third, the introduction of wild-type p53 by using a replication defective adenovirus vector universally sensitizes cancer (or normal) cells to killing by a number of cytotoxic DNA-damaging anticancer agents *(10)*.

In order to understand the relationship between p53 and chemosensitivity, it becomes crucial to understand the complexities of p53 regulation and function (Fig. 1). The pathway upstream of p53 is being elucidated, and it appears to involve a number of very interesting genes, some of which are clearly tumor-suppressor genes in their own right. The targets of p53 continue to be uncovered and links between p53 and cell execution are being made. For many years, *p53* has been known as a major tumor-suppressor gene with no close homologs; however, this is rapidly changing, with the identification of p73 *(11,12)*, p42 (the p53-competing protein) *(13)*, and p63 (KET) *(14)*. However, no p53 homolog has been reported to be frequently inactivated in human cancer, including p73, which may be inactivated rarely in neuroblastomas *(11)*.

REGULATORS OF p53

Because p53 may be the most important suppressor of human cancer, there has been an intense effort to identify its mechanism of action. One of the basic approaches to this question has been to identify what proteins interact with p53, and how the interactions modulate its function. Historically, the first group of p53-interacting proteins discovered were several transforming viral oncoproteins. In fact, p53 was discovered as an SV40 large T-antigen interacting protein. It is also targeted for inactivation by HPV E6, which, through E6-associated protein, an E3-ubiquitin ligase, resulted in its degradation. It appears that the p53 homolog, p73, unlike p53, is resistant to HPV E6-mediated degradation, and could suppress growth through the induction of apoptosis in E6-overexpressing cancer cells *(15)*. Other p53-interacting viral proteins include the adenovirus E1B 55-kDa protein, which binds to p53 and inhibits its transcriptional activity, and the hepatitis B X protein, which sequesters p53 in the cytoplasm. p53 function in transcription is also inhibited indirectly by adenovirus E1A, through its p300/(CBP)-interacting domain *(16)*, as well as by the adenovirus E4ORF6 protein *(17)*, and its function as an inducer of apoptosis is

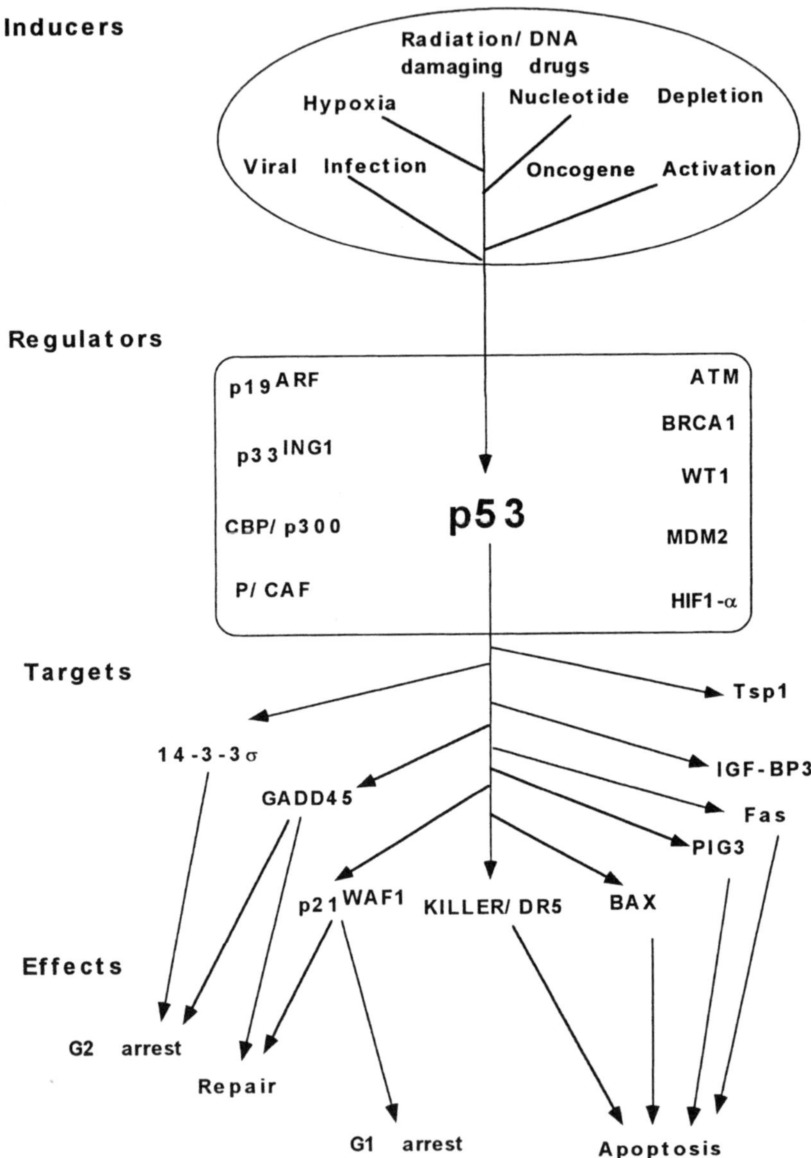

Fig. 1. The p53 pathway in 1998. Multiple different cellular stresses stabilize the p53 protein through a mechanism that, at least for DNA damage, involves phosphorylation and acetylation, leading to enhanced DNA binding and transactivation of target genes. Phosphorylated p53 (serine 15) is stabilized by virtue of decreased MDM-2 binding and degradation. p53 is subject to positive and negative regulation by a number of interacting proteins, some of which are known tumor suppressor genes. Targets of p53-dependent transcriptional activation mediate many of its biological effects, as indicated.

blocked indirectly by the Adenovirus E1B 19-kDa protein, which is a homolog of the antiapoptotic protein Bcl-2 *(18)*.

As the list of cellular proteins that interact with p53 grows, it becomes possible to classify them into at least four groups. These include proteins involved in control of gene expression, such as TAFs, p300, or CBP; proteins involved in replication or repair such as ERCC3, RAD51, or RPA; tumor-suppressor proteins, such as p33[ING1] *(19)*, breast cancer 1 susceptibility gene (BRCA1) *(20,21)*, WT1, p19[ARF] *(22,23)*; and the negative regulator of p53, Mdm-2. The hypoxia-inducible factor 1α (HIF-1α) was recently found to interact with p53, stabilize its expression, and stimulate p53-dependent transcription *(24)*.

Mdm-2, which targets p53 for degradation and also binds its N-terminal region, thereby concealing its transactivation domain, confers resistance to killing by certain DNA-damaging agents, such as cisplatin *(25)*. Strategies to bypass Mdm-2-dependent inhibition of p53-mediated apoptosis include the use of p53 mutants incapable of interaction with Mdm-2, drugs that block the p53–Mdm-2 interaction, or the use of adenoviruses that express growth-inhibitory targets of p53, such as p21, which is not sensitive to Mdm-2 *(26)*. Although it may be expected that tumor suppressors that interact with p53 may enhance chemosensitivity, there is little evidence of this at present. It has been shown, however, that the p33[ING1] candidate tumor suppressor can enhance the p53-dependent killing of human skin fibroblasts by the DNA-damaging topoisomerase II inhibitor, etoposide *(19)*. There is some evidence that BRCA1 may enhance p53-dependent transcription, leading to increased apoptosis *(20)*, and, although BRCA1 phosphorylation is altered following exposure to a variety of DNA-damaging agents *(27)*, it is not yet clear whether BRCA1 can sensitize cells to cytotoxicity caused by exposure to DNA-damaging agents that induce cell killing through the p53 pathway. This will be of particular interest, given that enhanced DNA repair capacity, a role proposed for the BRCA1 and BRCA2 gene products, may in some cases confer a survival advantage. At least in the case of BRCA2, loss of BRCA2 leads to radiation hypersensitivity *(28)*. It is well known that loss of the ataxia telangiectasia mutated gene (ATM) protein, which functions upstream of p53, leads to radioresistant DNA synthesis, i.e., loss of G1 and S-phase checkpoints manifested by continued DNA synthesis, despite DNA damage by ionizing radiation, and a severe hypersensitivity to the killing induced by such radiation. It should be emphasized that the loss of checkpoint control in AT cells does not account for the radiation hypersensitivity *(29)*. In fact, this hypersensitivity to ionizing radiation is further increased by loss of p21, the prototype cyclin-dependent kinase (CDK) inhibitor, which is a target of p53 in G1 checkpoint control; loss of both ATM and p21 has been correlated with delayed lymphomagenesis *(30)*.

REGULATION OF p53
BY PHOSPHORYLATION AND ACETYLATION

It has been known since the early 1980s that p53 protein is stabilized following exposure of mammalian cells to ultraviolet light. Work by Kastan and colleagues in the early 1990s revealed that p53 protein is stabilized following exposure to ionizing radiation, leading to cell-cycle arrest in G1 by a posttranscriptional mechanism. Although the p53 protein has numerous phosphorylation sites and could be phosphorylated by a number of kinases in vitro, including S-phase and G2/M CDKs, casein kinase II, and DNA-dependent protein kinase (DNA-PK), elucidation of the role of phosphorylation events and the mechanism of stabilization has remained enigmatic until recently *(31,32••,33••)*. Recent experiments by Shieh et al. *(32)* and Siliciano et al. *(33)* has revealed that, following exposure of mammalian cells to ionizing radiation, the p53 protein becomes rapidly phosphorylated at serine 15. DNA-PK was shown to be capable of phosphorylating p53 in vitro at serine 15 and 37, which reside within its transactivation/Mdm-2-interacting domain. The phosphorylated p53 interacts poorly with Mdm-2, which normally binds to and inhibits the transactivation function of p53, and also targets its degradation by the proteasome. Thus, it is currently believed that, through DNA-damage-induced phosphorylation, p53 becomes stabilized by inactivation of the pathway, which normally makes it a labile protein. In addition to regulation of its stability, there is potential for regulation of p53 function through phosphorylation. The phosphorylation of p53 by CDKs has been shown to selectively modulate its DNA-binding function *(31)*. The crystallization of the p53–Mdm-2-binding pocket has provided a strategy for drug development for Mdm-2-overexpressing cancers. It may be predicted, based on the new information on serine 15 phosphorylation, that drugs that enhance this phosphorylation, or that block dephosphorylation at serine 15, may have potent cytotoxic effects. It will be of interest to determine if p53 stabilization, which follows oncogene activation, nucleotide-depleted, or hypoxic conditions, also involves serine 15 phosphorylation and the same or different kinases.

New evidence obtained by Shiloh *(34)* and Kastan et al. *(35)* suggests that the p53 and ATM proteins co-immunoprecipitate, that ATM-dependent kinase activity is increased following γ-irradiation, and that p53 becomes phosphorylated by this ATM-dependent kinase activity. The mechanism of increased ATM kinase activity remains unclear.

Evidence obtained within the past year has linked acetylation events to regulation of p53 by p300/CBP and p300/CBP-associated factor (P/CAF) *(36•)*. Acetylation of p53 on lysine residues 382 and 320 within its C-terminal domain has been found to disrupt interactions between the C-terminal domain and the core DNA-binding domain, thus allowing the core domain to adopt an active conformation. Thus, phosphorylated and acetylated p53 becomes activated in terms of DNA binding and activation of gene expression. Evidence obtained by Saka-guchi et al. *(37)* suggests that acetylation of p53 may be driven by prior phosphorylation following DNA damage, i.e., full-length p53 containing its N-terminal region, which is phosphorylated following DNA damage was found to be a much better substrate for acetylation by p300 or P/CAF. These observations suggest that there may be several possible sequential steps in p53 activation following DNA damage, including: activation of a kinase, such as ATM or DNA-PK; phosphory-lation of the N-terminal region of p53 at serine 15 and 37; inhibition of interaction between phosphorylated p53 and Mdm-2, leading to decreased p53 degradation, (this step is subject to further regulation by other proteins, such as the p19^{ARF} tumor suppressor); acetylation of the C-terminal region of p53 at lysines 382 and 320; activation of latent DNA-binding activity of p53; and transactivation of downstream target genes. This upstream pathway of p53 regulation provides new targets for drug discovery to modulate chemosensitivity.

DOWNSTREAM TARGETS OF p53

The best-characterized function of p53 is in transcriptional activation, although p53 could also repress gene expression and signal through protein–protein interactions *(1)*. Most tumor-derived mutants of p53 are defective in DNA binding and transactivation. All targets directly upregulated by p53 contain consensus DNA-binding sites, and in part explain the biological effects of p53, including cell-cycle arrest in G1 (p21^{WAF1}) or G2 (14-3-3σ *[38]*, growth arrest and DNA damage induc-ible gene 45 [GADD45] *[39]*), repair (GADD45, p21^{WAF1}), inhibition of growth factor signaling insulin growth factor-binding protein 3 (IGF-BP3), inhibition of angiogenesis thrombospondin 1 (Tsp1), and apoptosis induction (Bax, Fas, p53-inducible gene 3 [PIG3], p53-induc-ible TRAIL death receptor 5 [KILLER/DR5]) *(40•,41•)*. Recently, p85, a regulator of the signaling protein, phosphatidylinositol-3 kinase has been identified in the p53-dependent apoptotic response to oxidative stress *(42)*. However, it is not yet clear whether p85 is a direct downstream target of p53-dependent transactivation.

Little is known about what determines which biological effect of p53 dominates, but it is clear that there are cell-type-specific differences in p53 function. For example, p53 tends to cause apoptosis in hematopoietic cells, but fibroblasts tend to arrest. However, overexpression of p53 at high levels or exposure of wild-type p53-expressing epithelial cancer cells to cytotoxic doses of DNA-damaging agents generally leads to apoptosis. It is likely that upstream regulators of p53 may direct activation of different transcriptional programs. For example, there is evidence that phosphorylation of p53 by CDKs could selectively affect its target specificity in DNA binding *(38)*. The precise mechanism that determines the outcome of p53 activation remains a mystery. However, it is likely that p53 is the favored target for inactivation in human cancer, because it activates multiple downstream effector target genes that mediate its effects. There is probably much redundancy in this pathway that is critical for generational health and cancer prevention.

DEATH RECEPTORS, p53, AND DNA DAMAGE

Studies in *Caenorhabditis elegans* and *Drosophila* have provided evidence for a genetic basis for programmed cell death. The pathway leading to death in higher organisms is being elucidated *(43••)*. Death can be triggered by extracellular signals that engage so-called death receptors, such as the Fas receptor or the tumor necrosis factor (TNF) receptor. Following binding of the ligand to the extracellular cysteine-rich repeats, members of the TNF receptor family trimerize within the cell membrane. Association of receptors is in part facilitated by their protein–protein interacting cytoplasmic death domains. The death domains of death receptors recruit so-called adaptor molecules that also contain death domains. The adaptor for the Fas receptor is known as Fas-associated death domain (FADD). This adaptor contains a death domain at its C-terminus and a second protein–protein interaction domain, known as the death-effector domain (DED) at its N-terminus. The DED of adaptors binds to the DED (also known as the prodomain) of caspases. The recruitment of caspases to the receptor-signaling complex generates an active entity (the death-inducing signaling complex), which signals subsequent proteolysis and endonucleolytic cleavage. Caspases have the property of autoactivation, which leads to cleavage into p10 and p20 subunits that can act on downstream caspases. The TNF receptor associates with TNF receptor-associated death domain (TRADD), which, through its death domain, binds to FADD and transmits a death signal. The TNF receptor can also trigger nuclear factor-κB activation, leading to decreased toxicity following exposure of cells to TNF.

A relatively new member of the cytotoxic cytokine family is the TNF-related apoptosis-inducing ligand (TRAIL) *(44)*. There are at least four TRAIL receptors (DR4, KILLER/DR5, decoy TRAIL receptor no. 3 lacking an intracellular domain [TRID], and decoy TRAIL receptor no. 4 containing a truncated death domain [TRUNDD]) that have been discovered (*see* Fig. 2) within the past year *(45••,46)*. TRID and TRUNDD are antiapoptotic decoy receptors that can inhibit TRAIL-induced apoptosis by competing for ligand binding. The two proapoptotic TRAIL receptors are DR4 and KILLER/DR5. The precise signaling pathway of TRAIL-mediated apoptosis remains unclear. Although FADD has been implicated in the apoptotic pathway of TRAIL *(47–49)*, FADD is not required for TRAIL-induced apoptosis *(50••)*.

A potential mechanism of p53-dependent apoptosis was suggested when the KILLER/DR5 receptor was identified as a candidate p53 target gene that is induced following exposure of wild-type p53-containing cells to a variety of DNA-damaging agents, including adriamycin, etoposide, and radiation *(41)*. KILLER/DR5 is an extremely potent growth-inhibitory gene through apoptosis induction. The mechanism of KILLER/DR5 induction by p53 appears to require new transcription, and may in part be mediated by p53 DNA-binding sites located within the genomic regulatory sequences of the KILLER/DR5 gene (R. Takimoto and W.S. El-Deiry, unpublished observations). Recent studies have uncovered a p53-independent pathway of KILLER/DR5 induction following exposure of cells to TNF-α *(51)*. Although the TNF-α induction appears to be transcriptional, the transducer(s) of the signal remains unknown.

Chemotherapeutic agents and radiation can induce p53-dependent expression of the death receptor Fas *(52);* (and *see* Chapter 12). There is also recent evidence that Fas ligand may be upregulated following exposure of cells to cytotoxic DNA-damaging agents *(53)*. It is possible that p53-mediated apoptosis induced by DNA damage involves upregulation of these receptors, which then signal to downstream caspases, leading to apoptosis. Although the Fas-signaling pathway may be involved in p53-dependent apoptosis, neither the Fas receptor nor the FADD adaptor are required for DNA-damage-induced p53-dependent death *(50,54;* G. S. Wu and W.S. El-Deiry, unpublished observations).

Some data have suggested that decoy receptors of TRAIL are expressed only in normal cells, but not in transformed and tumor cells *(45••)*. Because TRAIL can preferentially kill tumor cells over normal

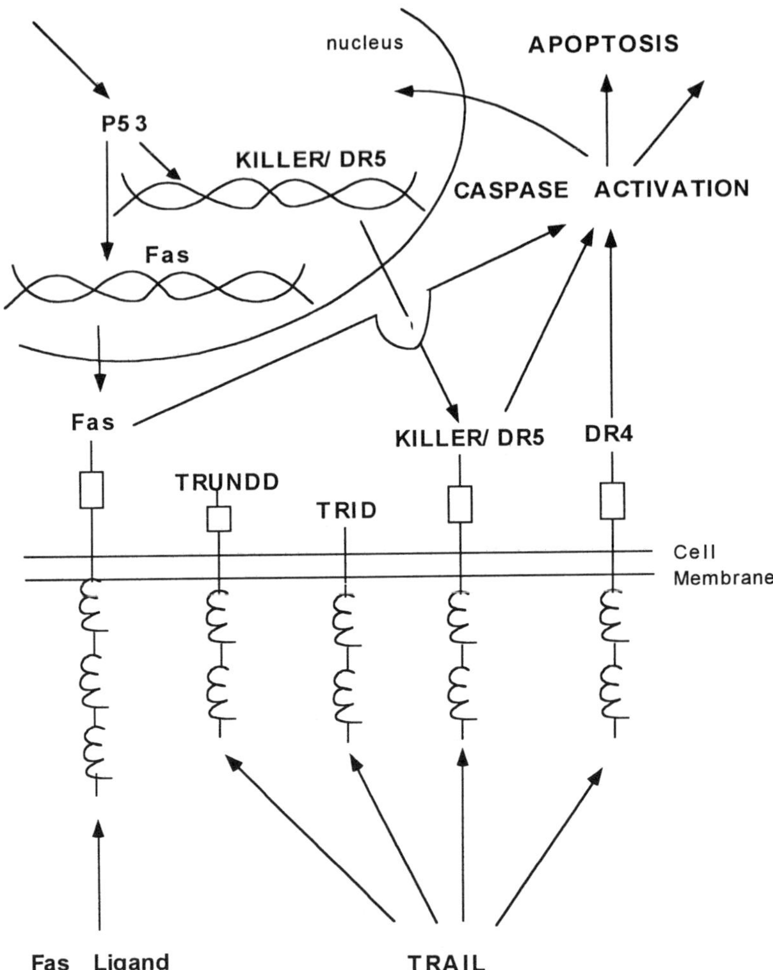

Fig. 2. p53 regulates the expression of death receptor genes in apoptosis induction. These include the Fas receptor and the TRAIL receptor KILLER/DR5. KILLER/ DR5 is part of a growing family of TRAIL death receptors that includes two decoys (TRID and TRUNDD) and the proapoptotic receptor DR4. The proapoptotic TRAIL receptors signal caspase activation through a mechanism that may involve the FADD adaptor. However, there is an unknown mechanism of caspase activation by these TRAIL receptors, because FADD is not required for TRAIL receptor death signaling. Although they may be involved in p53-dependent cell death, neither Fas nor FADD are required for DNA damage-induced p53-dependent death. It is not yet known if TRAIL receptors are required for p53-dependent death. It may be that multiple p53 targets, such as Fas, *bax*, KILLER/DR5, and PIG3, serve partly redundent roles in p53-dependent apoptosis, so that inactivation of any one of them alone is not suffi- cient to completely disable the p53 pathway. Such a hypothesis fits well with the fact that p53 is a highly favored target for mutation in human cancer, far beyond any of its targets, which are rarely, if ever, mutated.

cells, it shows promise for future use in the treatment of human cancers. However, there is no current information about the efficacy of combining TRAIL and classical cytotoxic agents. It is also not yet clear if decoy receptors can inhibit p53-dependent or DNA-damage-induced cell death, or whether their expression modulates the cytotoxicity of antineoplastic drugs.

All the known TRAIL receptor genes localize to human chromosome 8p21, a hotspot for translocations and a site where tumor suppressors have been mapped by allelic losses in head and neck, colon, breast, and prostate cancer. Thus, TRAIL receptor genes represent candidate tumor-suppressor genes, not only based on their genomic location, but also because of their function as signaling molecules involved in apoptosis induction. However, as with other targets of p53, such as *bax*, p21, or Fas, it is expected that mutations in KILLER/DR5 would be rare in human cancer, because the more common p53 mutations disable multiple targets simultaneously. At least one death-domain-truncating loss of function mutation in the KILLER/DR5 gene has been identified in a head and neck squamous cell carcinoma that contained wild-type p53 *(55)*. However, it is not clear if such a mutation contributed to the emergence of this head and neck cancer, and whether it affected its response to therapy.

INHIBITORS OF DEATH AND CHEMOSENSITIVITY

As the pathways of cell death have unfolded, so have a number of mechanisms for blocking apoptosis. Viruses have been useful in identifying some of these mechanisms (e.g., viral FLICE-inhibitory protein (FLIPs), crmA, and the Bcl-2 homolog 19-kDa, E1B protein, from adenovirus). Cellular proteins also function as potent inhibitors of apoptosis, and such molecules can lower the sensitivity of cancer cells to cytotoxic agents. It is well known, for example, that Bcl-2 is an inhibitor of apoptosis, and the overexpression of Bcl-2 has been linked to chemoresistance *(56); and other chapters in this book).* Recently, it has been shown that Bcl-2 is able to inhibit Bax-induced apoptosis, as well as cytochrome-*c* release *(57),* a process that occurs in cells undergoing apoptosis *(58).* Bcl-X$_L$, another Bcl-2 family member, identified as an antiapoptotic molecule, inhibits cell death induced by a variety of apoptotic stimuli, and its expression level has been correlated with chemoresistance in certain tumor types. Thus, there is evidence that elevated expression of Bcl-2 and Bcl-X$_L$ confers resistance to apoptosis and chemosensitivity *(55).* Bcl-2 has been previously reported as a target for transcriptional repression by p53 (*see* Chapter 7).

The cellular homolog of a viral FLIP was recently found to be overexpressed in human melanoma *(59••)*. Cellular FLIPs (both the short form, $FLIP_S$, and the long form, $FLIP_L$) contain two DEDs that interact with the DEDs of FADD and FLICE (also known as caspase-8 or Mch-5), thereby blocking transmission of the Fas-induced death signal. In addition, $FLIP_L$ contains a defective caspase-like domain, because of a substitution of the active-site cysteine residue by a tyrosine. It is not yet clear whether FLIPs inhibit TRAIL receptor signaling, but it appears that FLIPs do not block apoptosis induced by γ-irradiation or growth factor withdrawal *(59••)*.

Another mechanism for attenuating p53-dependent apoptosis and chemosensitivity involves the Mdm-2 oncoprotein. Mdm-2 is amplified or overexpressed in a significant number of human tumors. Overexpression of Mdm-2 has been correlated with increased resistance of human glioblastoma cells to the DNA-damaging agent, cisplatin *(25)*. Mdm-2 not only inhibits endogenous p53-dependent apoptosis, but also inhibits growth suppression following infection of Mdm-2-overexpressing cancer cells by p53-expressing adenovirus *(26)*.

CONCLUSIONS

p53 is a major determinant of chemosensitivity and radiosensitivity. There is a beginning understanding of the mechanism that signals its stabilization after DNA damage, and the targets that mediate its biological effects. The relationship between multiple-target gene expression and death vs cell-cycle arrest is currently under investigation. Much opportunity exists for novel cancer therapeutics aimed at modulating various aspects of the p53 pathway, to enhance cell killing.

ACKNOWLEDGMENTS

Thanks are extended to members of the El-Deiry lab for critical comments. G. S. Wu is also acknowledged for assistance with the artwork. Apologies are made to colleagues whose references were omitted because of space limitations; please refer to the cited reviews for the less recent references. W.S.E.-D. is an Assistant Investigator of the Howard Hughes Medical Institute.

REFERENCES

• of special interest
•• of outstanding interest

••1. Levine AJ. p53, the cellular gatekeeper for growth and division. *Cell* 1997; 88: 323–331.
   Outstanding comprehensive review of the p53 field until early 1997.

 2. Ko LJ, Prives C. p53: puzzle and paradigm. *Genes Dev* 1996; 10: 1054–1072.

 3. Velculescu VE, El-Deiry WS. Biological and clinical importance of the p53 tumor suppressor gene. *Clin Chem* 1996; 42: 858–868.

•4. Greenblatt MS, Bennett, WP Hollstein M, Harris CC. Mutations in the p53 tumor suppressor gene: clues to cancer etiology and molecular pathogenesis. *Cancer Res* 1994; 54: 4855–4878.
   Very useful and fairly up-to-date review of the involvement of p53 in human cancer.

 5. Weinstein JN, Myers TG, O'Connor PM, Friend SH, Fornace Jr AJ, Kohn KW, et al. An information-intensive approach to the molecular pharmacology of cancer. *Science* 1997; 275: 343–349.

•6. O'Connor PM, Jackman J, Bae I, Myers TG, Fan S, Mutoh M, et al. Characterization of the p53 tumor suppressor pathway in cell lines of the National Cancer Institute anticancer drug screen and correlations with the growth-inhibitory potency of 123 anticancer agents. *Cancer Res* 1997; 57: 4285–4300.
   Most recent summary of the results from the NCI cell screen regarding p53 and chemosensitivity. Information about World Wide Web access to the data, and links to other molecular characteristics, can be found in this article.

 7. Slickenmeyer WJ, Nelson WG, Slebos RJ, Kastan MB. Loss of a p53-associated G1 checkpoint does not decrease cell survival following DNA damage. *Cancer Res* 1993; 53: 4164–4168.

 8. Wu GS, El-Deiry WS. Apoptotic death of tumor cells correlates with chemosensitivity, independent of p53 or Bcl-2. *Clin Cancer Res* 1996; 2:623–633.

 9. Wu GS, El-Deiry WS. p53 and chemosensitivity. *Nature Med* 1996; 2: 255–256.

10. Blagosklonny MV, El-Deiry WS. Acute overexpression of wt p53 facilitates anticancer-induced death of cancer and normal cells. *Int J Cancer* 1998; 75: 933–940.

11. Kaghad M, Bonnet H, Yang A, Creancier L, Biscan JC, Valent A, et al. Monoallelically expressed gene related to p53 at 1p36, a region frequently deleted in neuroblastoma and other human cancers. *Cell* 1997; 90: 809–819.

12. Jost CA, Marin MC, Kaelin WG, Jr. p73 is a human p53-related protein that can induce apoptosis. *Nature* 1997; 389: 191–194.

13. Bian JH, Sun Y. p53CP, a putative p53 competing protein that specifically binds to the consensus p53 DNA binding sites: a third member of the p53 family. *Proc Natl Acad Sci USA* 1997; 94: 14,753–14,758.

14. Schmale H, Bamberger C. Novel protein with strong homology to the tumor suppressor p53. *Oncogene* 1997; 15: 1363–1367.

15. Prabhu NS, Somasundaram K, Satyamoorthy K, Herlyn M, El-Deiry WS. p73β, unlike p53, suppresses growth and induces apoptosis of human papillomavirus E6-expressing cancer cells. *Int J Oncol* 1998; 13: 5–9.

16. Somasundaram K, El-Deiry, WS. Inhibition of p53-mediated transactivation and cell cycle arrest by E1A through its p300/CBP-interacting region. *Oncogene* 1997; 14: 1047–1057.

17. Dobner T, Horikoshi N, Rubenwolf S, Shenk T. Blockage by adenovirus E4orf6 of transcriptional activation by the p53 tumor suppressor. *Science* 1996; 272: 1470–1473.

18. Han JH, Sabbatini P, Perez D, Rao L, Modha D, White E. E1B 19K protein blocks apoptosis by interacting with and inhibiting the p53-inducible and death promoting Bax protein. *Genes Dev* 1996; 10: 461–477.

19. Garkavtsev I, Grigorian IA, Ossovskaya VS, Chernov MV, Chumakov PM, Gudkov AV. Candidate tumour suppressor p33[INGl] cooperates with p53 in cell growth control. *Nature* 1998; 391: 295–298.

20. Zhang H, Somasundaram K, Peng Y, Tian H, Zhang H, Bi D, Weber BL, El-Deiry WS. BRCA1 physically associates with p53 and stimulates its transcriptional activity. *Oncogene* 1998; 16: 1713–1721.

21. Ouchi T, Monteiro AN, August A, Aaronson SA, Hanafusa H. BRCA1 regulates p53-dependent gene expression. *Proc Natl Acad Sci USA* 1998; 95: 2302–2306.

22. Pomerantz J, Schreiber-Agus N, Liegeois NJ, Silverman A, Alland L, Chin L, et al. INK4a tumor suppressor gene product, p19[Arf,] interacts with MDM2 and neutralizes Mdm-2's inhibition of p53. *Cell* 1998; 92: 713–723.

23. Zhang YP, Xiong Y, Yarbrough WG. ARF promotes Mdm-2 degradation and stabilizes p53: ARF-INK4a locus deletion impairs both the Rb and p53 tumor suppression pathways. *Cell* 1998; 92: 725–734.

24. An WG, Kanekal M, Simon MC, Maltepe E, Blagosklonny MV, Neckers LM. Stabilization of wild-type p53 by hypoxia-inducible factor 1α. *Nature* 1998; 392: 405–408.

25. Kondo S, Barnett GH, Hara H, Morimura T, Tekeuchi J. Mdm-2 protein confers the resistance of human glioblastoma cell line to cisplatin-induced apoptosis. *Oncogene* 1995; 10: 2001–2006.

26. Meng RD, Shih H, Prabhu NS, George DL, El-Deiry WS. Bypass of abnormal MDM2 inhibition of p53-dependent growth suppression. *Clin Cancer Res* 1998; 4: 251–259.

27. Scully R, Chen J, Ochs RL, Keegan K, Hoekstra M, Feunteun J, Livingston DM. Dynamic changes of BRCA1 subnuclear location and phosphorylation state are initiated by DNA damage. *Cell* 1997; 90: 425–435.

28. Patel KJ, Yu VPCC, Lee H, Corcoran A, Thistlethwaite FC, Evans MJ, et al. Involvement of Brca2 in DNA repair. *Mol Cell* 1998; 1: 347–357.

29. Rotman G, Shiloh Y. Ataxia-telangiectasia: Is ATM a sensor of oxidative damage and stress? *BioEssays* 1997; 19: 911–917.

30. Wang YA, Elson A, Leder P. Loss of p21 increases sensitivity to ionizing radiation and delays the onset of lymphoma in atm-deficient mice. *Proc Natl Acad Sci USA* 1997; 94: 14,590–14,595.

31. Wang Y, Prives C. Increased and altered DNA binding of human p53 by S and G2/M but not G1 cyclin-dependent kinases. *Nature* 1995; 376: 88–91.

••32. Shieh S-Y, Ikeda M, Taya Y, Prives C. DNA damage-induced phosphorylation of p53 alleviates inhibition by MDM2. *Cell* 1997; 91: 325–334.
These authors identified serine 15 and serine 37 of p53 as targets for phosphorylation after DNA damage. They further showed that phosphorylation of p53 at these serines impairs the ability of p53 to interact with MDM2, thereby rendering it more stable.

••33. Siliciano JD, Canman CE, Taya Y, Sakaguchi K, Appella E, Kastan MB. DNA damage induces phosphorylation of the amino terminus of p53. *Genes Dev* 1997; 11: 3471–3481.
These authors identified serine 15 as one of two serines within the human p53 protein that get phosphorylated following exposure to ionizing radiation or UV

radiation. They further found that, following ionizing irradiation of AT cells, but not UV irradiation, serine 15 phosphorylation on p53 is significantly (but not indefinitely) delayed.

34. Shiloh Y. ATM: a caretaker of the genome wearing many hats. *9th p53 Workshop.* 1998; http://athens.wistar.upenn.edu/p53.

35. Kastan MB, Siliciano JD, Canman CE, Lim D-S, Morgan SE. Cellular responses to irradiation: p53 and ATM. *9th p53 Workshop.* 1998; http://athens.wistar.upenn.edu/p53.

•36. Gu W, and Roeder RG. Activation of p53 sequence-specific DNA binding by acetylation of the p53 C-terminal domain. *Cell* 1997; 90: 595–606.
These authors found that the C-terminus of p53 is a substrate for acetylation, thereby enhancing its DNA-binding activity.

37. Sakaguchi K, Saito S, Herrera JE, Apella E, Anderson CW. DNA damage activates p53 through acetylation of the C-terminus: a phosphorylation–acetylation cascade. *9th p53 Workshop.* 1998; http://athens.wistar.upenn.edu/p53.

38. Hermeking H, Lengauer C, Polyak K, He TC, Zhang L, Thiagalingam S, Kinzler KW, Vogelstein B. 14-3-3 σ is a p53-regulated inhibitor of G2/M progression. *Mol Cell* 1997; 1: 3–11.

39. Wang XW, Hussain P, Zhou X, Zhan O, Fornace A, Wolkowicz R, et al. Etiology and pathobiological consequences of p53 mutations in human cancer. *9th p53 Workshop.* 1998; http://athens.wistar.upenn edu/p53.

•40. Polyak K, Xia Y, Zwier JK, Kinzler KW, Vogelstein B. Model for p53-induced apoptosis. *Nature* 1997; 389: 300–305.
Using serial analysis of gene expression (SAGE) technology, these authors uncovered evidence for p53-dependent induction of a number of genes involved in the generation of reactive oxygen species. At least one of these genes, PIG3, appears to be directly regulated by p53 at the level of transcription.

•41. Wu GS, Burns TF, McDonald III ER, Jiang W, Meng R, Krantz ID, et al. KILLER/DR5 is a DNA damage-inducible p53-regulated death receptor gene. *Nature Genet* 1997; 17: 141–143.
Using a subtractive hybridization approach comparing the gene expression patterns in chemosensitive vs chemoresistant isogenic human cancer cell lines, these authors identified the KILLER/DR5 as a downstream target of p53. This provided a link between DNA damage stabilization of p53 and the expression of a death receptor capable of caspase activation.

42. Yin Y, Terauchi Y, Solomon GG, Aizawa S, Rangarajan PN, Yazaki Y, Kadowaki, T, Barrett JC. Involvement of p85 in p53-dependent apoptotic response to oxidative stress. *Nature* 1998; 391: 707–710.

••43. Nagata S. Apoptosis by death factor. *Cell* 1997; 88: 355–365.
Outstanding comprehensive review of death signaling, until early 1997.

44. Wiley SR, Schooley K, Smolak PJ, Din WS, Huang CP, Nicholl JK, et al. Identification and characterization of a new member of the TNF family that induces apoptosis. *Immunity* 1995; 3: 673–682.

••45. Golstein P. Cell death: TRAIL and its receptors. *Curr Biol* 1997; 7: R750–R753.
Outstanding review of the rapidly moving TRAIL receptor field. This review contains many recent primary references that were omitted from this chapter because of space limitations.

46. Pan GH, Ni J, Yu G, Wei Y, Dixit VM. TRUNDD, a new member of the TRAIL receptor family that antagonizes TRAIL signalling. *FEBS Lett* 1998; 424: 41–45.

47. Schneider P, Thome M, Burns K, Bodmer JL, Hofmann K, Kataoka H, Holler N, Tschopp J. TRAIL receptor 1 (DR4) and 2 (DR5) signal FADD-dependent apoptosis and activate NF-kappa B. *Immunity* 1997; 7: 831–836.

48. Chaudhary PM, Eby M, Jasmin A, Bookwalter A, Murray J, Hood L. Death receptor 5, a new member of the TNFR family, and DR4 induce FADD-dependent apoptosis and activate the NF-κB pathway. *Immunity* 1997; 7: 821–830.

49. Walczak H, Degli-Esposito MA, Johnson RS, Smolak PJ, Waugh JY, Boiani N, et al. TRAIL-R2: a novel apoptosis-mediating receptor for TRAIL. *EMBO J* 1997; 16: 5386–5397.

••50. Yeh WC, De La Pompa JL, McCurrach ME, Shu HB, Elia AJ, Shahinian A, et al. FADD: essential for embryo development and signaling from some, but not all, inducers of apoptosis. *Science* 1998; 279: 1954–1958.
Using a mouse knockout approach, these authors probed the importance of the FADD death-signaling adaptor in development and death induced by a variety of different signals. FADD is required for Fas, TNFR1, and DR3 death signaling, but not for DR4, c-myc, E1A, or adiamycin-induced cell death.

51. Sheikh MS, Burns TF, Huang Y, Wu GS, Amundson S, Brooks KS, Fornace AJ, El-Deiry WS. p53-dependent and -independent regulation of the death receptor KILLR/DR5 gene expression in response to genotoxic stress and tumor necrosis factor alpha. *Cancer Res* 1998; 58: 1593–1598.

52. Sheard MA, Vojtesek B, Janakova L, Kovarik J, Zaloudik J. Upregulation of Fas (CD95) in human p53 (wild-type) cancer cells treated with ionizing radiation. *Int J Cancer* 1997; 73: 757–762.

53. Kasibhatla S, Brunner T, Genestier L, Echeverri F, Mahboubi A, Green DR. DNA damaging agents induce expression of Fas ligand and subsequent apoptosis in T lymphocytes via the activation of NF-κB and AP-1. *Mol Cell* 1998; 1: 543–551.

54. Fuchs EJ, McKenna KA, Bedi A. p53-dependent DNA damage-induced apoptosis requires Fas/APO-1-independent activation of CPP32β. *Cancer Res* 1997; 57: 2550–2554.

55. Pai SI, Wu GS, Ozoren N, Wu L, Jen J, Sidransky D, El-Deiry WS. Rare loss-of-function mutation of a death receptor gene in head and neck cancer. *Cancer Res* 1998; 58: 3513–3518.

56. El-Deiry, WS. Role of oncogenes in resistance and killing by cancer therapeutic agents. *Curr Opin Oncol* 1997; 9: 79–87.

57. Rosse T, Olivier R, Monney L, Rager M, Conus S, Fellay T, Jansen, B, Borner C. Bcl-2 prolongs cell survival after Bax-induced release of cytochrome c. *Nature* 1998; 391: 496–499.

58. Yang J, Liu X, Bhalla K, Kim CN, Ibrado AM, Cai J, et al. Prevention of apoptosis by Bcl-2: release of cytochrome c from mitrochondria blocked. *Science* 1997; 21: 1081,1082.

••59. Irmler M, Thome M, Hahne M, Schneider P, Hofman K, Steiner V, et al. Inhibition of death receptor signals by cellular FLIP. *Nature* 1997, 388: 190–194.
These authors identified a potentially important mechanism of cell death inhibition that may contribute to the neoplastic phenotype. The mechanism involves overexpression of a DED-containing molecule that can interact with FADD and FLICE, thereby blocking transmission of the death signal. Furthermore, the long form of FLIP contains a defective caspase-like domain, incapable of caspase activation.

Function of the p53 Gene Family
A Holistic Perspective

Sandra J. Campbell, BSc, PhD,
Mary O'Neill, BSc,
and Peter A. Hall, MD, PhD, FRCPath

CONTENTS

ABSTRACT

The plethora of biochemical and genetic data on p53 is chiefly based on in vitro and cell culture systems or the analysis of tumors late in their life history. Translating this data into truly physiological systems reveals many levels of complexity, and draws into question many of the dogmas and assumptions that litter the field. To advance knowledge of the p53 response pathway and utilize the information rationally in the clinic will require a concerted attempt to broaden understanding of the physiological biochemistry of the p53 family of proteins in vivo.

THE p53 RESPONSE: OVERVIEW

A large body of data places the transcription factor p53 at the focus of converging pathways from diverse cellular insults that can elicit coordinated cellular responses that result in adaptation to the insult *(1)*.

From: *Apoptosis and Cancer Chemotherapy*
Edited by: J. A. Hickman and C. Dive © Humana Press Inc., Totowa, NJ

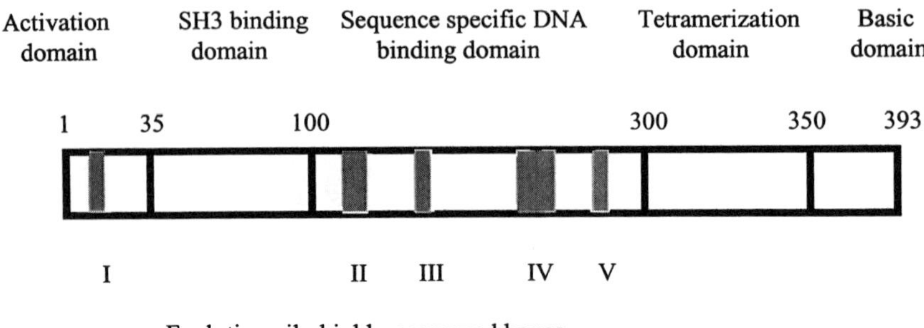

Fig. 1. A representation of the structure of mammalian p53. In humans p53 is composed of 393 residues. There are five highly conserved boxes and five identifiable regions subserving different functions. However, it should be recognized that the functions are interdependent, and regulation in one domain can profoundly influence other domains. Interactions with other macromolecules are of great significance *(see text)*.

The history of the p53 field and the complex biochemistry of this critical tumor suppressor have been well reviewed elsewhere, and readers are directed to these sources (*2–4,5••; see* Chapter 3). The purpose of this chapter is to give a perspective based not on neoplasia *per se,* but on the current view of the physiological role of the p53 pathway. Particularly, the chapter highlights the heterogeneity of p53 responses seen in vivo and reviews data suggesting that p53 has roles over and above simply being a tumor suppressor. The expansion of the p53 family to include other members is discussed. Finally, consideration is given the implications of the (as yet incomplete) view of the homeostatic role of p53 for therapeutic developments.

The current model of p53 function shows this 393-amino-acid nuclear phosphoprotein being composed of several critical domains (Fig. 1). The C-terminus contains an oligomerization domain and residues that can be modified by phosphorylation, RNA binding, O-glycosylation, and acetylation. As a consequence, the DNA-binding properties of homo tetramers of p53 may be modified, and conversion between active and inactive forms regulated. The central DNA-binding domain contains critical residues that either contact DNA or are essential for the stability of the protein backbone, which in turn is needed for the correct orientation of those contact residues. It is in this domain, and principally in those critical residues, that missense mutations lead to the loss of function phenotype so characteristic of neoplasia. N-terminal of the DNA-binding domain is a proline-rich region with striking similarity to SH3-binding proteins, and which appears to be involved in physical interaction with elements of signal transduction pathways, for example, c-*abl*. At the very N-terminus is an acidic domain involved in transcriptional activation by the recruitment of the basal transcription machinery, once p53 has bound to consensus p53-responsive elements within promoters of target genes. In addition, the N-terminus has a region that interacts with Mdm-2, and by so doing causes regulation of both p53 activity and level. This is achieved by the Mdm-2 targeting p53 for ubiquitin-mediated proteolysis, as well as acting as a steric block to interaction of p53 with the transcriptional apparatus. Data has recently accrued that posttranslational modification of p53 by phosphorylation of specific residues at the N-terminus can further regulate these interactions and activities. In general, it is argued that transcriptional regulation of p53 mRNA expression is not important; however, whether this is true in vivo remains an open question. Finally, it should be highlighted that, as well as DNA-sequence-specific (via the central DNA binding domain) binding, p53 can interact with DNA in a nonsequence-specific manner, and also with a very large number of cellular proteins. In many cases, the biological relevance of such interactions is not entirely clear.

COMPLEXITY OF THE p53 RESPONSE

An In Vivo Perspective

The widespread availability of well-characterized anti-p53 antibodies allowed pathologists in the early 1990s to provide a detailed compendium of p53 expression profiles in human tumors, and, in many cases, this correlated well with mutation of the *p53* gene *(6)*. Much less attention had been paid to the issue of the normal expression of p53 protein and the induction of p53 expression by homeostatic mechanisms. The work of Kastan et al. *(7)* and Lu and Lane *(8)* led to the investigation of the p53 response to DNA damage in vivo *(9)*. It was shown that low-dose (recreational) exposure to solar-simulated light was a potent inducer of p53 protein in normal human cells, without inducing discernible growth arrest or apoptosis. These studies were followed by collaborative investigation of the p53 response to ionizing radiation in the mouse, which included painstaking quantitation *(10)*. Taking these two studies together, it became very clear that, although p53 protein accumulation could be seen after genotoxic insult, the response was heterogeneous: Not all cells responded. This was essentially at odds with in vitro cell culture studies. Moreover, on the basis of conventional radiobiological target theory or (in the case of UV irradiation) direct experiment, it was clear that all cells had received very significant DNA damage *(11)*. Hence, even within a single tissue type (skin or gastrointestinal epithelium), there was quantitative variation in the p53 response.

These observations led to a systematic examination of the murine response to ionizing radiation in terms of p53 protein accumulation *(12•,13•)*, when ionizing radiation was employed because it gives identical dosage to all cells, without the vagaries of pharmacokinetics and metabolism seen with chemical agents. The critical observation was that "the coupling between ionizing radiation, p53 accumulation, and apoptosis (here a surrogate index for p53 activity) is dependent on cell type in vivo" *(12•)*. This was validated by biochemical assays as well as by immunostaining *(13•)*. Some populations of cells (e.g., gut, spleen, choroid plexus, bone marrow) were exquisitely sensitive to ionizing radiation, rapidly accumulating p53 protein and showing downstream events. Other populations (e.g., myocardium, adrenal, kidney, osteocytes of bone) showed accumulation of p53 protein, but there was no evidence of apoptosis. Finally, other tissues, such as brain, skeletal muscle, and liver, showed minimal induction of p53 protein. It should be clearly recognized that, in both of the first two groups, there was often extreme cell-to-cell variation within histological identical cellular populations. There is a rough correlation between these groupings and the categories of proliferative architecture defined by Leblond *(see* ref. *14)*, but this is by no means absolute. Although some data does suggest that the p53 response is cell-cycle-regulated *(15,16)*, this is not an entirely satisfactory explanation for the systematic heterogeneity observed: For example, consider the marked induction in the myocardium. Furthermore, when additional experiments relating to p53 activity (as opposed simply to level) were performed, even more heterogeneity was observed.

In order to evaluate p53 activity in vivo, as opposed simply to p53 level, three groups generated similar transgenic mice in which p53-responsive elements were linked to a reporter gene, β-galactosidase, so that the activity of p53 could be assayed on a cell-by-cell basis in vivo *(13•,17•,18•)*. In these mice, the heterogeneity of the p53 protein response was mirrored in terms of β-galactosidase activity. For example, in the gut, p53 protein accumulation and β-galactosidase correlated well. Similarly, the absence of radiation-induced p53 response in the liver was reflected by the absence of β-galactosidase activity. Elsewhere, however, the situation was more complex. For example, in the spleen, essentially all cells accumulate p53 protein *(12•)*, but there is considerable variation in the β-galactosidase response *(13•)*. In the myocardium, where there is accumulation of biochemically characterized p53 protein, there is no induction of β-galactosidase activity *(13•)*.

A further level of complexity is seen when different cellular insults are considered. For example, Bellamy et al. *(19)* have shown clearly that, in cultured hepatocytes, ionizing radiation does not induce p53 protein accumulation, but UV light does, and the liver in vivo can be shown to accumulate p53 protein after insult with certain chemical agents *(20)*. These data strikingly demonstrate that the heterogeneity manifest in biochemical studies is amplified when one begins to consider the in vivo perspective.

A Developmental Perspective

When considering the cell-type variation of the p53 response, the authors considered how the heterogeneity was established in development. For example, was the variation set up from a subset of cells, progressively acquiring the response from an unresponsive starting point, or was the reverse true, with all cells responding initially, and there being progressive restriction of the response during development? Early studies had suggested that p53 was regulated at either the mRNA or protein level during rodent development, and that it may have a role in development *(21)*. However, the generation of p53-null mice by several groups strongly argued against such a view, particularly given their very mild phenotype, but with high penetrance of neoplasia (proof of p53 having a tumor-suppressor role). However, several studies did demonstrate that the targeted expression of p53 in development could cause significant architectural and cellular defects. For example, using an MMTV promoter, Godley et al. *(22)* showed that there was profound failure of renal organogenesis. Similarly, expression in the developing lens could have dramatic consequences *(23)*. It has also shown that the absence of mdm-2 was not compatible with gestation beyond 6 d unless p53 was also absent, suggesting that the tight control of p53 and Mdm-2 in development was essential *(24)*. Finally, the groups that generated the original p53-null mice subsequently reported that there was a low, but very significant, level of developmental abnormality in the p53-null mice, with evidence of neural tube defects, particularly exencephaly *(25•,26•)*. Taken together, the above data suggest that, not only is p53 regulated in development, but it has some important roles to play. Direct evidence for this comes from the marked perturbation of *Xenopus* development by the overexpression of wild-type p53 *(27)*, or inactivation of p53 by the use of either dominant-negative p53 mutants or the *Xenopus* homolog of Mdm-2 *(28)*.

Examination of p53 protein expression in normal mouse development reveals that detectable protein is only seen in the neuraxis, and that this protein is biochemically active, as judged by activation of β-galactosidase in an identical distribution in the p53 reporter mice *(13•,17•,18•).* This profile of expression may be correlated with the abnormalities of the neuraxis reported in p53-null embryos *(25•, 26•).* When the authors examined the expression of p53 after irradiation of developing mice, a clear picture emerged: In the earliest times examined (d 8.5), all cells showed marked accumulation of p53 protein, but with progressive gestational age there was restriction to the profile seen in the adult. The kinetics of this change mirrored the well-established periods of particular radiosensitivity of mammalian development. How is the p53 response developmentally regulated? In both ES cells *(29,30)* and F9 teratocarcinoma cells *(31),* there are stockpiles of p53 protein present, but these appear to be inactive. In the case of ES cells, the protein is mostly cytoplasmic, and there appears to be a relative defect in nuclear transport. It is only on the induction of differentiation that activation of the pathway is effected, with resultant downstream sequelae. Such data are consonant with previous studies, suggesting that there is an effect of p53 on differentiation in several systems *(32–34).* The possible role of other transcription factors in regulating these properties is suggested by the potential role of PAX6 in controlling p53 expression in vivo *(35).* Another potential regulator of p53 function is Mdm-2 (and its homologs), and regulation of Mdm-2 expression in development has been reported in the rat *(36);* similar regulation is seen in mouse development (S.J. Campbell and P.A. Hall, unpublished). However, at present, there is no clear picture of how these different elements functionally interact.

A further level of p53 complexity has come from the recent recognition that there exists at least two family members exhibiting striking homology over the activation, DNA-binding, and oligomerization domains *(see* Fig. 2). They differ very much at the C-terminus, with both p73 *(37••)* and p63 *(38••)* having very long extensions, and p63 has a 40-amino-acid N-terminal extension. They appear to be closely related to mollusc sequences obtained by Friend (cited in ref. *39),* and it may be that p53 is the more recently evolved form. The existence of p53 homologs, of course, may explain the relatively mild phenotype of the p53-null mice (for example, with no detectable alteration of differentiation pathways known in cell culture to involve p53), because there are many examples in biology of at least partial functional complementation between family members. There is extreme conservation of critical

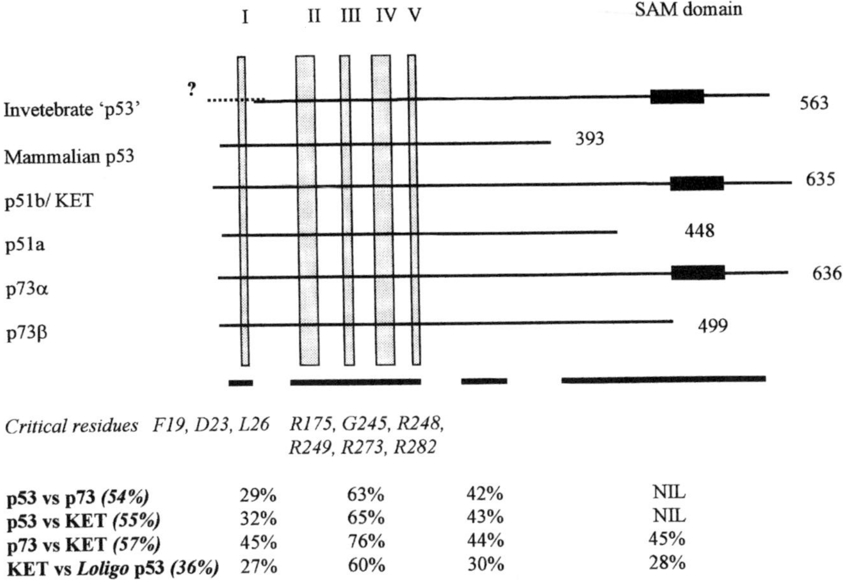

p53 vs p73 *(54%)*	29%	63%	42%	NIL
p53 vs KET *(55%)*	32%	65%	43%	NIL
p73 vs KET *(57%)*	45%	76%	44%	45%
KET vs *Loligo* p53 *(36%)*	27%	60%	30%	28%

Fig. 2. p53 homologs: similarities and differences. The recognition of possible homologs of p53 in the mammalian genome and the identification of a potential mollusk (invertebrate: squid, *Loligo*) homolog raise the notion of a family of p53 proteins. All have similarities in the conserved boxes, including conservation of critical contact residues in the N-terminal transactivation domain and in the DNA-binding domain. It is now clear that the *p73* and *p63* genes result in diverse polypeptides as a result of alternative promoter usage and alternate splicing as described by Yang et al. *(36b)*. Some of these species are illustrated here (adapted from ref. *1*), but other isoforms certainly exist. For example, alternate promoter usage gives rise to N terminal deletions of p73 in a similar manner to that shown for p63. Similarities in the oligomerization domain exist, with the figures representing percent identity. In the long C-terminal extension of p73α, p51b (KET), and squid sequence, there is a region of homology *(thick bar),* denoted a sterile alpha mutant (SAM) domain *(36a).* This is absent from the splice variants of p73 and p63.

residues in the Mdm-2 binding region, the SH3 domain, the DNA-binding domain, and the oligomerization domain. So much so that it seems probable that there will indeed be profound similarities in the biochemistry and biology of p53 family members. However, at present, very little is known about the properties of p73 or p63, but certainly p73 appears to be able to bind DNA and transactivate $p21^{WAF1}$, as well as induce apoptotic death *(40).* In the C-terminal extensions of both p63 and p73α, there is a SAM domain typical of protein–protein interaction regions of a number of developmentally regulated proteins. In p73-β, this domain is lost by alternate splicing. Indeed, as information accrues about p73 and p63 in man and other mammals, it seems that diverse isoforms are generated by alternate splicing. Nevertheless, one might speculate (based on the available data, and analysis of the sequences of p73 and p63, because of the crystal structure of p53 domains) that there may be potential for hetero-oligomerization, and a similar, but not necessarily identical, profile of DNA binding. It may be that this simply reflects a small fraction of the p53 family. Until more is known about these intriguing genes, there must be great caution in interpreting much of the data on p53 that has accrued over the past two decades. For example, it may be that the rather confusing data on the role of p53 in differentiation gleaned from overexpression studies in culture (and which should be contrasted with the absence of differentiation effects in p53-null mice) might be explained by influence of nonphysiological overexpression of p53, affecting the properties of p73 and/or p63 by the formation of hetero-oligomers, which are not normally seen.

BIOCHEMICAL PERSPECTIVES
ON IN VIVO COMPLEXITY

Recent rapid progress provides new insight into the possible mechanistic basis of the heterogeneity of the p53 response. The absolute level of p53 within a cell may be a critical control point. Low levels of p53 protein may, at least in culture systems, be actively antiapoptotic *(41••)*. At higher levels of protein growth, arrest phenomena appear to predominate, but at high levels apoptotic responses become manifest *(42••)*. The physiological effect of overexpression or inactivation is highlighted by the manipulations of *Xenopus* development by Hoever et al. *(27)* and Wallingford et al. *(28)*. How levels of p53 are regulated is currently a very active area. Certainly, the interaction with Mdm-2 is a critical point of regulation *(43)*. There is a general view that p53 is not controlled by the transcription of the p53 gene, but it has been reported that p53 is a target for its own transcriptional regulation *(44),* and, as is so often the case, the in vivo physiological relevance of this is uncertain. Another level of control on p53 seems to be translational regulation of nascent p53 mRNA *(45,46)*. The functional level of p53 protein may also be regulated by control of its subcellular localization, as emphasized by Moll et al. *(47),* and this is regulated in a cell-cycle-dependent manner *(15)*. In addition, the possibility that both p53 *(48)* and Mdm-2 *(49)* are actively shuttled in and out of the nucleus adds further levels of control to the system. Over and above the issue of p53 level and location, its effects are clearly cell-type-dependent. For example, there is increasing recognition that the profile of genes that p53 can transcriptionally regulate is cell-type-specific. For example, the authors have shown that the tyrosinase gene that encodes the rate-limiting step in melanogenesis is regulated by p53 (K. Nylander et al. 1999, submitted), and Nishimori et al. *(50)* have reported a brain-specific p53-regulated gene, *BAI1*. It makes sense that the adaptive responses that need to be elicited in different tissues (and cell types within tissues) will differ both quantitatively and qualitatively. In addition, it would seem that there is differential activity of wild-type and mutant p53 on target-gene promoters *(51,52):* Certainly there are many precedents for this in the field of transcriptional control. The recognition that p53 activates gene expression as part of a complex with p300 *(53),* and that interactions with human menopausal gonadotropin proteins can influence DNA binding *(54),* open up considerable scope for the further regulation of the p53 response. Furthermore, cofactors that can regulate p53 function, such as p33[ING1] *(55)* and JMY (N. LaThangue et al., submitted), have recently been reported. Finally, the upstream kinases that regulate p53 (and Mdm-2 *[56]*) activity are clearly of considerable potential utility for the cell to use as fine tuners of the p53 pathways function; these are not yet fully defined, but include DNA-dependent protein kinase *(57••,58••),* and possibly ataxia telangiectasia-mutated gene. The in vivo or physiological relevance of these burgeoning mechanisms of p53 regulation can only at present be speculated upon.

CLINICAL IMPLICATIONS OF p53 BIOLOGY

A large body of data has accrued suggesting that assays of p53, whether mutation analysis or expression studies, may provide clinically useful information, ranging from enhancing diagnostic precision *(59)* to allowing enhanced prognostication *(60)* or predicting therapeutic response. This area remains complex to interpret, with many studies of inadequate size or poorly designed or methodologically unsound. Nevertheless, there is a clear trend that, in at least some tumor types (for example, breast cancer), abnormal p53 function is associated with an adverse prognosis. Other studies, beginning with the seminal observations of Lowe *(61,62),* suggest that there may be a relationship between p53 function and sensitivity to therapeutic agents. This is reviewed in Chapters 2 and 3 in this book. Again, as they describe, the literature has become complex and difficult to interpret, perhaps in part because of the conflation of various tumor types and systems and assays. However, from the above discussion of the biological complexity of the normal p53 pathway, it can be seen why this might be so. The heterogeneity of the response between tissues and within cellular populations makes extrapolation difficult: Each cell type and, consequently, tumor type will have to be considered separately, until there is sufficient information to understand the mechanistic basis of the heterogeneity of the responses. Furthermore, it is well established that there are differing patterns of p53 overexpression in tumors, and there can be heterogeneity of mutation in tumor subclones. The former phenomena may reflect physiological responses to stresses acting on the tumor (hypoxia, metabolic disturbance, DNA damage, and so on) or pathological events, such as protein stabilization, contingent on p53 mutation. There may also be differentiation-related effects or reactivated developmental processes. At present, there is little mechanistic understanding of these phenomena in vivo, nor any understanding of how they impact on tumor behavior and response to therapy. Wynford-Thomas and Blaydes *(63)* have postulated that the differing involvement of p53 in histological subtypes of tumor may indeed reflect the diverse physiological roles of p53 in cell types and tissues. This may also be influenced by microanatomically regulated patterns of the expression profile of other p53 related genes: Certainly this should be investigated.

Many experiments have demonstrated that the introduction of functional p53 protein into cells can induce growth arrest and/or cell death. Indeed, this provides good evidence for the tumor-suppressor activities of p53, although the studies usually employ considerable overexpression of p53, and not the modest physiologically relevant levels normally seen in vivo. Consequently, there is considerable excitement and interest in the concept that manipulating the p53 pathway may be of therapeutic utility, and there are advocates of several strategies. For example, the use of adeno- and retroviral delivery systems to introduce wild-type p53 has been proposed. In addition, there has been interest in manipulation of mutant p53 protein, with the goal of reactivating wild-type activity. Certainly, in biochemical and cell culture assays, there is evidence that this approach can work. An alternative approach is to make use of the lack of function of p53 in tumor cells, and the need for certain viruses to inactivate p53 in order to proceed through a lytic cycle. In such circumstances, a mutant virus defective in the function of p53 inactivating ORFs will only replicate if p53 is already inactive. There is evidence that this approach may be clinically useful. An alternate strategy may be to target the normal cells in the body that limit the effectiveness of radio- and chemotherapy (bone marrow, gut, mucosa, and so on) and inhibit the p53-dependent apoptotic response here, hence, enhancing the potential therapeutic ration of current regimes. A limitation on this may be the existence of apoptotic pathways independent of p53. For example, even in p53-null mice, in which there is a clear relationship between apoptosis and p53 function *(10)*, at later time points, apoptosis is seen in the gut in the absence of p53 function *(64)*.

Despite the importance of p53 in human neoplasia and evidence that overexpression of p53 in cell lines in vitro can reverse the transformed phenotype, it may be quite inappropriate to conclude that restoration of p53 activity will necessarily be beneficial. A particularly striking example of this comes from the elegant studies of Ewald et al. *(65••)*. Using a tissue-specific regulatable promoter, p53 function was inactivated in vivo by the expression of the p53-binding domain of SV40 large T-antigen in salivary gland epithelium. When T-antigen is turned on (and hence p53 inactivated), there is progressive cellular hyperplasia and atypia. Turning T-antigen off (and hence restoring p53 function) leads to regression of these morphological lesions. In separate experiments, if T-antigen is left on for longer, there are more severe changes, which appear to be overtly neoplastic but benign. Again, these will regress if p53 function is restored by turning T-antigen off. However, if T-antigen is left on for longer, then overtly malignant tumors arise, which persist even in the absence of T-antigen expression: Presumably, other molecular events have occurred leading to the tumors persisting, irrespective of p53 function. Remembering that the clinical phase of tumors is short, compared with the life-span of the tumor, any manipulation of p53 will be late, and presumably in the milieu of multiple genetic changes and heterogeneity derived from the existence of multiple clonal sublines within the tumor.

One final perspective on the issue of the complex biology of p53: The p53 pathway has essential roles without its critical place as a tumor suppressor, which may be a more recent evolutionary development *(21)*. This broader perspective of p53 function emanates from the important observation of the role of p53 as a suppressor of teratogenesis *(66••,67••)*. Potential teratogens (ionizing radiation or benzapyrene) were shown to elicit far more embryopathy in the absence of a functioning p53 pathway than when it is functional, suggesting that p53 has an important (and possibly critical) role as a teratological suppressor. A fuller understanding of these issues will inevitably require a better understanding of p53 family physiology, and it may be that the p53 family members have important modulating roles in development and differentiation. The message of this chapter is that, although there has been tremendous progress in the understanding of the biochemistry of p53, and although there is a good overall picture of the pathway(s) in which it is involved, understanding of the biology of p53 in real physiological systems is remarkably crude and, until there is, it will be difficult to bridge the gap between bench and bedside. This must be the goal, and renewed effort should be placed on in vivo studies, because that may resolve the many apparent paradoxes that exist in the areas that form the focus of this book.

ACKNOWLEDGMENTS

The authors thank the members of the Hall Laboratory and, in particular, David MacCallum for invaluable comment and discussion. Work in the Hall Laboratory is supported by the Cancer Research Campaign, the Association for International Cancer Research, the Department of Health, the European Community, and the Pathological Society of Great Britain and Ireland. The authors apologize to the many hundreds of workers in the field whose work was not cited because of space constraints.

REFERENCES

- of special interest
- - of outstanding interest

1. Prives C, Hall PA. The p53 pathway. *J Pathol* 1999; 187: 98–116.
2. Ko LJ, Prives C. p53: puzzle and paradigm. *Genes Dev* 1996; 10: 1054–1072.
3. Levine AJ. p53, the cellular gatekeeper for growth and division. *Cell* 1997; 88: 323–331.
4. Agarwal ML, Taylor WR, Chernov MV, Chernova OB, Stark GR. The p53 network. *J Biol Chem* 1998; 273: 1–4.
••5. Harris CC. p53 tumor suppressor gene: from the basic research laboratory to the clinic—an abridged historical perspective. *Carcinogenesis* 1996; 17: 1187–1198. *Very valuable historical perspective.*
6. Hall PA, Lane DP. p53 in tumour pathology: can we trust immunohistochemistry. . . revisited! *J Pathol* 1994; 172: 1–4.
7. Kastan MB, Onyekwere O, Sidransky D, Vogelstein B, Craig RW. Participation of p53 protein in the cellular response to DNA damage. *Cancer Res* 1991; 51: 6304–6311.
8. Lu X, Lane DP. Differential induction of transcriptionally active p53 following UV or ionizing radiation: defects in chromosome instability syndromes? *Cell* 1993; 75: 765–778
9. Hall PA, McKee PH, du P Menage H, Dover R, Lane DP. High levels of p53 protein in UV-irradiated normal human skin. *Oncogene* 1993; 8: 203–207.
10. Merritt A, Potten CS, Kemp CJ, Hickman JA, Balmain A, Lane DP, Hall PA. Role of p53 in spontaneous and radiation induced apoptosis in the gastrointestinal tract of normal and p53 deficient mice. *Cancer Res* 1994; 54: 614–617.
11. Coates PJ, Save V, Ansari B, Hall PA. Demonstration of DNA damage/repair in individual cells using *in situ* end labelling: association of p53 with sites of DNA damage. *J Pathol* 1995; 176: 19–26.

•12. Midgley CA, Owens B, Briscoe CV, Thomas DB, Lane DP, Hall PA. Coupling between gamma-irradiation, p53 induction and the apoptotic response depends upon cell type in vivo. *J Cell Sci* 1995; 108: 1843–1841.
Important in highlighting the striking differences in the in vivo responses of cells to genotoxic insult, and demonstrating that the accumulation of p53 protein is cell-type-dependent.

•13. MacCallum DE, Hupp TR, Midgley CA , Stuart D, Campbell SJ, Harper A, et al. Te p53 response to ionising radiation in adult and developing murine tissues. *Oncogene* 1996; 13: 2575–2587.
Took the observations of Midgley et al. (12) a step further, in two ways. First, extending the evidence for heterogeneity of response by defining differential p53 activity in a transgenic mouse system with lacZ regulated by a p53-responsive element, and, second defining the developmental regulation of the p53 response. Similar studies were reported by Komarova et al. (17) and Gottlieb et al. (18)

14. Ansari B, Hall PA. Kinetic organisation of tissues. In: *Assessment of Cell Proliferation in Clinical Practice,* Hall PA, Levison DA, Gottlieb et al (18).Wright NA eds. Springer Verlag: Berlin. 1992; pp. 45–62.

15. Komarova EA, Zelnick CR, Chin D, Zeremski M, Gleiberman AS, Bacus SS, Gudkov AV. Intracellular localization of p53 tumor suppressor protein in gamma-irradiated cells is cell cycle regulated and determined by the nucleus. *Cancer Res* 1997; 57: 5217–5220.

16. Pitkanen K, Haapajarvi T, Laiho M. U.V.C.-induction of p53 activation and accumulation is dependent on cell cycle and pathways involving protein synthesis and phosphorylation. *Oncogene* 1998; 16: 459–469.

17. Komarova EA, Chernov MV, Franks R, Wang K, Armin G, Zelnick CR, et al. Transgenic mice with p53-responsive lacZ: p53 activity varies dramatically during normal development and determines radiation and drug sensitivity in vivo. *EMBO J* 1997; 16: 1391–1400.

•18. Gottlieb E, Haffner R, King A, Asher G, Gruss P, Lonai P, Oren M. Transgenic mouse model for studying the transcriptional activity of the p53 protein: age- and tissue-dependent changes in radiation-induced activation during embryogenesis. *EMBO J* 1997; 16: 1381–1390.

19. Bellamy CO, Prost S, Wyllie AH, Harrison DJ. UV but not gamma-irradiation induces specific transcriptional activity of p53 in primary hepatocytes. *J Pathol* 1997; 183: 177–181.

20. van Gijssel HE, Van Gijlswijk RP, De Haas RR, Stark C, Mulder GJ, Meerman JH. Immuno-histochemical visualization of wild-type p53 protein in paraffin-embedded rat liver using tyramide amplification: zonal hepatic distribution of p53 protein after N-hydroxy-2-acetylaminofluorene administration. *Carcinogenesis* 1998; 19: 219–222.

21. Hall PA, Lane DP. p53—a developing role? *Curr Biol* 1997; 7: 144–147.

22. Godley LA, Kopp JB, Eckhaus M, Paglino JJ, Owens J, Varmus HE. Wild-type p53 transgenic mice exhibit altered differentiation of the ureteric bud and possess small kidneys. *Genes Dev* 1996; 10: 836–850.

23. Nakamura T, Pichel JG, Williams-Simons L, Westphal H. Apoptotic defect in lens differentiation caused by human p53 is rescued by a mutant allele. *Proc Natl Acad Sci USA* 1995; 92: 6142–6146.

24. Montes DE Oca Luna R, Wagner DS, Lozano G. Rescue of early embryonic lethality in mdm2-deficient mice by deletion of p53. *Nature* 1995; 378: 203–206.

•25. Armstrong JF Kaufman MH, Harrison DJ, Clarke AR. High-frequency developmental abnormalities in p53-deficient mice. *Curr Biol* 1995; 5: 931–936.

•26. Sah VP, Attardi LD, Mulligan GJ, Williams BO, Bronson RT, Jacks T. Subset of p53-deficient embryos exhibit exencephaly. *Nat Genet* 1995; 10: 175–180. *These two papers provide evidence that p53 does indeed have a role in development as a significant proportion of p53 null mice do have developmental abnormalities.*

27. Hoever M, Clement JH, Wedlich D, Montenarh M, Knochel W. Overexpression of wild-type p53 interferes with normal development in Xenopus laevis embryos. *Oncogene* 1994; 9: 109–120.

28. Wallingford JB, Seufert DW, Virta VC, Vize PD. p53 activity is essential for normal development in Xenopus. *Curr Biol* 1997; 7: 747–757.

29. Sabapathy K, Klemm M, Jaenisch R, Wagner EF. Regulation of ES cell differentiation by functional and conformational modulation of p53. *EMBO J* 1997; 16: 6217–6229.

30. Aladjem MI, Spike BT, Rodewald LW, Hope TJ, Klemm M, Jaenisch R, Wahl GM. ES cells do not activate p53-dependent stress responses and undergo p53-independent apoptosis in response to DNA damage. *Curr Biol* 1998; 8: 145–155.

31. Lutzker SG, Levine AJ: Functionally inactive p53 protein in teratocarcinoma cells is activated by either DNA damage or cellular differentiation. *Nature Med* 1996; 2: 804–810.

32. Eizenberg O, Faber-Elman A, Gottlieb E, Oren M, Rotter V, Schwartz M. p53 plays a regulatory role in differentiation and apoptosis of central nervous system-associated cells. *Mol Cell Biol* 1996; 16: 5178–5185.

33. Soddu S, Blandino G, Scardigli R, Coen S, Marchetti A, Rizzo MG, et al. Interference with p53 protein inhibits hematopoietic and muscle differentiation. *J Cell Biol* 1996; 134: 193–204.

34. Ferreira A, Kosik KS. Accelerated neuronal differentiation induced by p53 suppression. *J Cell Sci* 1996; 109: 1509–1516.

35. Stuart ET, Haffner R, Oren M, Gruss P. Loss of p53 function through PAX-mediated transcriptional repression. *EMBO J* 1995; 14: 5638–5645.

36. Ibrahim AP, Gallimore PH, Grand RJ. Expression of MDM2 and other p53-regulated proteins in the tissues of the developing rat. *Biochim Biophys Acta* 1997; 1350: 306–316.

36a. Schultz J, Ponting CP, Hofmann K, Bork P. SAM as a protein interaction domain involved in developmental regulation. *Protein Sci* 1997; 6: 249–253.

36b. Yang A, Kaghad M, Wang Y, Gillett E, Fleming MD, Dotsch V, Andrews NC, Caput D, McKeon F. p63, a p53 homolog at 3q27–29 encodes multiple products with transactivating, death inducing, and dominant negative activities. *Mol Cell* 1998; 2: 305–316.

••37. Kaghad M, Bonnet H, Yang A, Creancier L, Biscan JC, Valent A, et al. Monoallelically expressed gene related to p53 at 1p36, a region frequently deleted in neuroblastoma and other human cancers. *Cell* 1997; 90: 809–819.

••38. Schmale H, Bamberger C. Novel protein with strong homology to the tumor suppressor p53. *Oncogene* 1997; 15: 1363–1367. *These two papers are important in defining the existence of p53 homologs in mammals, although, at present, we are ignorant of their functions and properties.*

39. Soussi T, May P. Structural aspects of the p53 protein in relation to gene evolution: a second look. *J Mol Biol* 1996; 260: 623–637.

40. Jost CA, Marin MC, Kaelin WG, Jr. p73 is a human p53-related protein that can induce apoptosis. *Nature* 1997; 389: 191–194.

••41. Lassus P, Ferlin M, Piette J, Hibner U. Anti-apoptotic activity of low levels of wild-type p53. *EMBO J* 1996; 15: 4566–4573.

••42. Chen X, Ko Lj, Jayaraman L, Prives C. p53 levels, functional domains, and DNA damage determine the extent of the apoptotic response of tumor cells. *Genes Dev* 1996; 10: 2438–2451.
 These two papers employ tetracycline regulated expression systems to titrate the level of p53 in cells, and to examine its functional consequences. Both elegantly demonstrate that p53 levels are critically important to its function.

43. Lane DP, Hall PA. MDM2: arbiter of p53's destruction. *Trends Biochem Sci* 1997; 22: 372–374.

44. Deffie A, Wu H, Reinke V, Lozano G. Tumor suppressor p53 regulates its own transcription. *Mol Cell Biol* 1993; 13: 3415–3423.

45. Mosner J, Mummenbrauer T, Bauer C, Sczakiel G, Grosse F, Deppert W. Negative feedback regulation of wild-type p53 biosynthesis. *EMBO J* 1995; 14: 4442–4449.

46. Fu L, Benchimol S. Participation of the human p53 3'UTR in translational repression and activation following gamma-irradiation. *EMBO J* 1997; 16: 4117–4125.

47. Moll UM, Ostermeyer AG, Haladay R, Winkfield B, Frazier M, Zambetti G. Cytoplasmic sequestration of wild-type p53 protein impairs the G1 checkpoint after DNA damage. *Mol Cell Biol* 1996; 16: 1126–1137.

48. Middeler G, Zerf K, Jenovai S, Thulig A, Tschodrich-Rotter M, Kubitscheck U, Peters R. Tumor suppressor p53 is subject to both nuclear import and export, and both are fast, energy-dependent and lectin-inhibited. *Oncogene* 1997; 14: 1407–1417.

49. Roth J, Dobbelstein M, Freedman DA, Shenk T, Levine AJ. Nucleo-cytoplasmic shuttling of the hdm2 oncoprotein regulates the levels of the p53 protein via a pathway used by the human immunodeficiency virus rev protein. *EMBO J* 1998; 17: 554–564.

50. Nishimori H, Shiratsuchi T, Urano T, Kimura Y, Kiyono K, Tatsumi K, et al. Novel brain-specific p53-target gene, BAI1, containing thrombospondin type 1 repeats inhibits experimental angiogenesis. *Oncogene* 1997; 15: 2145–2150.

51. Friedlander P, Haupt Y, Prives C, Oren M. Mutant p53 that discriminates between p53-responsive genes cannot induce apoptosis. *Mol Cell Biol* 1996; 16: 4961–4971.

52. Ludwig RL, Bates S, Vousden KH. Differential activation of target cellular promoters by p53 mutants with impaired apoptotic function. *Mol Cell Biol* 1996; 16: 4952–4960.

53. Lill NL, Grossman SR, Ginsberg D, Decaprio J, Livingston DM. Binding and modulation of p53 by p300/CBP coactivators. *Nature* 1997; 387: 823–827.

54. Jayaraman L, Moorthy NC, Murthy KGK, Manley JL, Bustin M, Prives C. High mobility group protein-1 (HMG-1) is a unique activator of p53. *Genes Dev* 1998; 12: 462–472.

55. Garkavtsev I, Grigorian IA, Ossovskaya VS, Chernov MV, Chumakov PM Gudkov AV. Candidate tumour suppressor p33ING1 cooperates with p53 in cell growth control. *Nature* 1998; 391: 295–298.

56. Mayo LD, Turchi JJ, Berberich SJ. Mdm-2 phosphorylation by DNA-dependent protein kinase prevents interaction with p53. *Cancer Res* 1997; 57: 5013–5016.

••57. Shieh Sy, Ikeda M, Taya Y, Prives C. DNA damage-induced phosphorylation of p53 alleviates inhibition by MDM2. *Cell* 1997; 91: 325–34.

••58. Siliciano JD, Canman CE, Taya Y, Sakaguchi K, Appella E, Kastan MB. DNA damage induces phosphorylation of the amino terminus of p53. *Genes Dev* 1997; 11: 3471–3481.
Although both these papers demonstrate, using phosphorylation site-specific antibodies, the importance of p53 phosphorylation in the regulation of its activity, the true physiological significance of this remains uncertain.

59. Dowell SP, Derias N, Wilson POG, Lane DP, Hall PA. Clinical utility of the immunocytochemical detection of p53 protein in cytological specimens. *Cancer Res* 1994; 54: 2914–2918.

60. Dowell SP, Hall PA. p53 tumour suppressor gene and tumour prognosis: is there a relationship? *J Pathol* 1995; 177: 221–224.

61. Lowe SW, Bodis S, Mcclatchey A, Remington L, Ruley HE, Fisher DE, Housman DE, Jacks T. p53 status and the efficacy of cancer therapy in vivo. *Science* 1994; 266: 807–810.

62. Lowe SW, Ruley HE, Jacks T, Housman DE. p53-dependent apoptosis modulates the cytotoxicity of anticancer agents. *Cell* 1993; 74: 957–967.

63. Wynford-Thomas D, Blaydes J. Influence of cell context on the selection pressure for p53 mutation in human cancer. *Carcinogenesis* 1998; 19: 29–36.

64. Merritt AJ, Allen TD, Potten CS, Hickman JA. Apoptosis in small intestinal epithelial from p53-null mice: evidence for a delayed, p53-independent G2/M-associated cell death after gamma-irradiation. *Oncogene* 1997; 14: 2759–2766.

••65. Ewald D, Li M, Efrat S, Auer G, Wall RJ Furth PA, Hennighausen L. Time-sensitive reversal of hyperplasia in transgenic mice expressing SV40 T antigen. *Science* 1996; 273: 1384–1386.
Elegant demonstration of the complexity of p53 function in vivo by the use of tissue-specific and inducible p53 expression in a transgenic mouse system. Raises important issues concerning the potential value of p53-based therapeutics.

••66. Nicol CJ, Harrison ML, Laposa RR, Gimelshtein IL Wells PG. Teratologic suppressor role for p53 in benzo[a]pyrene-treated transgenic p53-deficient mice. *Nat Genet* 1995; 10: 181–187.

••67. Norimura T, Nomoto S, Katsuki M, Gondo Y, Kondo S. p53-dependent apoptosis suppresses radiation-induced teratogenesis. *Nat Med* 1996; 2: 577–580.
These two papers provide clear evidence for the potential of p53 in physiological systems that are not related to tumor suppression. They both raise important issues, and one must question the assumption that the primary role of p53 is tumor suppression. This area is discussed in a recent review (see ref. 21).

Mismatch Repair Deficiency, Apoptosis, and Drug Resistance

Robert Brown, PhD

CONTENTS

ABSTRACT

The most clearly understood function of mismatch repair (MMR) is its role in the correction of mismatches occurring during DNA replication or DNA recombination. However, expression of MMR proteins is also now associated with sensitivity of mammalian cells to an ever-increasing range of DNA-damaging agents, and loss of MMR has been correlated with reduced ability of cells to undergo apoptosis, induced by a wide variety of clinically important chemotherapeutic drugs. The mechanisms by which MMR affects DNA damage-induced apoptosis are unclear. It has been proposed that certain types of DNA damage induce futile MMR activity and/or

From: *Apoptosis and Cancer Chemotherapy*
Edited by: J. A. Hickman and C. Dive © Humana Press Inc., Totowa, NJ

MMR-dependent replication stalling, which could lead to cell death. Alternatively, MMR may modulate the probability of bypassing DNA lesions by recombination-dependent mechanisms. Experimental models have implicated MMR as being necessary to activate (c-Jun NH2-terminal kinase), c-Abl nonreceptor tyrosine kinase, p53-dependent apoptosis, and G2 arrest in response to DNA damage. Thus, loss of MMR in cells may affect many signaling pathways, leading to important cellular responses to DNA damage, including apoptosis, although the relative contribution of any given pathway will depend on the cell type being examined. Correlation of MMR status with prognosis and response of tumors to treatment are beginning to emerge; however, larger prospective studies need to be done. If reduced apoptosis of patients' tumors is mediated by loss of MMR, this will provide new challenges for how to overcome drug resistance.

INTRODUCTION

Mismatch repair (MMR) plays an important role in maintaining the integrity of the genome, and in repairing mispaired bases in DNA *(1•)*. In addition, expression of MMR proteins is now associated with sensitivity of mammalian cells to an ever-increasing range of DNA-damaging agents *(2•)*, and loss of MMR has been correlated with reduced ability of certain tumor cells to undergo drug-induced apoptosis *(3•)*. Thus, loss of expression of MMR proteins is correlated with resistance, or tolerance, to methylating agents, 6-thioguanine, cisplatin (and carboplatin), doxorubicin, etoposide, and ionizing radiation. Clearly, many of these agents are clinically important in the treatment of cancer. How loss of MMR leads to drug resistance is not understood. Inactivation of MMR genes in mice causes cellular resistance to methylating agent and cisplatin *(4••,5)*. Restoration of MMR activity by chromosome transfer into MMR-defective human tumors leads to increased sensitivity to a range of drugs *(6)*. These observations argue against resistance being caused by increased mutations at other drug-resistance genes in these mutator cells, and supports a direct involvement of MMR proteins in sensitivity to DNA damage (Fig. 1). Thus, in mammalian cells, it has been proposed that certain types of DNA damage induce futile MMR activity *(7•)* or MMR-dependent replication stalling *(8•)*, which could lead to cell death. Recent data from this laboratory has raised a third possibility: MMR may modulate the probability of bypassing DNA lesions by recombination-dependent mechanisms *(60••)*. The absence of MMR in human ovarian tumor models correlates with loss of p53-dependent apoptosis *(3)*, although how MMR-generated signals lead to apoptosis is unclear. In MMR-deficient colon and endo-

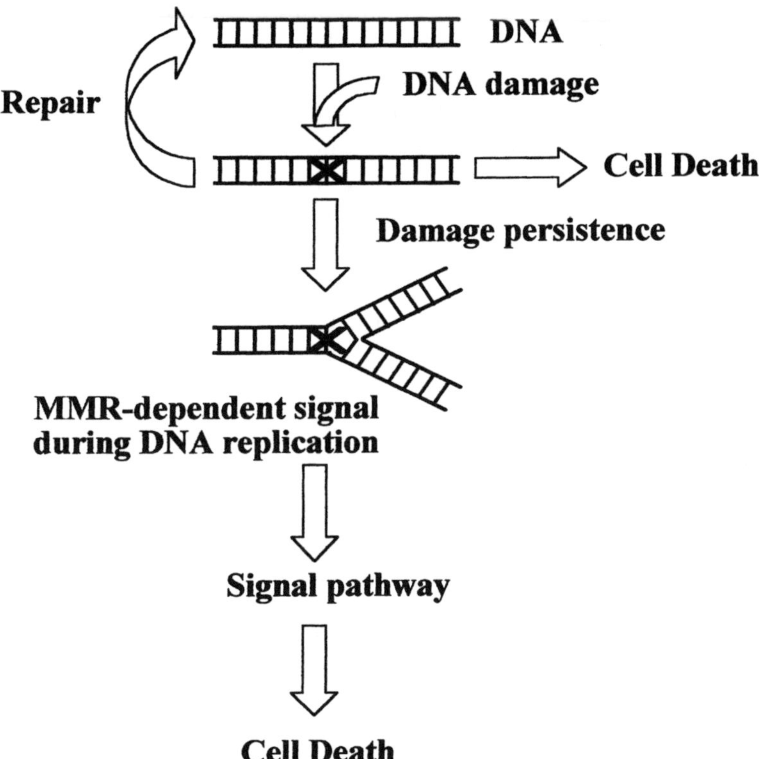

Fig. 1. MMR-dependent and -independent cell death. DNA damage can have a number of consequences affecting cell sensitivity. The damage can be rapidly repaired, with loss of repair leading to hypersensitivity. The damage can lead directly to cell death, such as double-strand break-induced apoptosis. Alternatively, certain types of DNA adducts are poorly recognized by repair pathways, and do not directly lead to cell death without DNA replication.

metrial tumor models, cisplatin activates c-Jun NH2-terminal kinase (JNK) and the c-Abl nonreceptor tyrosine kinase less efficiently than MMR-proficient cells *(9•)*.

Mutations in MMR genes occurs in the cancer-susceptibility syndrome, hereditary nonpolyposis colorectal carcinoma (HNPCC) *(10–12)*, which results in a predisposition to colorectal carcinoma, as well as to a number of other tumors, including adenocarcinomas of the endometrium, stomach, and ovary *(13)*. Mutations in the MMR genes *hMLH1, hMSH2, hPMS2,* and *hPMS1* have all been found to be associated with HNPCC; the vast majority of mutations are seen in either *hMLH1* or *hMSH2 (14)*. Transgenic mice with genetic inactivation of MMR genes have confirmed the tumor susceptibility associated with defects in *Mlh1, Msh2,* and *Pms2 (15),* although no tumor susceptibility was observed in *Pms1* knockouts *(16••).* One of the striking features of the resulting MMR-defective tumors is their greatly increased rate of mutation at microsatellite sequences, known as microsatellite instability (MIN+) *(17).* For many tumor types, the MIN+ phenotype has also been detected in sporadically occurring disease *(18):* For example, in ovarian cancer, about 15% of sporadic tumors exhibit the MIN+ phenotype. This suggests that these tumors are also MMR-defective, but mutations of MMR genes have only been observed at low frequency in MIN+ sporadic tumors *(19).* Recent evidence suggests that methylation of the promoter of MMR genes may play an important role in transcriptional silencing of MMR genes, either during acquisition of drug resistance *(61•)* or during tumorogenesis *(20•).*

MECHANISM OF HUMAN
STRAND-SPECIFIC MISMATCH REPAIR

The most clearly understood function of MMR is its role in the correction of mismatches occurring during DNA replication or DNA recombination *(1•).* The proposed model for postreplication MMR is based on work with DNA adenine methylation-instructed MMR pathway of *Escherichia coli (21••).* Thus, it is proposed that MutS protein recognizes and binds mispaired nucleotides that result from polymerase misincorporation errors. Initiation of the bacterial repair reaction occurs via mismatch-dependent incision of the newly synthesized, unmethylated strand at a hemimethylated d(GATC) site, and it is this nick in the new strand that actually directs repair. The nature of the strand signal has not been defined in any eukaryotic organism, but DNAs containing a nick are subject to mismatch-dependent, strand-specific repair by human cell extracts in vitro *(22).*

MutS and MutL homologs in mammalian cells exist primarily as heterodimeric proteins (Fig. 2). Thus, hMSH2 protein associates with hMSH3 and hMSH6. These complexes are referred to as MutSα (hMSH2/hMSH6) and MutSβ (hMSH2/hMSH3). Both complexes recognize short insertions of bases, but G/T mismatches appear to be primarily recognized by MutSα *(23).* Adenine nucleotide binding and hydrolysis by MutSα has been suggested to act as a molecular switch that determines the timing of downstream MMR events *(24•).* Thus, hMSH2–hMSH6 complex is ON (binds mismatched nucleotides) in the ADP-bound form and OFF in the ATP-bound form. A central role

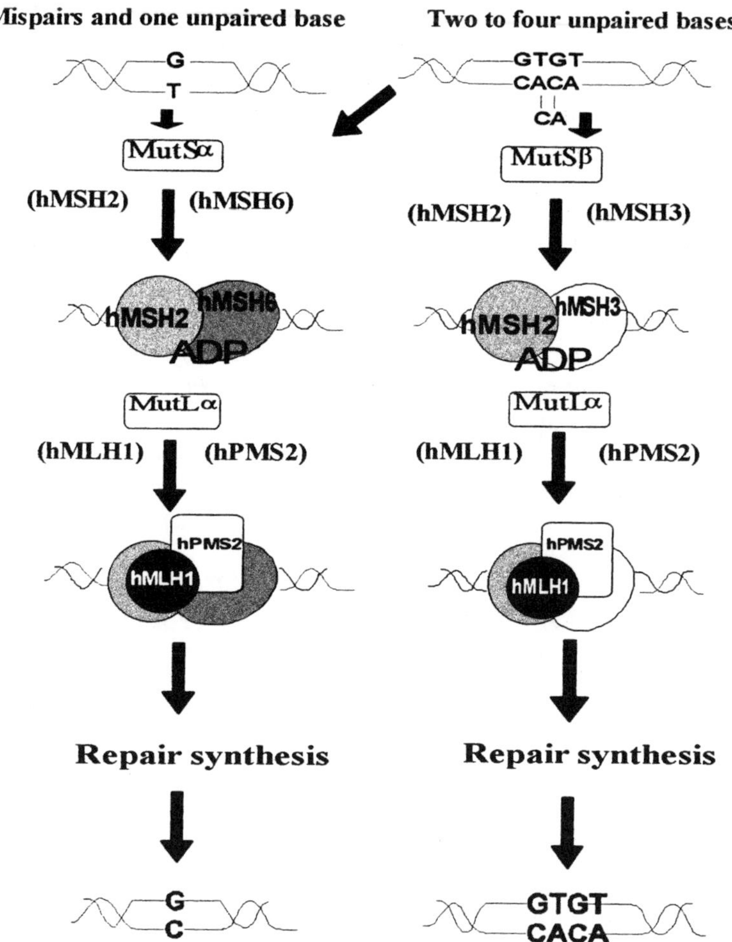

Fig. 2. Model of human MMR. MutSα (a heterodimer of hMSH2 and hMSH6) and MutSβ (a heterodimer of hMSH2 and hMSH3) recognize short insertions of bases, but G/T mismatches appear to be primarily recognized by MutSα. The hMSH2–hMSH6 complex binds mismatched nucleotides in the ADP-bound form, and not in the ATP-bound form. ATP hydrolysis has not yet been shown to be necessary for MutSβ binding. The hMLH1 protein forms a complex with hPMS2 referred to as MutLα. It has been proposed that MutS homologs binding to mismatched DNA allow recruitment of MutLα, which then allows MMR to proceed. Studies implicate Polδ, and possibly Polε, as well as PCNA in the repair DNA synthesis step of MMR.

for the adenine nucleotide-binding domain is consistent with the ATP-dependent translocation model of MMR proposed by Allen et al. *(25)*.

The MutL homologs also form heterodimers; thus, the hMLH1 protein forms a complex with hPMS2, referred to as MutLα. Defects in either MutL homolog leads to loss of mismatch correction at, or prior to, the excision stage of repair *(26)*. It has therefore been proposed that MutS homologs binding to mismatched DNA allows recruitment of MutL homologs, which then allows MMR to proceed. How this occurs is unclear.

Studies so far implicate the DNA polymerase Polδ, and possibly Polε, in the repair DNA synthesis step of MMR *(27)*. In *Saccharomyces cerevisiae*, proliferating cell nuclear antigen (PCNA) has also been suggested to participate in MMR *(28)*. PCNA is a required component of the eukaryotic replication apparatus, forming a homotrimeric sliding clamp around the helix, and increasing the processivity of DNA Polδ and Polε *(29)*.

HOW DOES MMR COUPLE TO APOPTOSIS AND LOSS OF MMR LEAD TO DRUG RESISTANCE?

Model 1: Futile Repair Cycles

The role of MMR in drug resistance was first identified in the alkylation tolerance phenotype observed in cells selected for resistance to monofunctional alkylating agents, such as *N*-methyl-*N*-nitro-*N*-nitro guanidine (MNNG) and *N*-methyl nitrosourea (MNU) *(30)*. The concept of tolerance is one in which the damaged base is not removed from the DNA, but appears unable to induce cytotoxic effects *(7)*. Agents that methylate DNA, such as MNNG and MNU, are mutagenic and cytotoxic as a result of the formation of O^6-methylguanine. Under normal circumstances, this lesion is directly repaired by the enzyme methyl transferase (MGMT). Functional loss of this enzyme, most commonly by epigenetic silencing, results in cellular hypersensitivity to killing and mutagenesis by such agents *(31)*. However, resistance (or tolerance) to methylating agents can occur because of inactivation of MMR *(32,33)*. Although the association between defective MMR and tolerance to certain methylating agents is clear, the mechanism underlying this phenomenon remains undefined. It has been proposed that these methylated bases are not a block to replication, but instead the DNA polymerase inserts the best-fitting base into the nascent strand opposite the modified guanine. It has been shown that O^6MeG·T introduces the least structural distortion to the DNA *(34•)*. In the absence of efficient methyltransferase function, this mismatched O^6MeG·T base pair is recognized by the MMR pathway as abnormal. Repair synthesis, which occurs in the newly synthesized strand opposite O^6MeG, is doomed to failure because of the inability to find a good complementary match for the methylated base. It is proposed that repeated, futile attempts to repair this mismatch eventually results in cell death, perhaps by the generation of strand breaks *(7)*. Tolerance would therefore arise when the MMR system could no longer initiate these aborted attempts at repair.

Model 2: Mismatch-Repair-Dependent Replication Stalling

The observation that loss of MMR correlated with acquisition of resistance to cisplatin opened up a much wider examination of the role of MMR in drug resistance beyond monofunctional methylating agents *(3)*. Loss of MMR is now associated with resistance to a diverse range of DNA-damaging agents *(2)*, and hence any model of resistance must take into account this wide variety of types of DNA damage. hMutSα recognizes 1,2-cisplatin crosslinks in a duplex DNA in which the complementary DNA strand contains two C residues opposite a 1,2-diguanyl crosslink *(35)*. However, this is a relatively poor substrate for hMutSα, and a duplex molecule, in which the platinated guanine residues are opposite noncomplementary bases, is bound with much greater affinity *(36)*. Such structures can arise in the cell if platinum-damaged DNA has undergone replication. Cellular proliferation, and hence, presumably, DNA replication is required for induction of apoptosis by cisplatin in sensitive cell types *(37)*. Certain cisplatin-resistant human ovarian cells, which have lost MMR protein expression, appear able to bypass cisplatin DNA intrastrand crosslinks during DNA replication *(38)*. The mechanisms leading to bypass are mostly unknown, but if loss of MMR leads to reduced replication stalling and increased bypass, then this could lead to resistance. In general, replicative bypass of 1,2-diguanyl cisplatin adducts has been considered inefficient, but the replicative polymerases δ and ε are indeed able to bypass these adducts in structures that resemble replication forks, although, so far, no role for MMR in this process has been demonstrated *(39)*. Thus, one possible mechanism of cisplatin toxicity is that inability to bypass this lesion in MMR-proficient sensitive cells has an intrinsic probability of being lethal, or of generating a signal that activates an apoptotic pathway. A direct corollary of this model is that cisplatin resistance may be acquired by reducing the probability of lethal events occurring, or of preapoptotic signals being generated during replication, allowing replication bypass and cell survival.

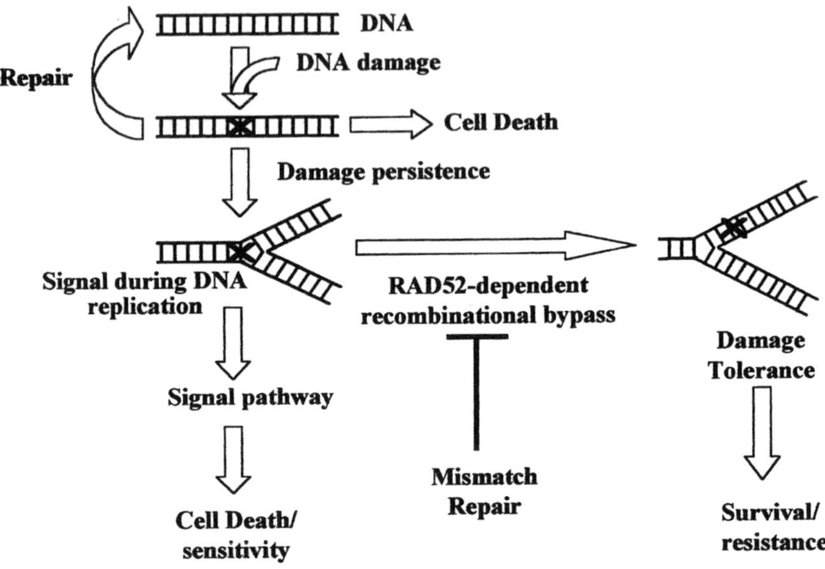

Fig. 3. Recombination-bypass model for drug resistance mediated by loss of MMR. Certain types of DNA damage induced by chemotherapeutic drugs such as cisplatin are poorly repaired, and may persist in the genome. The author proposes that signals are generated during DNA replication of this unrepaired damage that could lead to cell death, but have the potential to be bypassed in a RAD52-dependent manner that will lead to damage tolerance and cell survival. This recombinational bypass can be inhibited by MMR expression; thus, loss of MMR leads to increased drug resistance caused by increased bypass.

Model 3: Modulation of Recombination Bypass

Recently, the author et al. has shown that *S. cerevisiae* acquire resistance to cisplatin and carboplatin if MMR genes are inactivated (*60••*). Thus, genetic inactivation of MMR genes (*MLH1, MLH2, MSH2, MSH3, MSH6,* but not *PMS1*) in isogenic strains of *S. cerevisiae* leads to increased resistance to the anticancer drugs cisplatin and/or carboplatin, but has no effect on UVC sensitivity. However, inactivation of *MLH1, MLH2,* or *MSH2* has no significant effect on drug sensitivities in *rad52* mutant strains. Thus, drug resistance mediated by loss of MMR appears to require expression of RAD52, a protein known to be required for recombination *(40)*. The author proposes a model (Fig. 3) in which MMR proteins can modulate levels of adduct tolerance during DNA replication by a recombination-dependent mechanism, such as a sister chromatid exchange (SCE). Support for this hypothesis in human ovarian tumor cells is provided by loss of hMLH1, correlating with acquisition of cisplatin resistance and increased cisplatin-induced SCEs, both of which are reversed by restoration of hMLH1 expression *(60••)*. Furthermore, elevated somatic recombination has been shown in HeLa cells that are methylation-tolerant and defective in MMR compared to the parental line *(41)*.

As already discussed, bypass of DNA lesions during DNA replication has been suggested as a mechanisms for cisplatin-adduct tolerance *(38)*. The author proposes that loss of MMR proteins can lead to increased *RAD52*-dependent recombinational bypass of adducts *(see* Fig. 3). The 1,2-intrastrand crosslink induced by cisplatin is poorly repaired, either because of not being recognized by nucleotide excision repair (NER) *(42)* or by inhibition of repair, for instance, by damage-recognition proteins *(43)*. DNA damage persistence or nonrepaired DNA lesions could lead to a cytotoxic signal being generated during DNA replication, perhaps by stalling of the replication complex. RAD52-dependent recombinational bypass during replication would lead to resistance, but inhibition of this process by MMR would lead to sensitivity. It is known that MMR proteins can inhibit levels of recombination in yeast and mammalian cells *(41,44)*. Therefore, resistance may be acquired by loss of MMR proteins, reducing the probability of lethal signals being generated during replication by allowing increased recombination-dependent bypass and cell survival.

The experimental data discussed earlier, and all three of the models discussed above, predict that MMR is necessary to engage apoptosis in response to certain types of DNA damage. However, the signal pathways leading from MMR to apoptosis are as yet unclear. Cisplatin activates JNK and the c-Abl nonreceptor tyrosine kinase less efficiently than in MMR-deficient colon and endometrial tumor models in MMR-proficient cells *(9)*. Furthermore, in tumor cell models selected for cisplatin resistance in vitro, loss of MMR occurs concomitantly with loss of p53 function and ability to undergo p53-dependent apoptosis *(3)*. However, restoration of MMR activity, although restoring drug sensitivity, does not restore p53 function *(45)*. Loss of MMR correlates with loss of G2 arrest induced by certain drugs *(46)*, although there is no evidence that the loss of G2 arrest is functionally linked with the loss of drug-induced apoptosis. Thus, loss of MMR in cells affects many signaling pathways, leading to important cellular responses to DNA damage, including apoptosis, although the relative contribution of any given pathway may depend on the cell type being examined.

RELEVANCE OF LOSS
OF MISMATCH REPAIR FOR DRUG
RESISTANCE OF HUMAN TUMORS

A microsatellite instability (MIN)+ phenotype has been shown to correlate with reduced survival and poor disease prognosis in breast cancer *(47•)*. Conversely, MIN+ correlates with good prognosis in colon cancer *(48)*. These differences may reflect the different impact of a mutator phenotype on tumor progression (in the case of colon cancer) vs lack of MMR on drug sensitivity (in the case of breast cancer). MIN+ has been suggested to occur in 15–20% of sporadic ovarian tumors, and two separate studies have raised the possibility that this may have prognostic significance, although possible correlations with response to chemotherapy were not explored *(49,50)*. An increase in ovarian tumors has been observed that is immunologically negative for the hMLH1 subunit of the MMR MutLα heterodimer in samples taken at second-look laparotomy after chemotherapy (36%), compared to untreated tumors (9.5%) *(8)*. Together with the data observed in cell line models *(3,51)*, these observations encourage larger studies to be done in ovarian tumors correlating MMR with response to chemotherapy.

The acquisition of an MIN+ phenotype will lead to greater levels of genomic instability and higher mutation rates at genes throughout the genome. This has the possibility of increasing the mutation rate at genes involved in drug resistance or in tumor progression *(52)*. Thus, if a tumor does recur with resistant disease because of defects in MMR, it may be more likely to progress to a more advanced, aggressive neoplasm. Chemotherapeutic drugs, such as cisplatin, can induce hypermethylation in DNA of tumor cells *(53)*. The increasing evidence for methylation of MMR gene promoters in silencing transcription would suggest that chemotherapeutic agents that induce hypermethylation may also lead to increased inactivation of genes involved in drug resistance or tumor progression *(20)*.

CIRCUMVENTION OF DRUG
RESISTANCE MEDIATED
BY LOSS OF MISMATCH REPAIR

If reduced apoptosis and drug resistance of patients' tumors is mediated by loss of MMR, can new methods be devised to circumvent drug resistance?

Utilization of Drugs That Do Not Require Mismatch Repair to Engage Cell Death

The ideal would be to find an agent to which MMR-defective tumors are hypersensitive. So far, no such agent has been identified, although it has been suggested that the sensitivity to some drugs is not affected by MMR status *(2)*. For instance, although resistance to cisplatin and carboplatin can be mediated by loss of MMR, the sensitivity of colon tumor cells to the platinum analog, Oxaliplatin, appears not to be affected by MMR status *(54)*. This needs to be examined in a wider variety of experimental models. Oxaliplatin is presently undergoing clinical trials in patients with colon and ovarian cancer.

Inhibition of Replicative Bypass

If loss of MMR mediates resistance by increasing replicative or recombinational bypass of DNA lesions, then inhibition of this bypass could lead to resistance. A specific role for DNA polymerase ζ has been suggested in bypass of DNA lesions *(34)*. The DNA polymerase inhibitor, aphidocolin, has been shown to increase the sensitivity of cisplatin-resistant cell line models *(55)*. The original proposal was that this sensitization was caused by inhibition of DNA repair in resistant cells, but it would be worth re-evaluating this drug in models of defined MMR status.

Given that genetic inhibition of RAD52 in yeast sensitizes cells to cisplatin and carboplatin, and abrogates the resistance mediated by loss of MMR, these would be interesting targets for new sensitizers of chemotherapeutic drugs. However, proof of principle (that loss of RAD52 conferred increased sensitivity) would have to be established in human cells.

Reversal of Loss of MMR

Reversing the loss of MMR expression in resistant cells should sensitize tumors to anticancer drugs such as cisplatin and doxorubicin. The author has recently shown that 5-azacytidine treatment of cisplatin-resistant derivatives of ovarian tumor cells resulted in decreased methylation of the *hMLH1* promoter and increased expression of the hMLH1 protein, with a corresponding increase in cisplatin sensitivity *(61•)*. Thus, tumors that have lost hMLH1 expression because of promoter methylation may be especially sensitive to combined treatment with cisplatin and 5-azacytidine.

Genetic Therapy

Genetic therapy approaches may provide a new means of targeting tumors with specific genetic defects. Tumor specific expression of suicide gene vectors would be one means of killing tumor cells without affecting normal cells *(56)*. For instance, it may be possible to design suicide gene vector systems that utilize defects in MMR to deliver increased levels of a suicide gene to tumor cells. The ability of MMR-defective cells to tolerate damage and allow DNA replication to proceed could be utilized to design vectors containing specific types of damage, that the suicide-gene vector would replicate and express preferentially in MMR-defective cells, but not in normal cells. An E1B-attenuated adenovirus that replicates preferentially in tumor cells defective in p53 *(57)* has preferential replication, and prevents tumor growth in nude mice of ovarian tumor cells that are cisplatin-resistant and defective in MMR *(58)*. However, viral replication and oncolysis correlates with the loss of p53 function observed in these models, rather than with the loss of MMR *(45)*.

CONCLUSIONS

The past few years has seen an explosion in the understanding of basic mechanisms of MMR in human cells, and an increasing awareness of the possible role of MMR proteins in determining the sensitivity of tumor cells to a wide variety of important chemotherapeutic drugs. The precise mechanisms through which loss of MMR leads to drug resistance remain unclear. Novel approaches to modulate or target MMR-defective tumors need validation in animal models. It is also now essential to properly evaluate the relevance of MMR protein expression to clinical drug resistance, both intrinsic and acquired. The MIN status of tumors can be readily assayed using PCR assays and measured from not only tumor biopsies, but also from serum samples from patients *(59•)*. Such noninvasive means of monitoring the molecular changes in a tumor provide the exciting possibility of being able to determine the appropriate therapy for a given patient based on molecular genetic determinants of disease progression.

ACKNOWLEDGMENTS

The author would like to thank all the members, past and present, of the Molecular Mechanisms of Drug Resistance Team in Medical Oncology, Glasgow, and numerous collaborators from other laboratories who have been involved in our work on mismatch repair. Professor

S. B. Kaye (Glasgow) for constructive comments on the manuscript. The Cancer Research Campaign (UK) for funding the majority of our research.

REFERENCES

• of special interest
•• of outstanding interest

•1. Modrich P. Strand-specific mismatch repair in mammalian cells. *J Biol Chem* 1997; 272: 24,727–24,730.
   Excellent brief overview of biochemistry of MMR.

•2. Fink D, Aebi S, Howell SB. Role of DNA mismatch repair in drug resistance. *Clin Cancer Res* 1998; 4: 1–6.
   Review of status of drugs associated with MMR-mediated sensitivity.

•3. Anthoney DA, McIlwrath AJ, Gallagher WM, Edlin ARM, Brown R. Microsatellite instability, apoptosis and loss of p53 function in drug resistant tumour cells. *Cancer Res* 1996; 56: 1374–1381.
   Study of ovarian tumor models showing an association between loss of MMR and reduced apoptosis induced by cisplatin.

••4. De Wind N, Dekker M, Berns A, Radman M, Te Riele H. Inactivation of the mouse MSH2 gene results in mismatch repair deficiency, methylation tolerance, hyperrecombination and predisposition to cancer. *Cell* 1995; 82: 321–330.
   Analysis of mice and cells with genetic inactivation of MSH2 having methylation tolerance. Demonstrates that in cells with inactivated MSH2, homologous recombination has lost dependence on complete identity between interacting DNA sequences, suggesting that msh2 is involved in safeguarding the genome from promiscuous recombination.

5. Fink D, Nebel S, Aebi S, Nehme A, Howell SB. Loss of DNA mismatch repair due to knockout of MSH2 or PMS2 results in resistance to cisplatin and carboplatin. *Int J Oncol* 1997; 11: 539–542.

6. Aebi S, Fink D, Gordon R, Kim HK, Zheng H, Fink JL, Howell SB. Resistance to cytotoxic drugs in DNA mismatch repair-deficient cells. *Clin Cancer Res* 1997; 3: 1763–1767.

•7. Karran P, Hampson R. Genomic instability and tolerance to alkylating agents. *Cancer Surveys* 1996; 28: 69–85.
   Discussion of role of futile repair in MMR-mediated cell death induced by alkylating agents.

•8. Brown R, Hirst GL, Gallagher WM, McIlwrath AJ, Margison GP, Van Der Zee AG, Anthoney DA. hMLH1 expression and cellular responses of ovarian tumour cells to treatment with cytotoxic anticancer agents. *Oncogene* 1997; 15: 45–52.
   Correlation of loss of hMLH1 expression and drug resistance in ovarian tumor samples. Discussion of role of replication stalling in MMR-dependent engagement of apoptosis.

•9. Nehme A, Baskaran R, Aebi S, Fink D, Nebel S, Cenni B, et al. Differential induction of c-Jun-NH2 terminal kinase and c-abl kinase in DNA mismatch repair proficient and deficient cells exposed to cisplatin. *Cancer Res* 1997; 57: 3253–3257.

Cisplatin activates JNK kinase and c-Abl more efficiently in MMR-proficient cells.

10. Bronner CE, Baker SM, Morrison PT, Warren G, Smith LG, Lescoe MK, et al. Mutation in the DNA mismatch repair gene homologue hMLH1 is associated with hereditary non-polyposis colon cancer. *Nature* 1994; 368: 258–261.
11. Leach FS, Nicolaides NC, Papadopoulos N, Liu B, Jen J, Parsons R, et al. Mutations of a mutS homolog in hereditary nonpolyposis colorectal cancer. *Cell* 1993; 75: 1215–1225.
12. Nicolaides NC, Papadopoulos N, Liu B, Wei Y, Carter KC, Ruben SM, et al. Mutations of two PMS homologues in hereditary nonpolyposis colon cancer. *Nature* 1994; 371: 75–80.
13. Lynch HT. Genetics, Natural history, tumor spectrum and pathology of hereditary nonpolyposis colorectal cancer: an updated review. *Gasteroenterology* 1993; 104: 1535–1549.
14. Liu B, Parsons R, Papadopoulos N, Nicolaides NC, Lynch HT, Watson P, et al. Analysis of the mismatch repair genes in hereditary non-polyposis colorectal cancer patients. *Nature Medicine* 1996; 2: 169–174.
15. Reitmair AH, Redston M, Chun J, Chuang TCY, Bjerkness M, Cheng H, et al. Spontaneous intestinal carcinomas and skin neoplasms in MSH2-deficient mice. *Cancer Res* 1996; 56: 3842–3849.
••16. Prolla TA, Baker SM, Harris AC, Tsao J, Yao X, Bronner E, et al. Tumour susceptibility and spontaneous mutation in mice deficient in MLH1, Pms1 and Pms2 DNA mismatch repair. *Nature Genet* 1997; 18: 276–279.
Transgenic mice with genetic inactivation of MMR genes have confirmed the tumor susceptibility associated with defects in Mlh1 *and* Pms2, *although no tumor susceptibility was observed in* Pms1 *knockouts.*

17. Aaltonen LA, Peltomaki P, Leach PS, Sistonen P, Pylkkanen L, Maklin J, et al. Clues to the pathogenesis of Familial Colorectal Cancer. *Science* 1993; 260: 812–815.
18. Eshleman JR, Markowitz SD. Microsatellite instability in inherited and sporadic neoplasms. *Curr Opin Oncol* 1995; 7: 83–89.
19. Liu B, Nicolaides N, Markowitz S, Wilson JKV, Parsons RE, Jen J, et al. Mismatch repair defects in sporadic colorectal cancers with microsatellite instability. *Nature Genet* 1995; 9: 48–54.
•20. Kane MF, Loda M, Gaida GM, Lipman J, Mishra R, Goldman H, Jessup JM, Kolodner R. Methylation of the hMLH1 promoter correlates with lack of expression of hMLH1 in sporadic colon tumors and mismatch repair-defective tumor cell lines. *Cancer Res* 1997; 57: 808–811.
Methylation of promoter of hMLH1 may be a frequent mechanism of MMR deficiency in sporadic tumors.

••21. Modrich P, Lahue R. Mismatch repair in replication fidelity, genetic-recombination and cancer biology. *Ann Rev Biochem* 1996; 65: 101–133.
Excellent detailed overview of biochemistry and role of MMR.

22. Holmes J, Clark S, Modrich P. Strand-specific mismatch correction in nuclear extracts of human and *Drosophila melanogaster* cell lines. *Proc Natl Acad Sci USA* 1990; 87: 5837–5841.
23. Drummond JT, Li G, Longley MJ, Modrich P. Isolation of a hMSH2-p160 heterodimer that restores DNA mismatch repair in tumour cells. *Science* 1995; 268: 1909–1912.

•24. Gradia S, Acharya S, Fishel R. The human mismatch recognition complex hMSH2–hMSH6 functions as a novel molecular switch. *Cell* 1997; 91: 995–1005.
Adenine nucleotide binding and hydrolysis by MutSα suggested to act as a molecular switch that determines the timing of downstream MMR events. Thus, hMSH2–hMSH6 complex is ON (binds mismatched nucleotides) in the ADP-bound form and OFF in the ATP bound form.

25. Allen DJ, Makhov A, Grilley M, Taylor J, Thresher R, Modrich P, Griffith JD. MutS mediates hetcroduplex formation by a translocation mechanism. *EMBO J* 1997; 16: 4467–4476.

26. Parsons R, Li GM, Longley MJ, Fang WH, Papadopoulos N, Jen J, et al. Hypermutability and mismatch repair deficiency in RER$^+$ tumor cells. *Cell* 1993; 75: 1227–1236.

27. Longley MJ, Pierce AJ, Modrich P. DNA polymerase δ is required for human mismatch repair *in vitro*. *J Biol Chem* 1997; 272: 10,917–10,921.

28. Johnson RE, Kovvaki GK, Guzder SN, Amin NS, Holm C, Habraken Y, et al. Evidence for involvement of yeast proliferating cell nuclear antigen in DNA mismatch repair. *J Biol Chem* 1996; 45: 27,987–27,990.

29. Krishna TS, Kong X, Gray S, Burgers P, Kuriyan J. Crystal structure of the eukaryotic DNA polymerase processivity factor PCNA. *Cell* 1994; 79: 1233–1243.

30. Branch P, Aquilina G, Bignami M, Karran P. Defective mismatch binding and a mutator phenotype in cells tolerant to DNA damage. *Nature* 1993; 362: 652–654.

31. Hampson R, Humbert O, Macpherson P, Aquilina G, Karran P. Mismatch repair defects and O-6-methylguanine-DNA methyltransferase expression in acquired resistance to methylating agents in human cells. *J Biol Chem* 1997; 272: 28,596–28,606.

32. Kat A, Thilly WG, Fang W, Longley MJ, Li G, Modrich P. An alkylating-tolerant, mutator human cell line is deficient in strand-specific mismatch repair. *Proc Natl Acad Sci USA* 1993; 90: 6424–6428.

33. Koi M, Umar A, Chauhan DP, Cherian SP, Carethers JM, Kunkel TA, Boland CR. Human chromosome 3 corrects mismatch repair deficiency and microsatellite instability and reduces N-methyl-N′-nitro-N-nitrosoguanidine tolerance in colon tumor cells with homozygous hMLH1 mutation. *Cancer Res* 1994; 54: 4308–4312.

•34. Wood RD, Shivji MKK. Which DNA polymerases are used for DNA-repair in eukaryotes? *Carcinogenesis* 1997; 18: 605–610.
Excellent review of the role of DNA polymerases in DNA repair and adduct bypass.

35. Duckett DR, Drummond JT, Murchie AIH, Reardon JT, Sancar A, Lilley DM, Modrich P. Human MutSalpha recognizes damaged DNA base pairs containing O6-methylguanine, O4-methylthymine or the cisplatin-d(GpG) adduct. *Proc Natl Acad Sci USA* 1996; 93: 6443–6446.

36. Yamada M, O'regan E, Brown R, Karran P. Selective recognition of a cisplatin-DNA adduct by human mismatch repair proteins. *Nucleic Acids Res* 1997; 25: 491–495.

37. Evans DL, Tilby M, Dive C. Differential sensitivity to the induction of apoptosis by cisplatin in proliferating and quiescent immature rat thymocytes is independent of the level of drug accumulation and DNA adduct formation. *Cancer Res* 1994; 54: 1596–1603.

38. Mamenta EL, Poma EE, Kaufmann WK, Delmastro DA, Grady HL, Chaney SG. Enhanced replicative bypass of platinum-DNA adducts in cisplatin-resistant human ovarian carcinoma cell lines. *Cancer Res* 1994; 54: 3500–3505.

39. Hoffmann J, Pillaire M, Maga G, Podust V, Hubscher U, Villani G. DNA polymerase β bypasses *in vitro* a single d(GpG)-cisplatin adduct placed on codon 13 of the HRAS gene. *Proc Natl Acad Sci USA* 1995; 92: 5356–5360.

40. Petes TD, Malone RE, Syminton LS. Recombination in yeast. In *The Molecular and Cellular Biology of the Yeast Saccharomyces,* Broach JR, Pringle JR, Jones EW, eds. Cold Spring Harbor Laboratory, Cold Spring Harbor, NY. 1991: 407–521.

41. Ciotta C, Ceccotti S, Aquilina G, Humbert O, Palombo F, Jiricny J, Bignami M. Increased somatic recombination in methylation tolerant human cells with defective DNA mismatch repair. *J Mol Biol* 1998; 276: 705–719.

42. Moggs JG, Szymkowski DE, Yamada M, Karran P, Wood RD. Differential human nucleotide excision repair of paired and mispaired cisplatin-DNA adducts. *Nucleic Acids Res* 1997; 25: 480–490.

43. Mca'nulty MM, Lippard SJ. HMG-domain protein Ixr1 blocks excision repair of cisplatin-DNA adducts in yeast. *Mutat Res* 1996; 362: 75–86.

44. Alani E, Reenan RA, Kolodner RD. Interaction between mismatch repair and genetic recombination in *Saccharomyces cerevisiae. Genetics* 1994; 137: 19–39.

45. Brown R, Ganley I, Illand M, Kim YT. Preferential replication of an E1B-attenuated adenovirus in drug resistant ovarian tumour lines with defective mismatch repair and loss of p53. *Proc Am Assoc Cancer Res* 1998; 39: 555 (abstract).

46. Hawn MT, Umar A, Carethers JM, Marra G, Kunkel TA, Boland CR, Koi M. Evidence for a connection between the mismatch repair system and the G2 cell cycle checkpoint. *Cancer Res* 1995; 55: 3721–3725.

•47. Paulson TG, Wright FA, Parker BA, Russack V, Wahl GM. Microsatellite instability correlates with reduced survival and poor disease prognosis in breast cancer. *Cancer Res* 1996; 56: 4021–4026.
     Clinical correlation between defective MMR and poor prognosis in breast tumors.

•48. Bubb VJ, Curtis LJ, Cunningham C, Dunlop MG, Carothers AD, Morris RG, et al. Microsatellite instability and the role of hmsh2 in sporadic colorectal-cancer. *Oncogene* 1996; 12: 2641–2649.
     Clinical correlation between defective MMR and good prognosis in colorectal tumors.

49. King BL, Carcangiu ML, Carter D, Kiechle M, Pfisterer J, Pfleiderer A, Kacinski BM. Microsatellite instability in ovarian neoplasms. *Br J Cancer* 1995; 72: 376–382.

50. Fujita M, Enomoto T, Yoshino K, Nomura T, Buzard GS, Inoue M. Microsatellite instability and alterations in the hMSH2 gene in human ovarian cancer. *Int J Cancer* 1995; 64: 361–366.

51. Drummond JT, Anthoney A, Brown R, Modrich P. Cisplatin and adriamycin resistance are associated with MutLalpha and mismatch repair deficiency in an ovarian tumour cell line. *J Biol Chem* 1996; 271: 19,645–19,648.

52. Markowitz S, Wang J, Myeroff L, Parsons R, Sun L, Lutterbaugh J, et al. Inactivation of the type II TGF-beta receptor in colon cancer cells with microsatellite instability. *Science* 1995; 268: 1336–1338.

53. Nyce J. Drug-induced DNA hypermethylation and drug resistance in human tumors. *Cancer Res* 1989; 49: 5829–5836.

54. Fink D, Nebel S, Aebi S, Zheng H, Cenni B, Nehme A, Christen RD, Howell SB. Role of DNA mismatch repair in platinum drug resistance. *Cancer Res* 1996; 56: 4881–4886.

55. O'Dwyer PJ, Moyer JD, Suffness M, Harrison SD, Cysyk R, Hamilton TC, Plowman J. Antitumor activity and biochemical effects of aphidocolin glycinate (NSC 303812) alone and in combination with cisplatin in vivo. *Cancer Res* 1994; 54: 724–729.

56. Vile RG, Sunassee K, Diaz RM. Strategies for achieving multiple layers of selectivity in gene therapy. *Mol Med Today* 1998; 4: 84–92.

57. Bischoff JR, Kirn DH, Williams A, Heise C, Horn S, Muna M, et al. An adenovirus mutant that replicates selectively in p53-deficient human tumour cells. *Science* 1996; 274: 373–376.

58. Heise C, Sampson-Johannes A, Williams A, Mccormick F, Von Hoff DD, Kirn DH. Onyx-015, an E1B gene attenuated adenovirus, causes tumour-specific cytolysis and antitumoural efficacy that can be augmented by standard chemotherapeutic agents. *Nature Med* 1997; 3: 639–645.

•59. Nawroz H, Koch W, Anker P, Stroun M, Sidransky D. Microsatellite alterations in serum DNA of head and neck cancer patients. *Nature Med* 1996; 2: 1035–1037. *Description of a noninvasive means of monitoring MMR status of tumors.*

••60. Durant ST, Morris MM, Illand M, McKay HJ, McCormick C, Hirst GL, Borts RH, Brown R. Dependence on RAD52 and RAD1 for anticancer drug resistance mediated by inactivation of mismatch repair genes. *Curr. Biol,* 1999; in press. *Genetic analysis in* S. cerevisiae *on the role of MMR genes in cisplatin, carboplatin, and doxarubicin sensitivity. Demonstrates possible involvement of RAD52 and RAD1 on adduct bypass and proposes recombination-dependent bypass mechanisms.*

•61. Strathdee G, MacKean MJ, Illad M, Brown R. A role for methylation of the hMLM1 promoter in loss of hMLH1 expression and during resistance in ovarian cancer. *Oncogene,* 1999; in press. *Inactivation of* hMLH1 *gene expression by methylation of the promoter in ovarian tumor cells.*

6 Involvement of c-Abl Tyrosine Kinase in Apoptotic Response to Anticancer Agents

Zhi-Min Yuan, MD, PhD,
Surender Kharbanda, PhD,
Ralph Weichselbaum, MD,
and Donald Kufe, MD

CONTENTS

ABSTRACT

The cellular response to anticancer agents that damage DNA includes cell-cycle arrest, activation of DNA repair, and induction of apoptosis. However, the signals that determine cell fate, that is, survival or apoptosis, are largely unknown. The c-Abl tyrosine kinase is activated by diverse types of DNA damage. The available information supports a model in which certain sensors of DNA lesions activate the c-Abl kinase. Other findings support a proapoptotic function for the activated c-Abl kinase that is mediated by at least several downstream effectors known to be associated with the induction of apoptosis.

From: *Apoptosis and Cancer Chemotherapy*
Edited by: J. A. Hickman and C. Dive © Humana Press Inc., Totowa, NJ

INTRODUCTION

The response of cells to DNA damage includes cell-cycle arrest, activation of DNA repair, and, in the event of irreparable damage, the induction of apoptosis. The decision by a cell to repair DNA lesions and progress through the cell cycle, or to undergo apoptosis is of importance to mutagenesis and carcinogenesis. In this context, incomplete repair of DNA damage before replication or mitosis can result in the accumulation of heritable genetic changes. In treatment with anticancer agents that damage DNA, the balance between induction of repair or a cell-death program is also central to therapeutic effect. The signals activated in response to the detection of spontaneous DNA lesions are conceivably similar to those induced by exposure to genotoxic anticancer agents. However, the nature of these signals and the basis for a survival or cell-death response are largely unknown.

For genotoxic anticancer agents, the upstream signals responsible for activating cellular response mechanisms can be distinguished from early events associated with exposure to drugs that damage other cellular components. For example, the detection of DNA lesions needs to be converted into informational signals that regulate cell-cycle progression and activation of DNA repair. Also, the sensing that DNA lesions are irreparable needs to be converted into signals that activate a cell-death response. The available evidence is insufficient to determine whether events in the DNA-damage response differ with respect to the induction of a repair-survival or a cell-death pathway. What is needed, at least in part, to address the mechanisms employed in determining cell fate is a better definition of signals activated by DNA damage, and how those signals affect cell behavior. This chapter describes a role for the c-Abl tyrosine kinase in the response of cells to genotoxic anticancer agents, and how c-Abl executes a proapoptotic function.

ACTIVATION OF c-Abl
IN RESPONSE TO DNA DAMAGE

The c-Abl tyrosine kinase is a ubiquitously expressed 145-kDa protein that contains SH3, SH2, and SH1 (catalytic) domains in the N-terminal region *(1)*. Also contained in the C-terminus are nuclear localization motifs *(2)*, a bipartite DNA-binding domain *(3)*, and F- and G-actin-binding domains *(4,5)*. Alternative splicing results in the expression of two c-Abl isoforms (1a and 1b) *(6)* which are both detectable in the nucleus and cytoplasm. c-Abl shares homology with the mammalian c-*abl*-related gene (Arg) product *(7)* and proteins expressed in the sea urchin (E-Abl), fruit fly (D-Abl), and nematode

(N-Abl) *(8–10)*. Targeted disruption of c-Abl in mice is associated with normal fetal development, but the pups are runted and die as neonates with defective lymphopoiesis *(11•,12•)*. Other studies demonstrating that c-Abl-deficient mice exhibit defects in spermatogenesis have supported a functional role for c-Abl in meiosis *(13)*.

c-Abl activity is induced in cells exposed to the alkylating agents, cisplatin and mitomycin C *(14••)*. Similar findings have been obtained after treatment with the antimetabolite 1-β-D-arabinofuranosylcytosine (ara-C), and ionizing radiation (IR) *(14–16)*. Cisplatin forms DNA intrastrand crosslinks *(17)*. Mitomycin C forms monofunctional and bifunctional DNA lesions *(18)*. By contrast, ara-C incorporates into DNA and inhibits strand elongation *(19,20)*, and IR induces DNA strand breaks *(21,22)*. These findings have collectively supported the activation of c-Abl by diverse forms of DNA damage. Because c-Abl contains a DNA-binding domain *(3)*, direct interaction with damaged DNA could contribute to the induction of c-Abl activity. However, the interaction of c-Abl with sensors of DNA damage, such as the DNA-dependent protein kinase (DNA-PK), has lent support to a model in which c-Abl is a target for activation by an upstream effector.

DNA-PK consists of the 470-kDa catalytic subunit DNA-PK$_{cs}$ and the Ku heterodimer. The serine/threonine protein kinase activity of DNA-PK$_{cs}$ is induced by recruitment of DNA-PK to sites of DNA lesions through the DNA-binding protein Ku *(23)*. Ku binds to DNA ends and also associates with the carboxyl terminus of DNA-PK$_{cs}$ near the protein kinase domain *(23)*. Recent studies have demonstrated that DNA-PK$_{cs}$ associates constitutively with c-Abl *(24••)*. Significantly, DNA damage induces association of the c-Abl–DNA-PK$_{cs}$ complex with Ku, and the activation of c-Abl by a DNA-PK$_{cs}$-dependent mechanism *(24)*. In this context, DNA-PK phosphorylates the C-terminal region of DNA-PK$_{cs}$, and thereby disassociates DNA-PK$_{cs}$ from Ku *(23,24)*. Thus, Ku and c-Abl exhibit opposing effects on DNA-PK$_{cs}$ activity. These findings indicate that DNA-PK contributes to activation of c-Abl in the response to DNA damage, and that the phosphorylation of DNA-PK$_{cs}$ by c-Abl represents a potential feedback mechanism.

DNA-PK$_{cs}$ is related to members of the phosphatidylinositol (PI) 3-kinase family, which includes the product of the ataxia telangiectasia mutated (ATM) gene and the FKBP12-rapamycin-binding protein (FRAP) *(25)*. Other studies have demonstrated that c-Abl associates with the ATM protein, and that ATM may also contribute to activation of c-Abl in the response to DNA damage *(26••,27••)*. Although the precise mechanisms responsible for the activation of c-Abl remain unclear, the potential consequence of the interactions between c-Abl and DNA-PK or ATM may contribute to cell fate in the response to DNA damage. In this context, cells defective in DNA-PK or ATM are hypersensitive to the lethal effects of IR *(28–31)*, whereas c-Abl-deficient cells are resistant to IR-induced cell death *(32••)*. c-Abl-deficient cells are also resistant to killing by other anticancer agents, such as ara-C, which damage DNA *(33)*.

ROLE OF c-Abl
IN DNA-DAMAGE-INDUCED APOPTOSIS

Transient transfection of wild-type, but not a kinase-inactive, c-Abl is associated with the appearance of cells with sub-G1 DNA *(32)*. The finding that overexpression of c-Abl is also lethal has provided support for the induction of apoptosis. To assess a role for c-Abl in apoptosis, MCF-7 cells were used that stably express the dominant-negative c-Abl(K-R). MCF-7/c-Abl(K-R) cells exhibited resistance to killing by ara-C *(33)* and by IR *(32)*. By contrast, the MCF-7/c-Abl(K-R) cells exhibited little resistance to treatment with the protein kinase inhibitor staurosporine. Terminal deoxynucleotidyl transferase nick-end labeling (TUNEL) assays were used to confirm that ara-C- and IR-induced cell death was by apoptosis *(32,33)*. Moreover, c-Abl (K-R) expression decreased the appearance of ara-C- and IR-treated cells with sub-G1 DNA content *(32,33)*. To confirm the link between c-Abl and apoptosis, mouse embryo fibroblasts (MEFs) were studied that are deficient in c-Abl expression (abl$^{-/-}$ mice with targeted disruption of the *c-abl* gene) *(12)*. The Abl$^{-/-}$ MEFs exhibited resistance to the killing effects of ara-C and IR *(32,33)*. The finding that induction of cells with sub-G1 DNA content is decreased in the Abl$^{-/-}$ MEFs, compared to wild-type cells, has further supported a role for c-Abl in DNA-damage-induced apoptosis *(32,33)*.

c-Abl FUNCTIONS UPSTREAM
TO PROAPOPTOTIC EFFECTORS

Interaction Between c-Abl and p53

Treatment of cells with ara-C is associated with binding of c-Abl and the p53 tumor suppressor *(34)*. Similar observations have been obtained with the topoisomerase inhibitors, camptothecin and etoposide, and with IR exposure *(34,35••)*. These findings indicate that diverse agents that damage DNA activate c-Abl and induce the interaction of c-Abl with p53. However, the demonstration that c-Abl binds to p53 following treatment of the MCF-7/c-Abl(K-R) cells with genotoxic agents indicates that activation of the c-Abl kinase is not necessary for the binding of these proteins *(34,35)*. Importantly, c-Abl binds directly to p53 and enhances the transactivation function of p53 *(36)*. Moreover, the binding of c-Abl to p53 is abolished by deletion of a proline-rich domain within the C-terminus of c-Abl (ΔProl) *(36)*. The functional significance of the c-Abl–p53 interaction is supported by studies demonstrating that c-Abl induces G1 phase growth arrest by a p53-dependent mechanism *(35–37••)*. Other studies have indicated that c-Abl induces apoptosis by an interaction with p53 *(32)*. Transfection of c-Abl into cells that stably express the human papillomavirus E6 protein to promote degradation of p53 is associated with a decreased apoptotic response, compared to that obtained with expression of c-Abl in wild-type cells *(32)*. Also, stable expression of both c-Abl(K-R) and E6 resulted in resistance to DNA damage-induced apoptosis that was more pronounced than that obtained with expression of c-Abl(K-R) or E6 alone *(32)*. Other studies in p53$^{-/-}$ cells indicate that c-Abl induces apoptosis, at least in part, by a p53-dependent mechanism *(32)*.

c-Abl Functions Upstream to Stress-Induced Protein Kinases

The stress-activated protein kinases (SAPKs) are induced in cells treated with ara-C, alkylating agents, topoisomerase inhibitors, and IR *(38–43)*. These findings have indicated that, like c-Abl, diverse types of DNA damage induce the activation of SAPK. Other studies have shown that c-Abl functions upstream to SAPK in the response to ara-C, alkylating agents, and IR *(14,15)*. In concert with these findings, expression of an activated form of Abl induces SAPK activity *(15,44–46)*. The available evidence indicates that c-Abl interacts directly with MEKK1, an upstream effector of the SEK1/ SAPK pathway, in the response to DNA damage (unpublished data). Activation of SAPK by agents such as tumor necrosis factor (TNF), which act by mechanisms not involving DNA damage, occurs in c-Abl-deficient cells, and thereby through c-Abl-independent pathways *(14)*. The finding that diverse classes of DNA-damaging agents, including ara-C, cisplatin, and IR, also activate p38 mitogen-activated protein kinase (MAPK) *(47)* by a c-Abl-dependent mechanism *(48)* has supported the involvement of c-Abl in at least two stress-activated pathways. Both SAPK and p38 MAPK have been implicated in the induction of apoptosis *(42,43,49,50)*, and thereby may contribute to the proapoptotic function of c-Abl in the response to DNA damage.

c-Abl Inhibits Phosphatidylinositol 3-Kinase Activity

Phosphatidylinositol-3 kinase PI-3K is activated by growth factor receptors and has been implicated in the transduction of survival signals *(51)*. Inositol lipids phosphorylated at the D3 position by PI-3K are necessary for the activation of the Akt (protein kinase B) serine/threonine protein kinase *(52–55)*. The PI-3K–Akt pathway inhibits induction of apoptosis by the c-Myc protein *(56)* and by serum withdrawal *(57)*. These findings have indicated that signals that downregulate PI-3K would be proapoptotic. Autophosphorylation of PI-3K on serine inhibits PI-3K activity *(58,59);* other signals that negatively regulate this kinase are unknown. In this context, the finding that c-Abl interacts with DNA-PK$_{cs}$ and ATM, members of the PI-3K family, prompted studies on the interaction, if any, with PI-3K. The results demonstrate that c-Abl interacts directly with the p85 subunit of PI-3K *(60)*. Activation of c-Abl by treatment with IR or ara-C is associated with c-Abl-dependent phosphorylation of PI-3K. The results also show that phosphorylation of p85 by c-Abl inhibits PI-3K activity in vitro and in treated cells *(60)*. These findings support the hypothesis that activation of c-Abl by DNA damage downregulates PI-3K by phosphorylation of p85. Whereas the Akt/protein kinase B is a downstream effector of PI-3K that participates in the suppression of apoptosis, c-Abl could contribute to the regulation of cell fate through downregulation of PI-3K.

c-Abl Activates Proapoptotic Protein Kinase C δ

The protein kinase C δ (PKCδ) isoform is involved in the induction of growth arrest *(61)* and has been found to exhibit a tumor suppressor function *(62)*. Other studies in cells treated with ara-C, alkylating agents, or IR have shown that PKCδ is activated by proteolytic cleavage *(63,64)*. PKCδ is cleaved in the V3 region by caspase-3 to a 40-kDa, catalytically active fragment *(65)*. Moreover, overexpression of activated PKCδ in cells is associated with chromatin condensation, nuclear fragmentation, induction of sub-G1 DNA, and lethality *(65)*. Of the known PKC isoforms *(66)*, the ubiquitously expressed PKCδ *(67)* appears to be unique as a substrate for tyrosine phosphorylation *(68)*. Recent work has shown that PKCδ is constitutively associated with c-Abl *(69)*. The SH3 domain of c-Abl interacts directly with PKCδ; c-Abl phosphorylates and activates PKCδ in vitro. Also, treatment of cells with IR or ara-C is associated with c-Abl-dependent phosphorylation of PKCδ and translocation of PKCδ to the nucleus *(69)*. The significance of this event is supported by the finding that activated

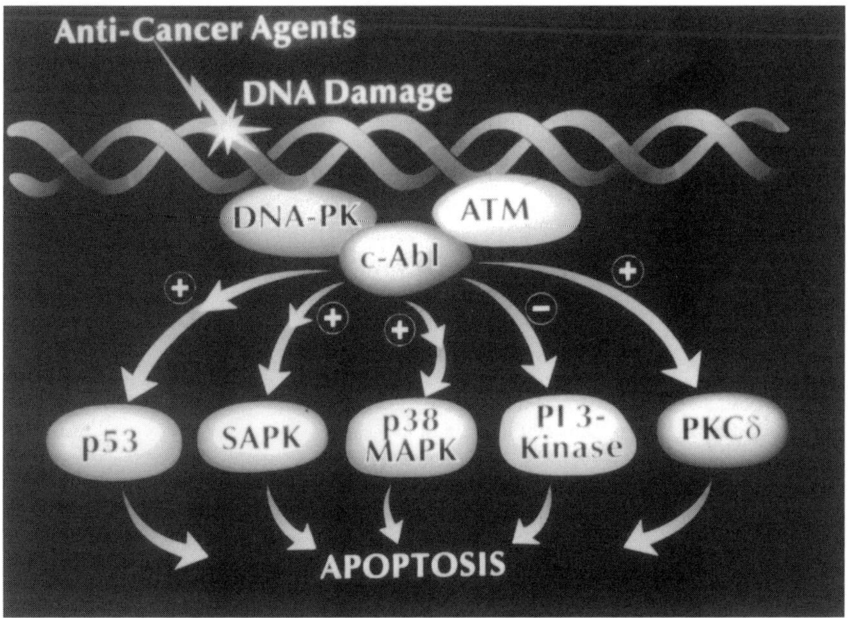

Fig. 1. Role for c-Abl as an upstream effector of proapoptotic signals.

PKCδ forms a nuclear complex with p53, and thereby may contribute to an apoptotic response (unpublished data). Further experimentation is needed to define the effects of nuclear PKCδ, the functional interaction between activated c-Abl and PKCδ in the response to DNA damage could potentiate the proapoptotic effects of c-Abl through induction of PKCδ activity.

CONCLUSIONS

An important question concerning treatment with genotoxic anticancer agents is how DNA damage is converted into intracellular signals that control cell fate. The available information supports a model in which sensors of DNA lesions, for example, DNA-PK and perhaps ATM, phosphorylate and thereby activate the c-Abl kinase (Fig. 1). The available data also supports a role for c-Abl in induction of an apoptotic response to DNA damage. Downstream effects of the activated c-Abl kinase include induction of p53-mediated transactivation, induction of SAPK and p38 MAPK, inhibition of PI-3K, and activation of PKCδ (Fig. 1). Because these downstream effectors of c-Abl have been associated with the induction of apoptosis, c-Abl may function in coordinating multiple events that each contribute to the cell-death response. Studies of c-Abl-dependent signaling mechanisms have been performed in cells treated with anticancer agents that act by damaging DNA. The physiologic function of c-Abl is therefore likely to reside in the elimination of cells that accumulate spontaneous, and irreparable, DNA lesions.

REFERENCES

• of special interest
•• of outstanding interest

1. Wang, JY: Abl tyrosine kinase in signal transduction and cell-cycle regulation. *Curr Opin Genet Dev* 1993; 3: 35–43.
2. Wen S-T, Jackson PK, Van Etten RA. Cytostatic function of c-Abl is controlled by multiple nuclear localization signals and requires the p53 and Rb tumor suppressor gene products. *EMBO J* 1996; 15: 1583–1595.
3. Kipreos ET, Wang JYJ. Cell cycle-regulated binding of c-Abl tyrosine kinase to DNA. *Science* 1992; 256: 382–385.
4. McWhirter JR, Wang JYJ. An actin-binding function contributes to transformation by the Bcr-Abl oncoprotein of Philadelphia chromosome-positive human leukemia. *EMBO J* 1993; 12: 1533–1546.
5. Van Etten RA, Jackson PK, Baltimore D, Sanders MC, Matsuddaira PT, Janmey PA. COOH terminus of the c-Abl tyrosine kinase contains distinct F- and G-actin binding domains with bundling activity. *J Cell Biol* 1994; 124: 325–340.
6. Ben-Neriah Y, Bernards A, Paskind M, Daley GQ, Baltimore D. Alternative 5′ exons in c-abl mRNA. *Cell* 1986; 44: 577–586.
7. Kruh GD, Perego R, Miki T, Aaronson SA. Complete coding sequence of arg defines the Abelson subfamily of cytoplasmic tyrosine kinases. *Proc Natl Acad Sci USA* 1990; 87: 5802–5806.
8. Henkemeyer MJ, Bennett RL, Gertler FB, Hoffmann FM. DNA sequence, structure, and tyrosine kinase activity of the Drosophila melanogaster Abelson proto-oncogene homolog. *Mol Cell Biol* 1988; 8: 843–853.
9. Goddard JM, Weiland JJ, Capecchi MR. Isolation and characterization of *Caenorhabditis elegans* DNA sequences homologous to the v-abl oncogene. *Proc Natl Acad Sci USA* 1986; 83: 2172–2176.
10. Moore KL, Kinsey WH. Identification of an abl-related protein tyrosine kinase in the cortex of the sea urchin egg: possible role at fertilization. *Dev Biol* 1994; 164: 444–455.
•11. Schwartzberg PL, Stall AM, Hardin JD, Bowdish KS, Humaran T, Boast S, et al. Mice homozygous for the ablT^{m1} mutation show poor viability and depletion of selected B and T cell populations. *Cell* 1991; 65: 1165–1175.
•12. Tybulewicz VLJ, Crawford CE, Jackson PK, Bronson RT, Mulligan RC. Neonatal lethality and lymphopenia in mice with a homozygous disruption of the c-abl proto-oncogene. *Cell* 1991; 65: 1153–1163.
13. Kharbanda S, Pandey P, Morris P, Whang Y, Xu Y, Sawantw S, et al. Functional role for the c-Abl tyrosine kinase in meiosis I. *Oncogene* 1998; 16: 1773–1777.

••14. Kharbanda S, Ren R, Pandey P, Shafman TD, Feller SM, Weichselbaum RR, Kufe DW. Activation of the c-Abl tyrosine kinase in the stress response to DNA-damaging agents. *Nature* 1995; 376: 785–788.
Provided the first evidence that c-Abl is activated by DNA-damaging agents.

15. Kharbanda S, Pandey P, Ren R, Feller S, Mayer B, Zon L, Kufe D. c-Abl activation regulates induction of the SEK1/stress activated protein kinase pathway in the cellular response to 1-β-D-arabinofuranosylcytosine. *J Biol Chem* 1995; 270: 30,278–30,281.

16. Liu Z-G, Baskaran R, Lea-Chou ET, Wood L, Chen Y, Karin M, Wang JYJ. Three distinct signalling responses by murine fibroblasts to genotoxic stress. *Nature* 1996; 384: 273–276.

17. Sherman SE, Lippard SJ. Structural aspects of platinum anticancer drug interaction with DNA. *J Chem Rev* 1987; 87: 1153–1181.

18. Tomasz M, Chawla AK, Lipman R. Mechanism of monofunctional and bifunctional alkylation of DNA by mitomycin C. *Biochemistry* 1988; 27: 3182–3187.

19. Kufe DW, Beardsley G, Major P, Egan EM. Correlation of cytotoxicity with incorporation of ara-C into DNA. *J Biol Chem* 1980; 255: 8997–9000.

20. Kufe D, Munroe D, Herrick D, Spriggs D. Effects of ara-C incorporation on eukaryotic DNA template function. *Mol Pharmacol* 1984; 26: 128–134.

21. Kemp LM, Sedgwick SG, Jeggo PA. X-ray sensitive mutants of Chinese hamster ovary cells defective in double-strand break rejoining. *Mutation Res* 1984; 132: 189–196.

22. Radford IR. Level of induced DNA double-strand breakage correlates with cell killing after X-irradiation. *Intl J Rad Biol* 1985; 48: 45–54.

23. Jin S, Kharbanda S, Mayer B, Kufe D, Weaver DT. Binding of Ku and c-Abl at the kinase homology region of DNA-dependent protein kinase catalytic subunit. *J Biol Chem* 1997; 272: 24,763–24,766.

••24. Kharbanda S, Pandey P, Jin S, Inoue S, Bharti A, Yuan Z-M, et al. Functional interaction of DNA-PK and c-Abl in response to DNA damage. *Nature* 1997; 386: 732–735.
Demonstrated that DNA-PK, a sensor of DNA damage, functions in the phosphorylation and partial activation of c-Abl.

25. Hartley KO, Gell D, Smith GCM, Zhang H, Divecha N, Connelly MA, et al. DNA-dependent protein kinase catalytic subunit: a relative of phosphatidylinositol 3-kinase and the ataxia telangiectasia gene product. *Cell* 1995; 82: 849–856.

••26. Shafman T, Khanna KK, Kedar P, Yen T, Spring K, Kozlov S, et al. Role of ATM protein in stress response to DNA damage: evidence for interaction with c-Abl. *Nature* 1997; 387: 520–523.
This paper and the following one, by Baskaran et al., demonstrate that the ATM protein interacts with, and contributes to, the activation of c-Abl. Thus, cells from AT patients exhibit defects in c-Abl activation in the response to ionizing radiation.

••27. Baskaran R, Wood LD, Whitaker LL, Xu Y, Barlow C, Canman CE, et al. Ataxia telangiectasia mutant protein activates c-abl tyrosine kinase in response to ionizing radiation. *Nature* 1997: 387: 516–519.

28. Jeggo PA, Taccioli GE, Jackson SP. Menage a trois: double strand break repair, V(D)J recombination and DNA-PK. *Bioessays* 1995; 17: 949–957.

29. Roth DB, Lindahl T, Gellert M. How to make ends meet. *Curr Biol* 1995; 5: 496–499.

30. McKinnon P. Ataxia-telangiectasia: an inherited disorder of ionizing-radiation sensitivity in man. *Hum Genet* 1987; 75: 197–208.

31. Taylor AMR, Harnden DG, Arlett CF, Harcourt AR, Lehmann AR, Stevens S, Bridges BA. Ataxia-telangiectasia: a human mutation with abnormal radiation sensitivity. *Nature* 1975; 258: 427–429.

32. Yuan Z-M, Huang Y, Ishiko T, Kharbanda S, Weichselbaum R, Kufe D. Regulation of DNA damage-induced apoptosis by the c-Abl tyrosine kinase. *Proc Natl Acad Sci USA* 1997; 94: 1437–1440.

33. Huang Y, Yuan ZM, Ishiko T, Nakada S, Utsugisawa T, Kato T, Kharbanda S, Kufe DW. Pro-apoptotic effect of the c-Abl tyrosine kinase in the cellular response to 1-β-D-arabinofuranosylcytosine. *Oncogene* 1997; 15: 1947–1952.

34. Yuan Z-M, Huang Y, Fan MM, Sawyers C, Kharbanda S, Kufe D. Genotoxic drugs induce interaction of the c-Abl tyrosine kinase and the tumor suppressor protein p53. *J Biol Chem* 1996; 271: 26,457–26,460.

••35. Yuan ZM, Huang Y, Whang Y, Sawyers C, Weichselbaum R, Kharbanda S, Kufe D. Role for the c-Abl tyrosine kinase in the growth arrest response to DNA damage. *Nature* 1996; 382: 272–274.
Showed that c-Abl contributes to the activation of p53 and G1 arrest in the response to DNA damage.

36. Goga A, Liu X, Hambuch TM, Senechal K, Major E, Berk AJ, Witte ON, Sawyers CL. p53 dependent growth suppression by the c-Abl nuclear tyrosine kinase. *Oncogene* 1995; 11: 791–799.

••37. Sawyers CL, McLaughlin J, Goga A, Havilik M, Witte O. Nuclear tyrosine kinase c-Abl negatively regulates cell growth. *Cell* 1994; 77: 121–131.
Demonstrated that c-Abl functions in cell-cycle arrest by a mechanism involving c-Abl kinase activity, nuclear localization, and an intact SH2 domain.

38. Saleem A, Datta R, Kharbanda S, Kufe D. Involvement of stress activated protein kinase in the cellular response to 1-β-D-arabinofuranosylcytosine and other DNA-damaging agents. *Cell Growth Differ* 1995; 6: 1651–1658.

39. Zanke BW, Boudreau K, Rubie E, Winnett E, Tibbles LA, Zon L, et al. Stress-activated protein kinase pathway mediates cell death following injury induced by cis-platinum, UV irradiation or heat. *Curr Biol* 1996; 6: 606–613.

40. Kharbanda S, Saleem A, Shafman T, Emoto Y, Weichselbaum R, Woodgett J, et al. Ionizing radiation stimulates a Grb2-mediated association of the stress activated protein kinase with phosphatidylinositol 3-kinase. *J Biol Chem* 1995; 270: 18,871–18,874.

41. Chen Y-R, Meyer CF, Tan T-H. Persistent activation of c-Jun N-terminal kinase 1 (JNK1) in γ radiation-induced apoptosis. *J Biol Chem* 1996; 271: 631–634.

42. Chen Y-R, Wang X, Templeton D, Davis RJ, Tan T-H. Role of c-Jun N-terminal kinase (JNK) in apoptosis induced by ultraviolet C and γ radiation. *J Biol Chem* 1996; 271: 31,929–31,936.

43. Verheij M, Bose R, Lin XH, Yhao B, Jarvis WD, Grant S, et al. Requirement for ceramide-initiated SAPK/JNK signaling in stress-induced apoptosis. *Nature* 1996; 380: 75–79.

44. Sanchez I, Hughes RT, Mayer BJ, Yee K, Woodgett JR, Avruch J, Kyriakis JM, Zon LI. Role of SAPK/ERK kinase-1 in the stress-activated pathway regulating transcription factor c-Jun. *Nature* 1994; 372: 794–798.

45. Raitano AB, Halpern JR, Hambuch TM, Sawyers CL. Bcr-Abl leukemia oncogene activates Jun kinase and requires Jun for transformation. *Proc Natl Acad Sci USA* 1995; 92: 11,746–11,750.

46. Renshaw MW, Lea-Chou E, Wang JYJ. Rac is required for v-Abl tyrosine kinase to activate mitogenesis. *Curr Biol* 1996; 6: 76–83.

47. Raingeaud J, Gupta S, Rogers JS, Dickens M, Han J, Ulevitch RJ, Davis RJ. Pro-inflammatory cytokines and environmental stress cause p38 MAP kinase activation by dual phosphorylation on tyrosine and threonine. *J Biol Chem* 1995; 270: 7420–7426.

48. Pandey P, Raingeaud J, Kaneki M, Weichselbaum R, Davis R, Kufe D, Kharbanda S. Activation of p38 MAP kinase by c-Abl-dependent and -independent mechanisms. *J Biol Chem* 1996; 271: 23,775–23,779.

49. Xia Z, Dickens M, Raingeaud J, Davis RJ, Greenberg ME. Opposing effects of ERK and JNK-p38 MAP kinases on apoptosis. *Science* 1995; 270: 1326–1331.

50. Johnson NL, Gardner AM, Diener KM, Lange-Carter CA, Gleavy J, Jarpe MB, et al. Signal transduction pathways regulated by mitogen-activated/extracurricular response kinase kinase kinase induce cell death. *J Biol Chem* 1996; 271: 3229–3237.

51. Kapeller R, Cantley LC. Phosphatidylinositol 3-kinase. *Bioassays* 1994; 16: 565–576.

52. Franke TF, Kaplan DR, Cantley LC, Toker A. Direct regulation of the *Akt* proto-oncogene product by phosphatidylinositol-3,4-bisphosphate. *Science* 1997; 275: 665–668.

53. Dudek H, Datta SR, Franke TF, Birnbaum MJ, Yao R, Cooper GM, et al. Regulation of neuronal survival by the serine–threonine protein kinase Akt. *Science* 1997; 275: 661–665.

54. Alessi DR, Andjelkovic M, Caudwell B, Cron P, Morrice N, Cohen P, Hemmings BA. Mechanism of activation of protein kinase B by insulin and IGF-1. *EMBO J* 1996: 15: 6541–6551.

55. Klippel A, Kavanaugh WM, Pot D, Williams LT. Specific product of phosphatidylinositol 3-kinase directly activates the protein kinase Akt through its pleckstrin homology domain. *Mol Cell Biol* 1997; 17: 338–344.

56. Kaufmann-Zeh A, Rodriguez-Viciana P, Ulrich E, Gilbert C, Coffer P, Downward J, Evan G. Suppression of c-Myc-induced apoptosis by Ras signalling through PI(3)K and PKB. *Nature* 1997; 385: 544–548.

57. Kennedy SG, Wagner AJ, Conzen SD, Jordan J, Bellacosa A, Tsichlis PN, Hay N. PI 3-kinase/Akt signaling pathway delivers an anti-apoptotic signal. *Genes Dev* 1997; 11: 701–713.

58. Carpenter CL, Auger KR, Duckworth BC, Hou WM, Schaffhausen B, Cantley LC. Tightly associated serine/threonine protein kinase regulates phosphoinositide 3-kinase activity. *Mol Cell Biol* 1993; 13: 1657–1665.

59. Dhand R, Hiles I, Panayotou G, Roche S, Fry JM, Gout I, et al. PI 3-kinase is a dual specificity enzyme: autoregulation by an intrinsic protein-serine kinase activity. *EMBO J* 1994; 13: 522–533.

60. Yuan ZM, Utsugisawa T, Huang Y, Ishiko T, Nakada S, Kharbanda S, Weichselbaum R, Kufe D. Inhibition of PI 3-kinase by c-Abl in the genotoxic stress response. *J Biol Chem* 1997: 272: 23,485–23,488.

61. Watanabe T, Ono Y, Taniyama Y, Hazama K, Igarashi K, Ogita K, Kikkawa U, Nishizuka Y. Cell division arrest induced by phorbol ester in CHO cells overexpressing protein kinase C-δ subspecies. *Proc Natl Acad Sci USA* 1992; 89: 10,159–10,163.

62. Lu Z, Hornia A, Jiang Y-W, Zang Q, Ohno S, Foster DA. Tumor promotion by depleting cells of protein kinase Cδ. *Mol Cell Biol* 1997: 17: 3418–3428.
63. Emoto Y, Manome G, Meinhardt G, Kisaki H, Kharbanda S, Robertson M, et al. Proteolytic activation of protein kinase C δ by an ICE-like protease in apoptotic cells. *EMBO J* 1995; 14: 6148–6156.
64. Emoto Y, Kisaki H, Manome Y, Kharbanda S, Kufe D. Activation of protein kinase C δ in human myeloid leukemia cells treated with 1-β-D-arabinofurano-sylcytosine. *Blood* 1996; 87: 1990–1996.
65. Ghayur T, Hugunin M, Talanian RV, Ratnofsky S, Quinlan C, Emoto Y, et al. Proteolytic activation of protein kinase C δ by an ICE/CED 3-like protease induces characteristics of apoptosis. *J Exp Med* 1996: 184: 2399–2404.
66. Nishizuka Y. Protein kinase C and lipid signaling for sustained cellular responses. *FASEB J* 1995; 9: 484–496.
67. Ogita K, Miyamoto S, Yamaguchi K, Koide H, Fujisawa N, Kikkawa U, et al. Isolation and characterization of δ-subspecies of protein kinase C from rat brain. *Proc Natl Acad Sci USA* 1992; 89: 1592–1596.
68. Li W, Li W, Chen X-H, Kelley CA, Alimandi M, Zhang J, et al. Identification of tyrosine 187 & nbsp as a protein kinase C-δ phosphorylation site. *J Biol Chem* 1996; 271: 26,404–26,409.
69. Yuan Z-M, Utsugisawa T, Ishiko T, Nakada S, Huang Y, Kharbanda S, Weichsel-baum R, Kufe D. Activation of protein kinase C δ by the c-Abl tyrosine kinase in response to ionizing radiation. *Oncogene* 1998; 16: 1643–1648.

7 Bcl-2 Family Proteins
Relative Importance as Determinants of Chemoresistance in Cancer

John C. Reed, MD, PhD

CONTENTS

ABSTRACT

Bcl-2 family proteins serve as critical regulators of pathways involved in apoptosis, acting to either inhibit or promote cell death. Because chemotherapeutic drugs typically exert their cytotoxic actions by inducing apoptosis, the ultimate efficacy of most anticancer drugs can be heavily influenced by the relative levels and state of activation of members of the Bcl-2 family. But how important are Bcl-2 family proteins in the overall complex picture of chemoresponses in cancer?

From: *Apoptosis and Cancer Chemotherapy*
Edited by: J. A. Hickman and C. Dive © Humana Press Inc., Totowa, NJ

Taking into consideration that such knowledge is far from complete, this chapter attempts to critically examine this question. It presents a discussion of some of the experimental and clinical observations that provide insights into the complexities of chemotherapeutic drug mechanisms and the role of Bcl-2 family proteins as modulators of tumor cell susceptibility to apoptosis.

Bcl-2 AS MULTIDRUG-RESISTANCE GENE

Defects in normal programmed cell death (PCD) mechanisms play a major role in the pathogenesis of tumors, allowing neoplastic cells to survive beyond their normally intended life-spans, subverting the need for exogenous survival factors, providing protection from hypoxia and oxidative stress as tumor mass expands, and allowing time for cumulative genetic alterations that deregulate cell proliferation, interfere with differentiation, promote angiogenesis, and increase cell motility and invasiveness during tumor progression. Though influenced by myriad genes, the core molecular machinery responsible for nearly all PCD, and its morphological equivalent, apoptosis, can probably be reduced to a few proteins that exist as multigene families in higher eukaryotes, such as mammals.

One of these key gene families is represented by Bcl-2 and its homologs, of which there are at least 16 in humans *(1•,2•)*. The apoptosis-suppressing Bcl-2 gene was discovered as a proto-oncogene found at the breakpoints of t(14;18) chromosomal translocations in low-grade B-cell lymphomas. These lymphomas represent quintessential examples of human malignancies in which the neoplastic cell expansion can be attributed primarily to failed PCD, rather than to rapid cell division. Initial gene transfection studies of Bcl-2 demonstrated that overproduction of the protein significantly prolongs cell survival in the face of classical apoptotic stimuli, including lymphokine deprivation from factor-dependent hematopoietic cells, glucocorticoid treatment of thymocytes and lymphoid leukemia cells, γ-irradiation of thymocytes, and nerve growth factor-deprivation from fetal sympathetic neurons *(1•)*. Conversely, antisense-mediated suppression of Bcl-2 expression was demonstrated to induce or accelerate cell death. It was thus that Bcl-2 emerged as the first example of an intracellular apoptosis-suppressor, and the first identified proto-oncogene that contributed to neoplasia through effects on cell life-span regulation, rather than cell division. Subsequent genetic studies of *Caenorhabditis elegans* mutants revealed an antiapoptotic Bcl-2 homolog, ced-9, implying that Bcl-2 governs an evolutionarily conserved step in the PCD machinery, a hypothesis further supported by experiments in which human Bcl-2 was found to be capable of rescuing, at least partially, ced-9-deficient worms *(3••,4••)*.

Not long after the antiapoptotic properties of Bcl-2 were recognized, cancer drug pharmacologists noted that the cell death induced by most chemotherapeutic agents in vitro resembled apoptosis, at least when the compounds were employed at clinically relevant concentrations *(5,6)*. These observations prompted explorations of the effects of Bcl-2 overexpression on tumor cell responses to anticancer drugs, revealing that Bcl-2 represents a novel type of multidrug-resistance protein that prevents or markedly delays apoptosis induction by most chemotherapeutic agents currently in clinical use *(7•–9•)*. Conversely, reductions in Bcl-2 achieved by antisense methods were shown to increase the susceptibility of cancer cells to apoptosis induction by multiple chemotherapeutic drugs *(10•,11•)*. This suppression of drug-mediated cytotoxicity, which is mediated by Bcl-2, is unrelated to rates of drug uptake or efflux, drug metabolism, nucleotide pools, amounts of drug-induced damage to macromolecules such as DNA, or rates of repair *(9•,12•)*. Rather, Bcl-2 suppresses steps that occur downstream of drug-induced damage that would otherwise result in apoptotic cell death. Subsequent analysis of other antiapoptotic Bcl-2 family proteins, such as Bcl-X_L, produced similar results, again demonstrating a unique multidrug-resistance phenotype unlike the MDR-family of drug-efflux proteins, glutathione and related endogenous antioxidants, DNA-repair enzymes, or other classical chemoresistance proteins.

Further evidence that Bcl-2 family proteins might be intimately connected to chemoresponses in cancer came from studies of the mechanisms by which the tumor suppressor p53 induces apoptosis. The p53 protein has been implicated by numerous studies in responses to genotoxic-stress injury, inducing cell-cycle arrest and apoptosis when DNA is damaged by anticancer drugs or radiation. Though p53 may use several cell type-specific mechanisms for inducing apoptosis, a direct connection to the Bcl-2 family was revealed when the promoter region of the proapoptotic *bax* gene was found to contain several typical p53-binding sites, and the *bax* gene was shown to be a direct transcriptional target of p53 *(13)*. Moreover, overexpression of Bax, as well as some other proapoptotic members of the Bcl-2 family, can render tumor cells relatively more sensitive to the cytotoxic actions of several anticancer drugs *(14)*; ablating *bax* reduces drug-induced apoptosis *(15)*, thus further supporting an important role for Bcl-2 family proteins as not only modulators of chemoresponses, but also as possible direct mediators of their ultimate cytotoxic actions.

CLINICAL CORRELATIVE STUDIES
PRESENT A CONFUSING PICTURE

With abundant data from in vitro gene-transfection studies, xenograph tumor models in animals, and transgenic mouse experiments providing the foundation for clinical investigations, a number of groups have set out to explore the potential prognostic significance of Bcl-2 or its family members regarding predicting survival and chemoresponses in patients *(16•–18•,19)*. In interpreting the results of these clinical-correlative studies, it is important to recognize that many are not well controlled, contain only small numbers of patients, or have no bearing on the question of Bcl-2 as a chemoresistance and radioprotective protein, because the treatment modality was primarily surgical. Moreover, none of the studies performed to date have used truly quantitative means to measure the relative amounts of Bcl-2 family proteins in cancer cells, relying instead on qualitative methods, such as immunohistochemistry or semiquantitative indirect immunofluorescence flow-cytometric assays. Nevertheless, several correlative studies of this type have provided support for the hypothesis that Bcl-2, or its close relatives Bcl-X_L and Mcl-1, confers a clinically important chemoresistant phenotype on cancer cells, including studies of patients with acute myelogenous leukemia (AML), acute lymphoblastic leukemia (ALL), chronic lymphocytic leukemia (CLL), non-Hodgkin's lymphomas, multiple myeloma, and prostate cancer *(16–19)*. In addition, attempts to compare the relative levels of Bcl-2 or other antiapoptotic Bcl-2 family proteins in leukemia or solid tumor cells before treatment, and again at the time of relapse, have demonstrated an apparent upregulation of Bcl-2, or a selective survival of those tumor cells that contain higher levels of Bcl-2 in some patients with ALL, AML, CLL, neuroblastomas, and prostate cancer. However, clinical correlative studies have not uniformly demonstrated an adverse prognostic role for higher levels of Bcl-2 protein production, and some have even shown that expression of Bcl-2 can bear a strong association with favorable clinical outcome *(20•–22•)*. What then explains these seemingly discrepant observations?

APOPTOSIS VS CLONOGENIC SURVIVAL

Without exception, when Bcl-2 has been overexpressed by gene transfection in any given tumor cell line, and then compared side by side with control-transfected, otherwise isogenic cells, enhanced resistance to anticancer-drug-induced apoptosis has been observed. This increased resistance imparted by Bcl-2 overexpression can typically be observed as a requirement for higher concentrations of drug to kill a given percentage of the tumor cells (e.g. 50%) within a defined amount of time (typically 2–3 d). In all cases, however, this protection imparted by Bcl-2 is relative, because, if enough drug is applied, the cells can be killed, which is consistent with the second rule of pharmacology (i.e., enough of anything will block [or kill] anything). Thus, some discrepancies between in vitro and in vivo observations could be attributable to dose issues, but this seems inadequate to account for everything. More at issue may be the question of whether short-term cytotoxicity assays of the sort described above (ED_{50} at 2–3 d) provide adequate information about what one can expect in the clinic.

Many in vitro studies of Bcl-2 have demonstrated a pronounced delay in drug-induced apoptosis, but not necessarily a true block of cell death, which implies that Bcl-2 may merely postpone the inevitable. This observation is not particularly damning if one considers the many differences between the in vitro experimental design and the in vivo clinical circumstance, such as maintenance of drugs at constant concentrations in vitro, compared to cyclical or bolus-type administration in vivo, in which drug is present only transiently. Thus, even a delay in cell death may produce an opportunity for rescue of sublethally damaged tumor cells, if the drug were removed. Moreover, to the extent that removal of drug is sufficient to rescue remaining viable cells from eventual death, it might also contribute to further genetic instability of drug-damaged cells. A particularly striking example has been reported with the antimicrotubule drugs in tumor cells protected by overexpression of $Bcl-X_L$, in which washing drug from the $Bcl-X_L$-protected cells after vincristine- or vinblastine-induced G_2/M-phase arrest prompted cells to re-enter G_1 and progress through S in the absence of cell division, resulting in tetraploidy *(23)*. Nevertheless, it does beg the questions of what defines the commitment step from which the cell cannot be rescued, and how important are Bcl-2 family proteins for regulating this step?

Examination of the effects of Bcl-2 overexpression on clonogenic survival of drug-treated cells may be more germane than short-term cytotoxicity studies, because clonogenic survival assays provide the additional benefit of determining not only whether cells survived, but also whether they retained replicative potential *(24)*. Given that clonogenic survival necessitates both the maintenance of cell survival and adequate preservation of chromosomal integrity for successful execution of the cell division cycle, however, it is perhaps not surprising that Bcl-2 overexpression has been shown to be sufficient for increasing clonogenic survival in some cases, but not in others *(25•,26•)*. Asking

Bcl-2 to do both may be requiring it to fulfill functions beyond its intended role as a regulator of cell life and death. Clearly, from the clinical standpoint, a viable tumor cell that cannot replicate is of little danger to the patient. But, what if millions of such cells remain viable in the body, perhaps for many days or a few weeks, and occasional ones experience genetic alterations that permit them to regain replicative competence? In that case, even preventing cell death could be sufficient to contribute to relapses.

Leaving issues of replicative integrity aside, how much is known about cell-death commitment concerning chemotherapeutic drugs? Recent studies have demonstrated that blocking apoptosis induced by anticancer drugs is not necessarily synonymous with blocking cell-death commitment. The effectors of apoptosis are a family of cysteine proteases that cleave peptide bonds just distal to aspartic acid residues occurring in specific amino-acid-sequence contexts. These proteases, known as caspases, are initially produced as single polypeptide chains representing inactive zymogens, but are converted to active proteases by proteolytic cleavage at specific aspartic acid residues, thus giving rise to the fully active heterotetrameric enzymes *(26•)*. Treatment of tumor cells with chemotherapeutic drugs uniformly results in proteolytic processing of certain caspases and cleavage of caspase substrates, such as poly ADP-ribose polymerase (PARP). Moreover, using broad-specificity, irreversible inhibitors of caspase, such as benzyloxycarbonyl-valinyl-alaninyl-aspartyl-fluoromethylketone (zVAD-fmk), it has been demonstrated that apoptosis can be partly, if not completely, prevented following exposure of tumor cells to etoposide, cisplatin, ara-C, ara-F, nitrosureas (bichloronitrosurea; BCNU), and other drugs. However, in many cases, despite blocking apoptotic proteases with zVAD-fmk or similar compounds , drug-damaged cells nevertheless die through what appears to be a delayed necrosis *(24, 27–30•)*. This delayed nonapoptotic cell death is preceded by, and indeed, is probably caused by, mitochondrial damage, resulting in release of certain mitochondrial proteins, including cytochrome-*c* from these organelles, interrupting electron-chain transport, causing generation of free radicals, loss of the electro-chemical gradient across the inner membrane, and presumably ATP depletion *(31, 32)*.

Though caspase inhibitors fail to prevent anticancer-drug-induced mitochondrial damage and the commitment to cell death, clonogenic survival studies indicate that Bcl-2 and Bcl-X$_L$ can prevent or reduce drug-induced cell death under the same circumstances *(32)*. Thus, at least when examined so far, Bcl-2 family proteins appear to govern a cell-death commitment step upstream of anticancer-drug-induced caspase activation and apoptosis. Though the mechanisms by which Bcl-2 accomplishes this remain debatable *(2•,31–33)*, it is noteworthy that many Bcl-2 family proteins, including Bcl-2 and Bcl-X$_L$, reside in mitochondrial membranes, chiefly the outer membrane. Moreover, Bcl-2 and Bcl-X$_L$ can prevent mitochondrial changes associated with drug-induced cell death, including cytochrome-*c* release, loss of $\Delta\psi$, and generation of reactive oxygen species (ROS), though caspase-independent mechanisms *(31,32)*. Conversely, the proapoptotic Bax protein can induce these changes in mammalian cells, and in yeasts which contain no apparent Bcl-2 homologs and no caspases *(31,32,34,35•)*.

BCL-2-INDEPENDENT PATHWAYS FOR APOPTOSIS

Although Bcl-2 clearly can govern a cell-death commitment step upstream of caspase activation, it has been shown that Bcl-2-independent pathways for caspase activation and apoptosis induction also exist *(36)*. For example, in some (but not all) types of cells, apoptosis triggered by Fas (CD95) and certain members of the tumor necrosis factor (TNF) family of death receptors is not blocked by overexpression of Bcl-2 or other antiapoptotic members of the Bcl-2 family (*see* Fig. 1). Fas and similar TNF-family receptors, including TNFR1, DR3/Weasle, DR4 (TRAIL-R1), DR5 (TRAIL-R2), and CAR, all contain death domains within their cytosolic tails, and directly induce caspase activation through ligand-induced recruitment of cytosolic procaspases, which can interact with specific adaptor proteins (FADD; TRADD) that bind to the death domains of these receptors *(37,38)*. Activation of the caspases, which interact with TNF-family receptor complexes (e.g., caspases-8, -10), can trigger a cascade of proteolysis, involving processing and activation of the zymogen forms of other downstream caspases (e.g., caspases-3, -6, -7), which serve as the final effectors (executioners) of apoptosis *(32)*. In many types of cells, this death receptor pathway for apoptosis runs via a Bcl-2-independent pathway, circumventing the participation of mitochondria or other organelles, where Bcl-2 and many of its homologs reside as integral membrane proteins.

How is this Bcl-2-independent process relevant to cancer chemotherapy? Some kinds of tumor cells apparently are induced to express TNF family receptors and/or ligands when damaged by anticancer drugs, including some T-cell leukemias, hepatocellular carcinomas, and colon cancers *(39–41•)*. The result is an autocrine (or paracrine) self-destruction *(39–41•)*. Moreover, p53 has been reported to upregulate transcrip-

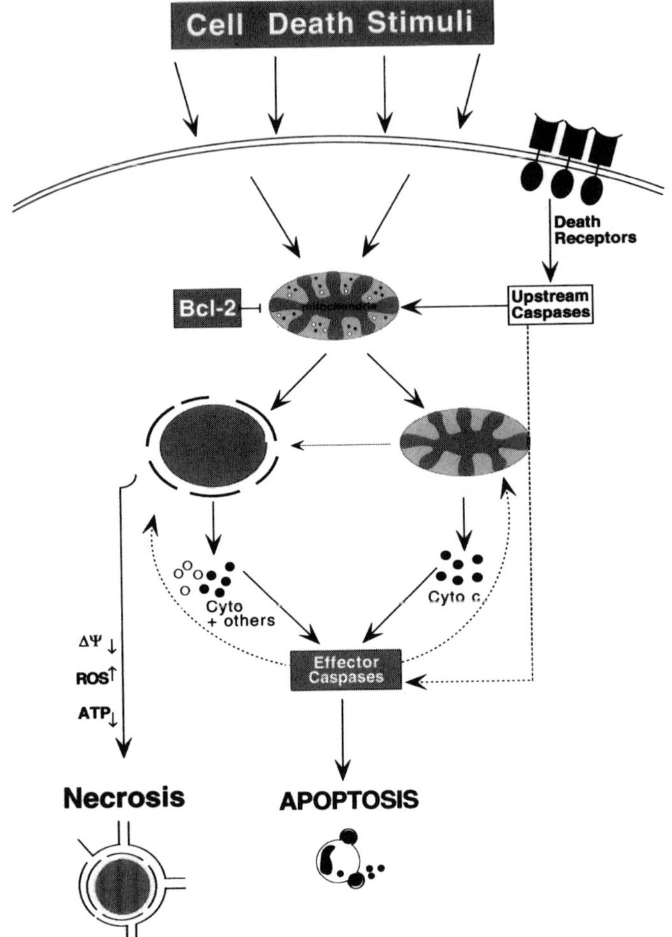

Fig. 1. Apoptosis Pathways. Apoptosis occurs when terminal-effector caspases are activated. Activation of these caspases can occur through several routes, including via mitochondria-dependent and -independent pathways. Two mechanisms of mitochondrial participation have been described: one involves opening of the mitochondrial permeability transition (PT) pore complex, with organelle swelling and rupture of the outer membrane *(left),* causing release of cytochrome-*c (dark circles),* and possibly other caspase activators *(open circles);* the other appears to allow for direct release of cytochrome-*c,* without evidence of PT pore opening *(right).* Bcl-2 blocks cytochrome-*c* release and PT pore opening. Some apoptotic stimuli involve TNF-family death receptors, which activate upstream caspases. These caspases can either directly cleave and activate downstream effector caspases (Bcl-2-independent), or they can indirectly activate effector caspases through effects on mitochondria (Bcl-2-dependent). Positive feedback loops exist with effector caspases, causing mitochondrial damage *(dashed lines).* Moreover, when sufficient cytochrome-*c* has been lost, resulting cessation of electron-chain transport can produce reactive oxygen species (ROS) that induce PT pore opening *(dashed line).* In the presence of caspase inhibitors, mitochondria damage can lead to necrosis-like cell death, typically involving generation of ROS, loss of membrane potential ($\Delta\psi$), and impaired ATP production.

tion of Fas and DR5 (reviewed here and in ref. *42*), linking these death receptors closely to this guardian of the genome. Thus, to the extent that this death receptor pathway is responsible for the cytotoxic actions of anticancer drugs in tumors, Bcl-2 overexpression may be irrelevant. Differences among tumors, even among tumors of the same lineage, however, make it difficult to predict how important this Bcl-2-independent pathway is within the context of clinical responses, because studies have shown that many tumors and leukemias do not rely on this autocrine mechanism for cell death. Further complicating attempts to predict chemoresponses, it has been shown that even apoptosis induced via Fas can be blocked substantially by Bcl-2 or Bcl-X_L overexpression in some cells, but not others. Recent comparisons of cell lines in which Bcl-2 overexpression was or was not Fas-protective suggest that two types of cellular contexts can be identified: one in which Fas ligation leads to abundant amounts of pro-caspase-8 processing that is Bcl-2-independent, and another in which Fas triggers only small amounts of caspase-8 activation that is Bcl-2-suppressible *(43)*. Again, even cells derived from the same lineage, such as T-cell leukemias, can differ in terms of which of these two Fas-signaling mechanism they rely on. Debatin discusses this in Chapter 12.

The basis for this difference in Bcl-2 sensitivity appears to reside in whether caspase-8 does or does not require a mitochondria-dependent amplification step to achieve sufficient activation of downstream effector caspases for apoptosis. For example, addition of recombinant caspase-8 at relatively high concentrations to cytosolic extracts results in activation of downstream caspases as well as apoptosis-like destruction of exogenously added nuclei. However, if lower concentrations of active caspase-8 are added, then downstream caspase activation and nuclear destruction are minimal, unless mitochondria are added to the cytosolic extracts *(44)*. Bcl-2 apparently can block caspase-8-mediated apoptosis only in cases in which a mitochondria-dependent amplication of the caspase cascade is required. Though caspase-8 may employ several mechanisms for bringing mitochondria into the picture, one intriguing pathway involves caspase-8-mediated cleavage of a cytosolic Bcl-2 family protein, Bid. Bid is a proapoptotic protein, which upon cleavage by caspase-8, translocates to mitochondrial membranes, binds to other Bcl-2 family proteins, and induces release of cytochrome-*c* from mitochondria (J. Yuan, X. Wang, personal communications). Once in the cytosol, cytochrome-*c* then binds to the ced-4-like protein, Apaf-1, inducing Apaf-1 to form complexes with pro-caspase-9, and triggering the dATP/ATP-dependent activation of caspase-9 *(45)*. Overexpression of Bcl-2 or Bcl-X$_L$ blocks both Bid-induced apoptosis and release of cytochrome-*c* from mitochondria.

Taken together, these observations concerning alternative pathways of Fas-induced apoptosis (i.e., Bcl-2-dependent/independent), and variations in the extent to which anticancer drugs utilize a death receptor-type mechanism for triggering tumor cell apoptosis, illustrate some of the complexities in attempting to predict clinical responses from analysis of Bcl-2 or Bcl-2 family protein levels.

MULTIPLE MECHANISMS OF BCL-2: HOW DOES IT WORK?

Perhaps it would be easier to correlate Bcl-2 expression with clinical responses of tumors to anticancer drugs and radiation, if one knew how the Bcl-2 protein works. Bcl-2 appears to defy the dictum: "One gene $\rightarrow$ One protein $\rightarrow$ One function." Rather, Bcl-2 is a multifunctional protein. At least three general functions for Bcl-2 and its antiapoptotic homologs have been identified: dimerization with other Bcl-2 family proteins; binding to nonhomologous proteins; and formation of ion-channels/pores *(2,31–33,36,46)*. A thorough discussion of the participation of Bcl-2 in these various biochemical functions is beyond the scope of this chapter, but the relevant point is that the relative importance of any of these three general mechanisms probably varies, depending on cell context (cell background) and the apoptotic stimulus involved. For example, Bcl-2 or Bcl-X$_L$ reportedly binds, or at least can be co-immunoprecipitated, not only with other several other members of the Bcl-2 family (both pro- and antiapoptotic), but also at least three proteins that can bind caspases (Apaf-1, Bap28, FLIP), a protein kinase (Raf-1), a protein phosphatase (calcineurin), a regulator of Hsc70/Hsp70 molecular chaperones (BAG-1), the spinal muscular atrophy (SMN) protein, a p53-binding protein (53BP-2), and others *(2,33)*. The extent to which any of these protein interactions with Bcl-2 and Bcl-X$_L$ is essential to antiapoptotic functions may differ greatly, depending on whether the host cell expresses some of these proteins, and whether the stimulus for apoptosis is p53 (e.g., 53BP-2), elevated Ca^{2+} (e.g., calcineurin), cytochrome-*c* release (e.g., Apaf-1), or something else. Similarly, some Bcl-2 family proteins are structurally similar to the pore-forming domains of bacterial toxins *(33,46)*, and have been reported to regulate channel/pore-like phenomenon in the membranes where they insert, including the mitochondria (e.g., opening/closing of the mitochondrial permeability transition pore complex; and release of cytochrome-*c*), endoplasmic reticulum (e.g., Ca^{2+} efflux from ER), and nuclear envelope (e.g., protein translocation through nuclear pore complexes) *(1,2,31–33)*. Thus, the ability to ascertain the relative level of Bcl-2 activity in a tumor cell is complicated by not knowing which of these various facets of Bcl-2 to measure.

REGULATION OF CELL DEATH AND MORE

Not only does Bcl-2 potentially employ multiple mechanisms for controlling cell life and death, but it also has an affect on cell proliferation. The antiapoptotic proteins Bcl-2, Bcl-X_L, and Bcl-W have been reported to suppress cell proliferation, probably by making it more difficult of cells to exit the quiescent (G_0) state and enter the cell cycle *(47•,48•)*. This function of Bcl-2 can be separated from its antiapoptotic actions. For example, site-directed mutagenesis of tyrosine 28 to phenylalanine in Bcl-2 ablates it ability to impede cell proliferation, but has no apparent affect on its antiapoptotic activity *(48•)*. Moreover, somatic point mutations in the *bcl-2* genes of tumors have been described, which appear to generate mutant Bcl-2 proteins that no longer suppress proliferation, but which still block cell death with efficiency indistinguishable from the wild-type protein *(49•)*. At present, few clues exist that might explain how Bcl-2, Bcl-X_L, and Bcl-W suppress cell proliferation, but binding to another protein(s) involved in regulating cell-cycle entry is probably the leading hypothesis.

It has been argued that the ability of Bcl-2 to inhibit cells from entering the cell-division cycle may contribute to its cytoprotective effects, for example, by helping to prevent cells with damaged DNA from attempting to replicate their genomes. Within the context of tumorigenesis, however, the inhibitory effects of antiapoptotic Bcl-2 family proteins on cell proliferation create a conflict of interest. On the one hand, tumors may rely on such proteins as Bcl-2 to prolong their lifespan, and to render them tolerant to hypoxia, growth factor deprivation, detachment from extracellular matrix, and the apoptotic drives of deregulated oncogene expression and genomic instability. On the other hand, suppression of proliferation is counterproductive to the clonal expansion of tumor cells, and thus could result in selective pressure for loss of Bcl-2 expression. At some point in the pathogenesis of most tumors, these competing and not necessarily compatible goals must be dealt with for maintaining cell survival while driving cell division. The resolution to that conflict could include: mutating Bcl-2, so that its cytoprotective actions are preserved, but its antiproliferative effects are lost; altering the expression or function of the putative protein(s) that interacts with Bcl-2 to impede cell cycle entry; or generating other lesions in the apoptotic machinery that render Bcl-2 superfluous, and thus allowing for loss of Bcl-2 expression. It is tempting to speculate that this conflict between the goals of tumor cell survival and cell division may explain the observations that Bcl-2 expression represents a favorable prognostic indicator for some kinds of solid tumors *(20–22)*.

HOW SHOULD BCL-2 FAMILY GENE EXPRESSION BE MEASURED IN TUMORS?

Ideally, attempts to delineate the relative importance of Bcl-2 family proteins with respect to tumor responses to chemotherapy or radiation would make use of highly sensitive and specific measures of Bcl-2 family protein activity. Instead, what is available is a variety of surrogate markers that merely assess the structural status of *bcl*-2-family genes or that assess qualitative changes in the levels of their encoded mRNAs or proteins. Some of the issues that ought to be considered in interpreting various measures of Bcl-2 family gene expression are presented here.

First, for historical reasons, many clinical-correlative studies have focused on Bcl-2 as a biomarker, but this is only because Bcl-2 was the first member of the family to be discovered. There is no *a priori* reason to suspect that Bcl-2 is any more or less important than any of the other approx 16 members of the human Bcl-2 protein family. Because the relative ratios of pro- and antiapoptotic Bcl-2 family proteins dictate the ultimate sensitivity or resistance of cells to myriad apoptotic stimuli, relying solely on measures of *bcl*-2 gene expression may be misleading or simply uninformative *(50)*. The advent of chip-based genomic and proteomic technologies should make it possible in the future to simultaneously assess the levels of mRNAs or proteins within the entire gene family.

Second, measures of *bcl*-2 family gene structure can be unreliable indicators of gene expression. For example, t(14;18) translocations are involved in the activation of the *bcl*-2 proto-oncogene in >85% of follicular non-Hodgkin's lymphomas (low-grade lymphomas), and in about one-third of diffuse large B-cell lymphomas (intermediate-grade lymphomas). Consequently, PCR, Southern blotting, and other DNA-based assays for detection of the t(14;18) have been explored as potential prognostic markers *(see* ref. *16•)*. Most of these studies, however, have failed to find a convincing correlation between the presence or absence of translocations and patient survival, responses to chemotherapy, or other measures of clinical outcome, probably in part because elevated Bcl-2 protein levels can commonly be found in lymphomas that do not contain t(14;18) translocations, either because of *bcl*-2 gene amplification, or other mechanisms that do not require structural changes to the *bcl-2* gene. In contrast to these DNA-based assays, immunohistochemical detection of Bcl-2 protein has been identified as an independent prognostic indicator of shorter disease-free or shorter overall survival in several studies of intermediate-grade lymphomas, in which simultaneous analysis of t(14;18) was uninformative *(51–53)*.

Third, measures of *bcl-2* family mRNA levels may not be indicative of protein production, representing a limitation to the application of new DNA-chip-based technologies. For example, *bcl-2* transcripts contain a region upstream of the open reading frame that has been shown to regulate translation *(54)*. Moreover, the relative levels of the Bcl-2, Bax, Mcl-1, and other possible Bcl-2 family proteins can be modulated by changes in their rates of protein turnover.

Fourth, measures of the levels of Bcl-2 family proteins, such as immunohistochemistry, immunofluorescence/flow cytometry, and immunoblotting, may be misleading if they fail to detect various posttranslational modifications that can alter the function of these proteins. For instance, cytosolic Bid may be inactive until cleaved by caspases, as mentioned above. Conversely, cleavage of Bcl-2 and Bcl-X_L by caspases has been reported to convert them from cytoprotective to cytodestructive proteins *(55)*. Moreover, phosphorylation of Bcl-2 has been associated with either enhanced or reduced antiapoptotic function, depending on the specific sites that become phosphorylated within the protein *(56,57)*. Likewise, phosphorylation of the proapoptotic BAD protein plays a critical role in the regulation of its function, with phosphorylation inactivating the protein by preventing it from dimerizing with Bcl-2/Bcl-X_L, and dephosphorylation restoring its cytotoxic activity *(58)*.

Fifth, the intracellular locations of some Bcl-2 family proteins are subject to regulation. In some types of cells, most of the Bax protein molecules are located in the cytosol, where they apparently have little influence on cell death, but, after apoptotic stimuli, these cytosolic Bax molecules translocate to mitochondrial membranes, where they induce cytochrome-*c* release and trigger apoptosis *(59)*. Thus, measures of Bax protein that fail to consider where the protein is located in the cell may not necessarily correlate with Bax activity, if it is discovered that this process of Bax translocation is differentially regulated in cells (i.e., permissive vs resistant to Bax translocation).

Sixth, the functional state of Bcl-2 family proteins may be dictated by their interactions with other proteins, and thus are not directly reflected by the levels of expression. Though controversial, for example, it has been argued that the apoptotic sensitivity of cells is controlled by the percentage of the total Bax protein molecules sequestered in dimeric complexes with Bcl-2, Bcl-X$_L$, or other antiapoptotic Bcl-2 family proteins *(60)*. Conversely, protection by Bcl-X$_L$ may correlate with binding to Apaf-1 or other proteins, rather than dimerization with Bax *(2,31–33,61)*. Thus, techniques that fail to measure these protein–protein interactions may not necessarily be reflective of the bioactivity of the proteins.

Seventh, conformational changes in Bcl-2 family proteins may determine whether they are inactive vs active. Evidence has been presented that Bcl-2, Bcl-X$_L$, and Bax, for example, can exist in two profoundly different conformational states, representing either a compact aqueous-soluble form or a membrane-integrated form associated with channel formation *(46)*. The events that regulate these conformational transitions remain mostly unknown, and techniques for readily measuring them in clinical specimens are not yet available.

CONCLUSIONS

With the large size of the Bcl-2 gene family and the enormous complexity in the regulation of Bcl-2 family gene expression and protein function, it is somewhat miraculous that immunohistochemical assessment of single members of the Bcl-2 family and some assays of *bcl-2* family gene structure have revealed strong correlations with clinical behavior in specific types of tumors. This implies that, although the gene family is large, certain members appear to play dominant roles in the pathogenesis or progression of particular cancers. Moreover, the striking observation that gene-transfer experiments performed in numerous laboratories, using multiple tumor cell lines, have almost uniformly demonstrated that elevations in Bcl-2 or Bcl-X$_L$ increase resistance to apoptosis induced by anticancer drugs, and provides a compelling argument that the relative amounts of these proteins are strong determinants of chemoresponses, despite the diverse mechanisms responsible for regulating their bioactivities. Probably the best determinants of the relative importance of Bcl-2 and its various homologs in controlling chemoresponses in cancers will come when potent antagonists of these proteins are finally available through drug screening, drug design, or other methods, and these compounds enter clinical trials. Until then, attempts to gain a greater understanding of the molecular mechanisms of action of Bcl-2 family proteins within tumor-specific contexts should proceed with hope of revealing optimal strategies for modulating the functions of these proteins in clinically useful ways for patients with fatal malignancies.

ACKNOWLEDGMENTS

The author thanks T. Brown and J. Waltz for manuscript preparation.

REFERENCES

• of special interest
•• of outstanding interest

•1. Reed JC. Bcl-2 and the regulation of programmed cell death. *J Cell Biol* 1994; 124: 1–6.

•2. Zamzami N, et al. Subcellular and submitochondrial mode of action of Bcl-2-like proteins. *Oncogene* 1998; 16: 2265–2282.
Two reviews on the mechanism of action of Bcl-2.

••3. Vaux DL, Weissman IL, Kim SK. Prevention of programmed cell death in Caenorhabditis elegans by human bcl-2. *Science* 1992; 258: 1955–1957.

••4. Hengartner MO., Horvitz HR. *C. elegans* cell survival gene ced-9 encodes a functional homolog of the mammlian proto-oncogene bcl-2. *Cell* 1994; 76: 665–676.
These two important papers provided evidence of evolutionarily conserved mechanism for Bcl-2 and ced-9.

5. Kaufmann SH. Induction of endonucleolytic DNA cleavage in human acute myelogenous leukemia cells by etoposide, campthothecin, and other cytotoxic anticancer drugs: a cautionary note. *Cancer Res* 1989; 49: 5870–5878.

6. Eastman A. Apoptosis: a product of programmed and unprogrammed cell death. *Toxicol Appl Pharmacol,* 1993; 121: 160–164.

•7. Miyashita T, Reed JC. bcl-2 gene transfer increases relative resistance of S49.1 and WEHI7.2 lymphoid cells to cell death and DNA fragmentation induced by glucocorticoids and multiple chemotherapeutic drugs. *Cancer Res* 1992; 52: 5407–5411.

•8. Miyashita T, Reed JC. Bcl-2 oncoprotein blocks chemotherapy-induced apoptosis in a human leukemia cell line. *Blood* 1993; 81: 151–157.

•9. Fisher TC, et al. Bcl-2 modulation of apoptosis induced by anticancer drugs: resistance to thymidylate stress is independent of classical resistance pathways. Cancer Res 1993, 53: 3321–3326.
Three examples of early papers that noted a correlation between Bcl-2 and chemoresistance.

•10. Kitada S, et al. Reversal of chemoresistance of lymphoma cells by antisense-mediated reduction of bcl-2 gene expression. *Antisense Res Dev* 1994; 4: 71–79.

•11. Campos L, et al. Effects of Bcl-2 antisense oligodeoxynucleotides on in vitro proliferation and survival of normal marrow progenitors and leukemic cells. *Blood* 1994; 84: 595–600.
Ref. 10 and 11 are examples of antisense Bcl-2 reversal of chemoresistance.

12. Bullock G, et al. Intracellular metabolism of Ara-C and resulting DNA fragmentation and apoptosis of human AML HL-60 cells possessing disparate levels of Bcl-2 protein. *Leukemia* 1996; 10: 1731–1740.

13. Miyashita T, Reed JC. Tumor suppressor p53 is a direct transcriptional activator of human BAX gene. *Cell* 1995; 80: 293–299.

14. Bargou RC, et al. Overexpression of the death-promoting gene bax-a which is downregulated in breast cancer restores sensitivity to different apoptotic stimuli and reduces tumor growth in SCID mice. *J Clin Invest* 1996; 97: 2651–2659.

15. Perez GI, et al. Apoptosis-associated signaling pathways are required for chemotherapy-mediated female germ cell destruction. *Nature Med* 1997; 3: 1228–1232.

•16. Reed JC. Bcl-2: Prevention of apoptosis as a mechanism of drug resistance. *Hematol/Oncol Clin N Am* 1995; 9: 451–474.

•17. Reed JC. Regulation of apoptosis by Bcl-2 family proteins and its role in cancer and chemoresistance. *Curr Opin Oncol* 1995; 7: 541–546.

•18. Strasser A, Huang DCS, Vaux DL: Role of the Bcl-2/ced-9 gene family in cancer and general implications of defects in cell death control for tumourigenesis and resistance to chemotherapy. *Biochim Biophys Acta* 1997; 1333: F151–F178.
Refs. 16–18 are recommended as reviews on Bcl-2 relation to clinical outcome and chemoresistance.

19. Reed J. Chronic lymphocytic leukemia: a disease of disregulated programmed cell death. *Clin Immunol Newslet* 1998; 17: 125–140.

•20. Silvestrini R, et al. bcl-2 protein: a prognostic indicator strongly related to p53 protein in lymph node-negative breast cancer patients. *J Natl Cancer Inst* 1994; 86: 499–504.

•21. Manne U, et al. Prognostic significance of Bcl-2 expression and p53 nuclear accumulation in colorectal adenocarcinoma. *Int J Cancer* 1997; 74: 346–358.

•22. Pezzella F, et al. bcl-2 protein in non-small-cell lung carcinoma. *N Engl J Med* 1993; 329: 690–694.
These three papers suggested that Bcl-2 expression associated with better clinical outcome in breast, colon, and non-small cell lung cancer.

23. Minn AJ, Boise LH, Thompson CB. Expression of Bcl-x$_L$ and loss of p53 can cooperate to overcome a cell cycle checkpoint induced by mitotic spindle damage. *Genes Devel* 1996; 10: 2621–2631.

24. Brunet CL, et al. Commitment to cell death measured by loss of clonogenicity is separable from the appearance of apoptotic markers. *Cell Death Diff* 1998; 5: 107–115.

•25. Yin D, Schimke R. Bcl-2 expression delays drug-induced apoptosis but does not increase clonogenic survival after drug treatment in HeLa cells. *Cancer Res* 1996; 55: 4922–4928.

•26. Walker A et al. Germinal center-derived signals act with Bcl-2 to decrease apoptosis and increase clonogenicity of drug-treated human B lymphoma cells. *Cancer Res* 1997; 57: 1939–1945.
Examples of papers in which Bcl-2 alone delayed apoptosis, but was insufficient for increasing clonogenic survival.

•27. McCarthy NJ, et al. *J Cell Biol* 1997; 136: 215–227.

•28. Ohta T, et al. Requirement of the caspase-3/CPP32 protease cascade for apoptotic death following cytokine deprivation in hemotopoietic cells. *J Biol Chem* 1997; 272: 23,111–23,116.

•29. Amarante-Mendes G, et al. Anti-apoptotic oncogenes prevent caspase-dependent and independent commitment for cell death. *Cell Death Diff* 1998; 5: 298–306.

•30. Hirsch T, et al. The apoptosis–necrosis paradox. Apoptogenic proteases activated after mitochondrial permeability transition determine the mode of cell death. *Oncogene* 1997; 15: 1573–1581.

Refs. 24 and 27–30 present evidence that anticancer drugs do not require caspases for inducing cell-death commitment.

31. Reed JC. Cytochrome C: Can't live with it; Can't live without it. *Cell* 1997; 91: 559–562.
32. Green DR, Reed JC. Mitochondria and apoptosis. *Science* 1998; 281: 1309–1312.
33. Reed JC. Double identity for proteins of the Bcl-2 family. *Nature* 1997; 387: 773–776.
•34. Xiang J, Chao DT, Korsmeyer SJ. BAX-induced cell death may not require interleukin 1b-converting enzyme-like proteases. *Proc Natl Acad Sci USA* 1996; 93: 14,559–14,563.
•35. Jurgensmeier JM, et al. Bax directly induces release of cytochrome c from isolated mitochondria. *Proc Natl Acad Sci USA* 1998; 5: 4997–5002.
   These two interesting papers show that Bax can cause mitochondrial dysfunction and cytochrome-c release without active caspases.
36. Vaux DL, Strasser A. Molecular biology of apoptosis. *Proc Natl Acad Sci USA* 1996; 93: 2239–2244.
37. Wallach D, et al. Cell death induction by receptors of the TNF family: towards a molecular understanding. *FEBS Lett* 1997; 410: 96–106.
38. Yuan J: Transducing signals of life and death. *Cur Opin Cell Biol* 1997; 9: 247–251.
•39. Friesen C, et al. Involvement of the CD95 (APO-1/Fas) receptor/ligand system in drug induced apoptosis in leukemia cells. *Nature Med* 1996; 2: 574–577.
•40. Muller M, et al. Drug-induced apoptosis in hepatoma cells is mediated by the CD95 (APO-1/Fas) receptor/ligand system and involves activation of wild-type p53. *J Clin Invest* 1997; 99: 403–413.
•41. Houghton JA, Harwood FG, and Tillman DM. Thymineless death in colon carcinoma cells is mediated via fas signaling. *Proc Natl Acad Sci USA* 1997; 94: 8144–8149.
   These three papers provided evidence of death receptor involvement in anticancer-drug-induced apoptosis.
42. Kastan M. On the TRAIL from p53 to apoptosis? *Nature Genet* 1997; 17: 130, 131.
43. Scaffidi C, et al. Two CD95 (APO-1/Fas) signaling pathways. *EMBO J* 1998; 17: 1675–1687.
44. Kuwana T, et al. Apoptosis induction by caspase-8 is amplified through the mitochondrial release of cytochrome c. *J Biol Chem* 1998; 273: 16,589–16,594.
45. Li P, et al. Cytochrome c and dATP-dependent formation of Apaf-1/Caspase-9 complex initiates an apoptotic protease cascade. *Cell* 1997; 91: 479–489.
46. Schendel S, Montal M and Reed JC. Bcl-2 family proteins as ion-channels. *Cell Death Differ* 1998; 5: 372–380.
•47. Borner C. Diminished cell proliferation associated with the death-protective activity of Bcl-2. *J Biol Chem* 1996; 271: 12,695–12,698.
•48. Huang DCS, et al. Anti-apoptosis function of Bcl-2 can be genetically separated from its inhibitory effect on cell cycle entry. *EMBO J* 1997; 16: 4628–4638.
•49. Reed JC and Tanaka ST. Somatic point mutations in translocated bcl-2 alleles of non-Hodgkin's lymphomas and lymphocytic leukemias: implications for mechanisms of tumor progression. *Leuk Lymphoma* 1993; 10: 157–163.
   These three papers provided examples of independent function of Bcl-2 as blocker of cell death and inhibitor of cell proliferation.

50. Reed JC. Balancing cell life and death: Bax, apoptosis, and breast cancer. *J Clin Invest* 1996; 97(11): 2403–2404.

51. Hermine O, et al. Prognostic significance of Bcl-2 protein expression in aggressive non-Hodgkin's Lymphoma. *Blood* 1996; 87: 265–272.

52. Hill ME, et al. Prognostic significance of BCL-2 expression and bcl-2 major breakpoint region rearrangement in diffuse large cell non-Hodgkin's lymphoma: a british national lymphoma investigation study. *Blood* 1996; 88: 1046–1051.

53. Gascoyne RD, et al. Prognostic significance of Bcl-2 protein expression and Bcl-2 gene rearrangement in diffuse aggressive non-Hodgkin's lymphoma. *Blood* 1997; 90: 244–251.

54. Harigai M, et al. A cis-acting element in the BCL-2 gene controls expression through translational mechanisms. *Oncogene* 1996; 12: 1369–1374.

55. Cheng EH-Y, et al. Conversion of Bcl-2 to a bax-like death effector by caspases. *Science* 1997; 278: 1966–1968.

56. Ito T, et al. Bcl-2 phosphorylation required for anti-apoptosis function. *J Biol Chem* 1997; 272: 11,671–11,673.

57. Haldar S, Jena N, Croce CM. Inactivation of Bcl-2 by phosphorylation. *Proc Natl Acad Sci USA* 1995; 92: 4507–4511.

58. Gajewski TF, Thompson CB. Apoptosis meets signal transduction: elimination of a BAD influence. *Cell* 1996; 87: 589–592.

59. Wolter KG, et al. Movement of bax from the cytosol to mitochondria during apoptosis. *J Cell Biol* 1997; 139: 1281–1292.

60. Oltvai ZN, Korsmeyer SJ. Checkpoints of dueling dimers foil death wishes. *Cell* 1994; 79: 189–192.

61. Cheng EH-Y, et al. Bax-independent inhibition of apoptosis by Bcl-XL. *Nature* 1996; 379: 554–556.

8

Bax, a Death Effector Molecule
Its Role in Development and Oncogenesis

Yi-Te Hsu, PhD and Richard Youle, PhD

CONTENTS

ABSTRACT

Bax is a member of the Bcl-2 family that regulates apoptosis by promoting cell death. Abnormal expression of this protein has been linked to pathological features such as colorectal cancer, enhanced tumorigenesis, and male sterility. Currently, the molecular mechanism by which Bax regulates apoptosis is unknown, despite intensive investigation of a variety of proposed models. Emerging evidence suggests that Bax can undergo differential conformational changes, and that its site of action appears to reside in mitochondria.

INTRODUCTION

Bax (Bcl-2-associated protein X) was first isolated by immunoprecipitation as a heterodimer with Bcl-2 *(1••)*. Overexpression of Bax in an IL-3-dependent cell line (FL 5.12) was shown to accelerate cell death

From: *Apoptosis and Cancer Chemotherapy*
Edited by: J. A. Hickman and C. Dive © Humana Press Inc., Totowa, NJ

induced by IL-3 withdrawal. Subsequently, it was demonstrated that overexpression of this protein promotes apoptosis induced by a variety of cellular insults *(2)*.

Three different transcripts of Bax (termed α, β, and γ), representing alternatively spliced forms, were identified from cDNA clones isolated by library screening. However, Baxα appears to be the only transcript expressed in cells. The gene for human Bax contains 6 exons and codes for a protein of 192 amino acids. In human, the *bax* gene has been mapped to chromosome region 19q13.3–q13.4 *(3)* and it appears that this gene can be transcriptionally activated by p53 *(4,5)*. Currently, Bax from mouse, rat, and human have been cloned. Sequence analyses of these proteins indicate that they share high degree of homology *(1••,6)*.

Sequence analysis of Bax indicates that this protein, like other members of the Bcl-2 family, has three conserved regions, known as the BH1 (amino acids 97–118), BH2 (amino acids 150–165), and BH3 (amino acids 59–73) domains. The BH3 domain of Bax (**LK/R**RIG-DELD) is believed to be required for dimerization to prosurvival factors Bcl-2 and Bcl-X$_L$ *(7•,8•)*. However, the importance of this region in the regulation of cell death remains controversial *(9•,10)*. Another notable feature of the Bax sequence is the presence of a predicted hydrophobic membrane-spanning segment at the carboxyl terminal end. In Bcl-2, a similar C-terminal hydrophobic segment is believed to be required for anchoring the protein to endoplasmic reticulum, mitochondria, and outer nuclear membranes *(11, 12)*.

ROLE OF Bax IN TISSUE
DEVELOPMENT AND ONCOGENESIS

Bax is widely distributed in many tissues within the body *(13)*. Physiologically, Bax is believed to be involved in maintaining cellular homeostasis in spleen and thymus, which are the two main organs responsible for lymphocyte production. In Bax knockout mice, selective hyperplasia of these lymphoid tissues was observed *(14•)*. Bax also appears to play a role in mammalian reproduction systems. In males, Bax plays a crucial role in spermatogenesis: Bax-deficient male mice were found to be sterile *(14•,15)*. In females, follicles in ovaries of knockout mice have excess granulosa cells *(14•)*, and their oocytes are highly resistant to treatment with the chemotherapeutic agent, doxorubicin *(16)*. Bax also appears to play a role in neuronal development: An increased number of neurons were found in Bax knockouts *(17)*. In a

Bax/Bcl-X$_L$ double knockout system, the absence of Bax counteracted the increased cell death in immature central nervous system neurons in Bcl-X$_L$-deficient mice, yielding relatively normal neuron levels *(18)*.

The involvement of Bcl-2 family members in cancer was first demonstrated for Bcl-2. Bcl-2 was found to be overexpressed in human follicular B-cell lymphoma caused by a t[14;18] interchromosomal translocation *(19–21)*. The connection between Bax and tumorigenesis was initially reported in human colon cancer of the microsatellite mutator phenotype, a cancer that is typified by deletion or insertion mutations in repeat sequences. For Bax, mutations were found in approx 50% of the MMP$^+$ tumors within the G8 tract of codons 38–41 of the third exon (ATG GGG GGG GAG) *(22,23••)*. In these tumors, mutations leading to the formation of either G9 or G7 tract cause a frameshift in the Bax gene, resulting in an inactive protein product. Similar findings were noted in hereditary nonpolyposis colorectal cancer (HNPCC). An increased incidence of Bax mutations in the (G)$_8$ tract of Bax was observed through the progression of colorectal adenoma to carcinoma *(24)*. Bax has also been reported to act, to some extent, like the tumor suppressor p53 *(25•)*. In a brain tumor model, an increase in tumor growth and a decrease in apoptosis were found in Bax-deficient mice.

SUBCELLULAR LOCALIZATION OF Bax

Because of the presence of a predicted C-terminal membrane-spanning domain and the propensity of Bax to form heterodimers with membrane-associated Bcl-2, it has been widely assumed that Bax is a membrane protein. However, by the conventional hypotonic lysis, Dounce homogenization, and differential centrifugation analyses, it was found that Bax in murine thymocytes and splenocytes was predominantly in the cytosolic fraction *(26••,27•)*. Bcl-2, in contrast, was found exclusively in the membranes. Bcl-X$_L$, which also has a predicted membrane-spanning domain, was found to exist in both soluble and membrane-bound forms *(26••,27•)*. In addition to thymus and spleen, Bax also appeared to be predominantly soluble in brain and liver (Hsu and Youle, unpublished data). The cytosolic localization of Bax was confirmed by tagging Bax to the green fluorescent protein (GFP). This enabled the visualization of biologically active Bax in living cells and in real time. GFP-Bax expressed in COS-7 cells was cytosolic, based on its diffuse pattern, and on its rapid recovery from fluorescent photobleaching *(28•)*.

Although Bax was cytosolic in healthy cells, upon induction of apoptosis in murine thymocytes with dexamethasone or γ-irradiation and in HL-60 promyelocytic leukemia cells treated with staurosporine, a significant shift in the subcellular localization of Bax from the cytosol to the membranes was observed *(26••)*. This shift in Bax localization was confirmed by GFP-Bax transiently expressed in COS-7 cells *(28•)*. Upon induction of apoptosis by staurosporine, GFP-Bax, which initially displayed a diffuse cytosolic state, changed to a punctate membrane-bound state. The site of Bax insertion overlaps with mitochondria, suggesting that this organelle is a primary site for Bax location during early stages of apoptosis.

More recently, several reports have suggested that Bax resides in mitochondria in healthy cells *(6,8•,29–31)*. These studies were carried out either by immufluorescence staining of cells overexpressing Bax or by the GFP-Bax visualization system. Overexpression of Bax, in many cases, is toxic to cells, even in the absence of apoptotic stimuli *(32)*. It is possible that the mitochondrial localization of Bax observed in these studies, in fact, represents the membrane-bound state of Bax found in cells undergoing apoptosis.

DIMERIZATION AND DIFFERENTIAL CONFORMATIONS OF Bax

One of the few activities associated with members of the Bcl-2 family is their propensity to form dimers. Bax has been shown by a number of investigators to be capable of forming either homodimers or heterodimers with Bcl-2 and Bcl-X_L by immunoprecipitation and yeast two hybrid select systems *(1••,33,34)*. Based on the tendency of Bax to promote cell death, it has been proposed that the relative ratio of Bax heterodimers to Bax homodimers serves as a switch that dictates the cell fate *(2,35)*. According to this model, exposure of the cell to cytotoxic insults may promote the formation of Bax homodimers. However, this death-initiation step may be blocked, if sufficient quantities of the prosurvival factors Bcl-2 or Bcl-X_L are present to form heterodimers with Bax. Thus, depending on the relative ratio of Bax heterodimers to Bax homodimers, the cell may proceed toward either cell survival or cell death.

However, several reports describing site-directed mutagenesis of Bax, Bcl-2, and Bcl-X_L have suggested that these proteins may function independent of dimer formation *(9•,10,36•–39)*. In these studies, mutations were introduced into the genes encoding these proteins. These mutations, in many instances, were able to abrogate the propensity of Bax, Bcl-2, and Bcl-X$_L$ to form dimers without affecting their biological activities. Thus, these observations raise the question of why members of the Bcl-2 family form dimers, and yet they can function in a dimer-independent manner.

The unexpected and intriguing finding of Bax being in a different subcellular compartment than Bcl-2 in murine thymocytes, with the former being a soluble protein and the latter being a membrane protein, provided the first indication that dimer formation between these two proteins may not normally occur in healthy cells *(26••,27•)*. Bax homo- and heterodimerizations were re-examined with the aid of several epitope-specific monoclonal antibodies generated against different species of Bax *(27•,40•)*. These studies took advantage of the fact that Bax and a significant fraction of Bcl-X$_L$ are soluble in murine thymocytes, and thus do not require the presence of detergents during the immunoprecipitation process. It was found that Bax readily forms homo- and heterodimers in the presence of nonionic detergents such as nonidet P-40, but it does not form dimers in the absence of detergent. In fact, the type of dimer formed by Bax appears to be highly dependent on the type of detergent used during the immunoprecipitation process (Table 1). This detergent-dependent dimerization of Bax appears to coincide with a detergent-induced exposure of an N-terminal epitope of Bax (amino acids 9–16), as detected by a conformation-sensitive monoclonal antibody, µBax 6A7. Taken together, it appears that the presence of nonionic detergents somehow caused a conformational change in Bax, to enable its homo- and heterodimer formation. It is interesting that detergents, which are normally thought to dissociate protein–protein interactions, in this case activate dimer formation. How this detergent-induced conformational change of Bax relates to the apoptosis-induced subcellular localization change remains to be explored.

FUNCTION OF Bax

The molecular cascade by which Bax regulates cell death is still unknown. Recently, it has also been reported that Bax-induced cell death may be blocked by Bax inhibitor-1 (B-1), an integral membrane protein with six transmembrane helices *(41)*. However, the mechanism of Bax regulation by B-1 is not known, because they do not appear to interact with one another.

Table 1

Bax Dimer Formation and the Exposure of the 6A7 Antibody Epitope in the Presence of Different Detergents

	None	NP-40	TX-100	TX-114	Polydoc.	W-1	Chaps	Octyl.	Dodec. malt.	Tween 20	Brij 35	Na Chol.
Heterodimerization												
Bcl-2[a]	N/A	++[c]	++	+++	−	−	−	+	−	N/A	N/A	N/A
Bcl-XL[a]	N/A	+++	+++	+++	+++	+++	−	++	++	N/A	N/A	N/A
Bcl-XL[b]	−	+++	+++	+++	+++	+++	−	+++	+++	+++	−	+
Homodimerization[b]	−	+++	+++	+++	++	−	−	−	++	−	−	−
6A7 binding[b]	−	+++	+++	++	++	−	−	−	+	−	−	++

[a]Results obtained from detergent-solubilized thymocytes.
[b]Results obtained from soluble protein extract of thymocytes.
[c]The relative extent of Bax dimerization and 6A7 antibody binding were assigned as strong (+++), medium (++), weak (+), or none (−).

The recent identification of the crystal structure of Bcl-X$_L$ provided a first clue about the potential mechanism played by members of the Bcl-2 family *(42•)*. The most prominent feature associated with Bcl-X$_L$, based on X-ray and NMR analyses, appears to be the presence of two central α-helices, termed α5 and α6. These two hydrophobic α-helices resemble the membrane translocation domain of diphtheria toxin and colicins. Because the similar domains of diphtheria toxin and colicins have ion channel activity in vitro, several studies were carried out to test the plausibility of Bcl-2 family members being ion-conducting channels *(43•–46)*. These studies measured ion conductance by recombinant Bcl-X$_L$, Bcl-2, and Bax, using the synthetic lipid vesicles and planar lipid bilayers. It was found that Bcl-2 and Bcl-X$_L$ are more selective for cations such as K$^+$, but Bax is more selective for anions such as Cl$^-$. The typical conductance of Bax is in the pS range and appears to be dependent on both pH and voltage. How this activity may regulate cell death remains unclear, but recent studies on changes in mitochondrial membranes during apoptosis suggest several areas for research.

Bax-dependent cell killing appears to be linked to caspase activation *(32, 47)*. Inclusion of a pan-caspase inhibitor, such as zVAD-fmk, has been shown to be effective in delaying cell death promoted by Bax *(32)*. Caspase activation is initiated by the release of cytochrome-*c* from mitochodria *(48)*. Several reports have suggested that overexpression of Bax leads to the loss of inner mitochondrial potential, a phenomenon also known as mitochondrial permeability transition *(47,49)*. It has also been reported that mitochondrial F$_0$F$_1$-ATPase proton pump may be required for Bax-induced killings *(50)*. These observations, taken together, have led to the proposal that Bax may form pores on mitochondrial surface, leading to the dissipation of mitochondrial potential and to the release of cytochrome-*c* *(51)*. One could envision that these series of processes (Fig. 1) would first require a signaling mechanism that switches Bax from its normal physiologically inactive state, possibly represented by its cytosolic form, to an active apoptotic state, possibly represented by its mitochondria-bound form. Interestingly, the 6A7 antibody binds the membrane-bound form of Bax and not the soluble form *(40)*. In addition, these processes would also require a specific targeting system that selectively directs Bax to the mitochondrial membranes, and not to other cellular organelle membranes. Identifying these key components of the apoptosis molecular cascade involving Bax will undoubtedly provide further insights into the intricate regulation of apoptosis by members of the Bcl-2 family.

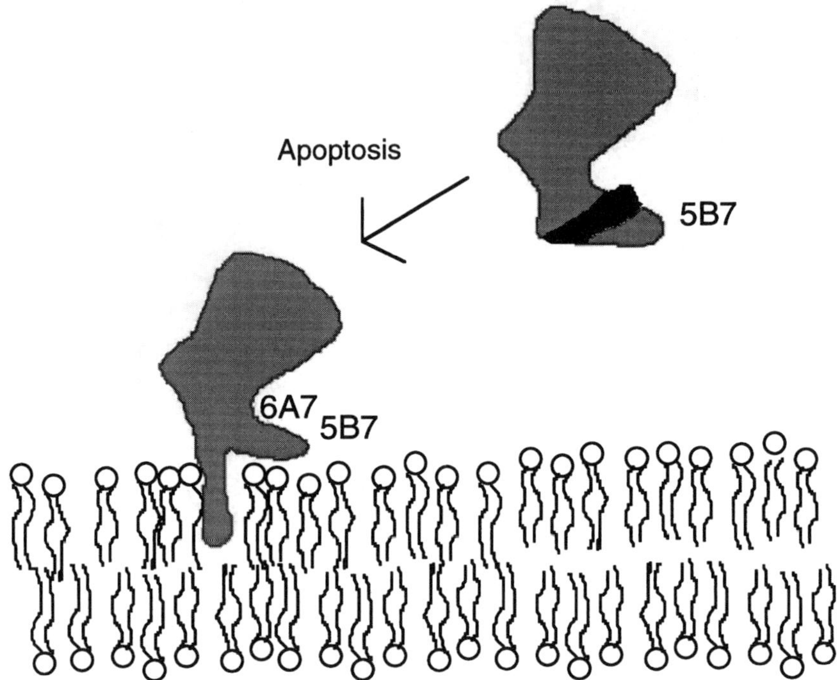

Fig. 1. Molecular regulation of Bax-induced cell death. In healthy living cells, Bax is a soluble protein, and probably exists in an inactive state. Upon induction of apoptosis, an unknown mechanism triggers the insertion of Bax into mitochondria. It is not known whether this mechanism acts directly on Bax or on mitochondria, or both. Insertion of Bax then causes the release of cytochrome-*c* and the activation of caspases, leading to cell death.

REFERENCES

- of special interest
- of outstanding interest

•• 1. Oltvai ZN, Milliman CL, Korsmeyer SJ. Bcl-2 heterodimerizes in vivo with a conserved homolog, Bax, that accelerates programmed cell death. *Cell* 1993; 74: 609–619.
Initial paper describing the discovery of Bax.

2. Yang E, Korsmeyer SJ. Molecular thanatopsis: a discourse on the Bcl-2 family and cell death. *Blood* 1996; 88: 386–401.

3. Apte SS, Mattei MG, Olsen BR. Mapping of the human Bax gene to chromosome 19q13.3–q13.4. *Genomics* 1995; 26: 592–594.

4. Miyashita T, Krajewski S, Krajewska M, Wang HG, Lin HK, Liebermann DA, Hoffman B, Reed JC. Tumor suppressor p53 is a regulator of bcl-2 and bax gene expression in vitro and in vivo. *Oncogene* 1994; 9: 1799–1805.

5. Miyashita T, Reed JC. Tumor suppressor p53 is a direct transcriptional activator of the human bax gene. *Cell* 1995; 80: 293–299.

6. Han J, Sabbatini P, Perez D, Rao L, Modha D, White E. E1B 19K protein blocks apoptosis by interacting with and inhibiting the p53-inducible and death-promoting Bax protein. *Gene Dev* 1996; 461–477.

•7. Zha H., Aime-Sempe C, Sato T, Reed JC. Proapoptotic protein Bax heterodimer-izes with Bcl-2 and homodimerizes with Bax via a novel domain (BH3) distinct from BH1 and BH2. *J Biol Chem* 1996; 271: 7440–7444.
   First paper describing the BH3 domain as an important regulatory element on Bax.

•8. Zha H, Fisk HA, Yaffe MP, Mahajan N, Herman B, Reed JC. Structure–function comparisons of the proapoptotic protein Bax in yeast and mammalian cells. *Mol Cell Biol* 1996; 16: 6494–6508.
   Describes a heterodimer-independent function of Bax.

•9. Simonian PL, Grillot DAM, Merino R, Nunez G. Bax can antagonize Bcl-XL during etoposide and cisplatin-induced cell death independently of its heterodim-erization with Bcl-XL. *J Biol Chem* 1996; 271: 22,764–22,772.
   Describes a homodimer-independent function of Bax.

10. Simonian PL, Grillot DAM, Andrews DW, Leber B, Nunez G. Bax homodimer-ization is not required for Bax to accelerate chemotherapy-induced cell death. *J Biol Chem* 1996; 271: 32,073–32,077.

11. Hockenbery D, Nunez G, Milliman C, Schreiber RD, Korsmeyer SJ. Bcl-2 is an inner mitochondrial membrane protein that blocks programmed cell death. *Nature* 1990; 348: 334–336.

12. Krajewski S, Tanaka S, Takayama S, Schibler MJ, Fenton F, Reed JC. Investiga-tion of the subcellular distribution of the bcl-2 oncoprotein: residence in the nuclear envelope, endoplasmic reticulum, and outer mitochondrial membranes. *Cancer Res* 1993; 53: 4701–4714.

13. Krajewski S, Krajewska M, Shabaik A, Miyashita T, Wang HG, Reed JC. Immunohistochemical determination of in vivo distribution of Bax, a dominant inhibitor of Bcl-2. *Am J Pathol* 1994; 145: 1323–1336.

•14. Knudson CM, Tung KSK, Tourtellotte WG, Brown GAJ, Korsmeyer SJ. Bax-deficient mice with lymphoic hyperplasia and male germ cell death. *Science* 1995; 270: 96–99.
    Shows that Bax is crucial for spermatogenesis.

15. Rodriguez I, Ody C, Araki K, Garcia I, Vassalli P. Early and massive wave of germinal cell apoptosis is required for the development of functional spermato-genesis. *EMBO J* 1997; 16: 2262–2270.

16. Perez GI, Knudson MC, Leykin L, Korsmeyer SJ, Tilly JL. Chemotherapy-mediated female germ cell destruction. *Nat Med* 1997; 3: 1228–1232.

17. White FA, Keller-Peck CR, Knudson CM, Korsmeyer SJ, Snider WD. Wide-spread elimination of naturally occurring neuronal death in Bax-deficient mice. *J Neurosci* 1998; 18: 1428–1439.

18. Shindler KS, Latham CB, Roth KA. Bax deficiency prevents the increased cell death of immature neurons in bcl-x-deficient mice. *J Neurosci* 1997; 17: 3112–3119.

19. Tsujimoto Y, Finger LR, Yunis J, Nowell PC, Croce CM. Cloning of the chromo-somes breakpoint of neoplastic B cells with the t(14;18) chromosome transloca-tion. *Science* 1984; 226: 1097–1099.

20. Bakhshi A, Jensen JP, Goldman P, Wright JJ, McBride OW, Epstein AL, Korsmeyer SJ. Cloning of the chromosomal breakpoint of t(14;18) human lymphomas: clustering around JH on chromosome 14 and near a transcriptional unit on 18. *Cell* 1985; 41: 899–906.

21. Cleary ML, Smith SD, Sklar J. Cloning and structural analysis of cDNAs from bcl-2 and the hybrid bcl-2/immunoglobulin transcript resulting from the (14;18) translocation. *Cell* 1986; 47: 19–28.

22. Yamamoto H, Sawai H, Perucho M. Frameshift somatic mutations in gastrointestinal cancer of the microsatellite mutator phenotype. *Cancer Res* 1997; 57: 4420–4426.

••23. Rampino N, Yamamoto H, Ionov Y, Li Y, Sawai H, Reed JC, Perucho M. Somatic frameshift mutations in the BAX gene in colon cancers of the microsatellite mutator phenotype. *Science* 1997; 275: 967–969.
First describes Bax mutations related to colon cancer.

24. Yagi OK, Akiyama Y, Nomizu T, Iwama T, Endo M, Yuasa Y. Proapoptotic gene Bax is frequently mutated in hereditary nonpolyposis colorectal cancers but not in adenomas. *Gastroenterol* 1998; 114: 268–274.

•25. Yin C, Knudson CM, Korsmeyer SJ, Dyke TV. Bax suppresses tumorigenesis and stimulates apoptosis in vivo. *Nature* 1997; 385: 637–640.
Shows that Bax exhibits tumor suppressor activity in mice.

••26. Hsu Y-T, Wolter KG, Youle RJ. Cytosol-to-membrane redistribution of Bax and Bcl-X_L during apoptosis. *Proc Natl Acad Sci USA* 1997; 94: 3668–3672.
First describes the cytosolic localization of Bax, and its redistribution from the cytosol to organelle membranes during apoptosis.

•27. Hsu Y-T, Youle RJ. Nonionic detergents induce dimerization among members of the Bcl-2 family. *J Biol Chem* 1997; 272: 13,829–13,834.
Describes unique detergent-induced conformational changes of Bax and Bax dimerization.

•28. Wolter KG, Hsu Y-T, Smith CL, Nechushtan A, Xi X-G, Youle RJ. Movement of Bax from the cytosol to mitochondria during apoptosis. J Cell Biol 1997; 139: 1281–1292.
Describes the translocation of GFP-Bax from the cytosol to mitochondria during apoptosis.

29. Shibasaki F, Kondo E, Akagi T, McKeon F. Suppression of signalling through transcription factor NF-AT by interactions between calcineurin and Bcl-2. *Nature* 1997; 386: 728–731.

30. Rosse T, Olivier R, Monney L, Rager M, Conus S, Fellay I, Jansen B, Borner C. Bcl-2 prolongs cell survival after Bax-induced release of cytochrome c. *Nature* 1998; 391: 496–499.

31. Otter I, Conus S, Ravn U, Rager M, Olivier R, Monney L, Fabbro D, Borner C. Binding properties and biological activities of Bcl-2 and Bax in cells exposed to apoptotic stimuli. *J Biol Chem* 1998; 273: 6110–6120.

32. Kitanaka C, Namiki T, Noguchi K, Mochizuki T, Kagaya S, Chi S, et al. Caspase-dependent apoptosis of COS-7 cells induced by Bax overexpression: differential effects of Bcl-2 and Bcl-X_L on Bax-induced caspase activation and apoptosis. *Oncogene* 1997; 15: 1763–1772.

33. Sato T, Hanada M, Bodrug S, Irie S, Iwama N, Boise LH, et al. Interactions among members of the Bcl-2 protein family analyzed with a yeast two-hybrid system. *Proc Natl Acad Sci USA* 1994; 91: 9238–9242.

34. Sedlak TW, Oltvai ZN, Yang E, Wang K, Boise LH, Thompson CB, Korsmeyer SJ. Multiple Bcl-2 family members demonstrate selective dimerizations with Bax. *Proc Natl Acad Sci USA* 1995; 92: 7834–7838.

35. Oltvai ZN, Korsmeyer SJ. Checkpoints of dueling dimers foil death wishes. *Cell* 1994; 79: 189–192.

•36. Cheng EH-Y, Levine B, Boise LH, Thompson CB, Hardwick JM. Bax-independent inhibition of apoptosis by Bcl-X. *Nature* 1996; 379: 554–556.
First describes a heterodimer-independent function of Bcl-X$_L$.

37. Knudson MC, Korsmeyer SJ. Bcl-2 and Bax function independently to regulate cell death. *Nat Genet* 1997; 16: 358–363.

38. Zha H, Reed JC. Heterodimerization-independent functions of cell death regulatory proteins Bax and Bcl-2 in yeast and mammalian cells. *J Biol Chem* 1997; 272: 31,482–31,488.

39. Clair EGS, Anderson SJ, Oltvai ZN. Bcl-2 counters apoptosis by Bax heterodimerization-dependent and -independent mechanisms in the T-cell lineage. *J Biol Chem* 1997; 272: 29,347–29,355.

•40. Hsu Y-T, Youle RY. Bax in murine thymus is a soluble monomeric protein that displays differential detergent-induced conformations. *J Biol Chem* 1998; 273: 10,777–10,783.
Describes the biochemical characterization of Bax.

41. Xu Q, Reed JC. Bax inhibitor-1, a mammalian apoptosis suppressor identified by functional screening in yeast. *Mol Cell* 1998; 1: 337–346.

•42. Muchmore SW, Sattler M, Liang H, Meadows RP, Harlan JE, Yoon HS, et al. X-ray and NMR structure of human Bcl-X$_L$, an inhibitor of programmed cell death. *Nature* 1996; 381: 335–341.
First shows that Bcl-X$_L$ shares a high degree of homology with the translocation domain of diphtheria toxin.

•43. Minn AJ, Velez P, Schendel SL, Liang H, Muchmore SW, Fesik SW, Fill M, Thompson CB. Bcl-X$_L$ forms an ion channel in synthetic lipid membranes. *Nature* 1997; 385: 353–357.
Shows that recombinant Bcl-X$_L$ can form ion-conducting channels in planar lipid bilayers.

•44. Antonsson B, Conti F, Ciavatta A, Montessuit S, Lewis S, Martinou I, et al. Inhibition of Bax channel-forming activity by Bcl-2. *Science* 1997; 277: 370–372.
Shows that recombinant Bax can form ion channels in lipid bylayers.

45. Schendel SL, Xie Z, Montal MO, Matsuyama S, Montal M, Reed JC. Channel formation by antiapoptotic protein Bcl-2. *Proc Natl Acad Sci USA* 1997; 94: 5113–5118.

46. Schlesinger PH, Gross A, Yin XM, Yamamoto K, Saito M, Waksman G, Korsmeyer SJ. Comparison of the ion channel characteristics of proapoptotic Bax and antiapoptotic Bcl-2. *Proc Natl Acad Sci USA* 1997; 94: 11,357–11,362.

47. Xiang J, Chao DT, Korsmeyer SJ. Bax-induced cell death may not be required interleukin 1 beta-converting enzyme-like proteases. *Proc Natl Acad Sci USA* 1996; 93: 14,559–14,563.

48. Liu X, Kim CN, Yang J, Jemmerson R, Wang X. Induction of apoptotic program in cell-free extracts: requirement for dATP and cytochrome c. *Cell* 1996; 86: 147–157.

49. Pastorino JG, Chen S-T, Tafani M, Snyder JW, Farber JL. Overexpression of Bax produces cell death upon induction of the mitochondrial permeability transition. *J Biol Chem* 1998; 273: 7770–7775.
50. Matsuyama S, Xu Q, Velours J, Reed JC. Mitochondrial Fo/F1-ATPase proton pump is required for function of the proapoptotic protein Bax in yeast and mammalian cells. *Mol Cell* 1998; 1: 327–336.
51. Manon S, Chaudhuri B, Guerin M. Release of cytochrome c and decrease of cytochrome c oxidase in Bax-expressing yeast cells, and prevention of these effects by coexpression of Bcl-X$_L$. *FEBS Lett* 1997; 415: 29–32.

9

Bax, a Proapoptotic Protein Forming Channels in Mitochondria

Jean-Claude Martinou, MD, PhD
and Bruno Antonsson, PhD

CONTENTS

ABSTRACT

Bax is a member of the Bcl-2 family of proteins, with proapoptotic activity. In cells, the protein is found in the cytosol and in the mitochondrial membrane. At least in some cell types, a translocation from the cytosol to the mitochondria has been detected during apoptosis. The protein contains a hydrophobic C-terminal domain presumably involved in the membrane attachment. Bax can form ion channels in artificial lipid membranes: These can be described as

From: *Apoptosis and Cancer Chemotherapy*
Edited by: J. A. Hickman and C. Dive © Humana Press Inc., Totowa, NJ

multiconductance, pH-sensitive, voltage-sensitive channels with rather poor ion selectivity. Overexpression of Bax in cells, or addition of recombinant Bax protein to isolated mitochondria, triggers loss of the mitochondrial membrane potential and release of cytochrome-c, both early events during apoptosis.

INTRODUCTION

Apoptosis or programmed cell death is an essential physiological process required for normal development and maintenance of tissue homeostasis *(1,2)*. However, apoptosis is also involved in a wide range of pathologic conditions, including acute neurological injuries, neurodegenerative diseases, immunological diseases, AIDS, and cancer *(3–5)*. Several signaling pathways appear to be involved in inducing apoptosis, depending on the initiating stimulus. Proteins of the Bcl-2 family, together with mitochondria, cytochrome-*c*, and caspases, have, among others, been identified as essential parts of the apoptotic signaling pathways *(6–9)*. Bcl-2 family members can be subdivided into two groups according to their function: the antiapoptotic members on one side and the proapoptotic members on the other side (Table 1).

PRIMARY STRUCTURE

The Bcl-2 family of proteins share homology within specific regions called Bcl-2 homology (BH) domains, which mediate protein interactions *(10–12)*. The proteins can interact in specific ways to form either homo- or heterodimers. The BH1 and BH2 domains are required for Bcl-2 and Bcl-X_L to dimerize with Bax, and to suppress apoptosis *(13)*. However, mutations in Bcl-X_L have been described that prevent heterodimerization with Bax or Bak, but still maintain antiapoptotic activity, suggesting that the antiapoptotic proteins can also function independently to regulate cell survival *(14)*.

The BH3 domain of proapoptotic proteins, such as Bax, Bak, or Bad, is sufficient, but are not required for their binding to Bcl-2 or Bcl-X_L, and to promote apoptosis *(15•)*. In fact, this domain seems to play a dominant role, because introduction of the BH3 domain of Bax into Bcl-2 is sufficient to convert Bcl-2 to a killer protein *(16)*. In addition, there are proteins, such as Bid, Bik, and Bim, which have BH3 domains, but lack identifiable BH1 and BH2 domains, and which are nevertheless efficient proapoptotic proteins, confirming the key role of the BH3 domain in triggering apoptosis. Mutagenesis studies demonstrate that the proapoptotic activity of the BH3 domain is mediated through heterodimerization

Table 1
Members of the Bcl-2 Protein Family, with Pro-
and Antiapoptotic Activity *(24,52–54)*

Antiapoptotic	*Proapoptotic*
	Bax
Bcl-2	Bax
Bcl-X$_L$	Bok
Bcl-w	Bcl-X$_S$
Mcl-1	Bad
A1	Bid
	Bik/Nbk
	Bim
	Krk
	Mtd

with either pro- or antiapoptotic proteins. For example, the BH3 domain of Bid can heterodimerize with either Bcl-2 or Bax. However, a Bid mutant that binds to Bax, but not to Bcl-2, is still apoptotic in FL5.12 cells *(17•)*. In contrast, the BH3 domain of Bad requires binding to Bcl-2 to kill the same cells *(18•)*. These results do not facilitate comprehension of the mechanism by which the BH3 domain triggers apoptosis.

A fourth domain, BH4, is found in the N-terminal region of antiapoptotic proteins only, except Bcl-X$_S$ which also contains a BH4 domain. This domain has been shown to bind to several proteins, including Raf1, Ced-4, and calcineurin *(19–21)*. Mutants of Bcl-2 lacking the BH4 domain not only lose their antiapoptotic activity, but behave like killer proteins *(22•)*. Such processed forms of Bcl-2 have been shown to be produced through caspase cleavage during apoptosis, and were found to accelerate the apoptotic process *(23)*.

Finally, some of these proteins display a hydrophobic amino-acid sequence at the C-terminus, which is required for their membrane interaction and localization, as described later.

Despite all this information concerning the primary structure of these proteins, little is known about the mechanism of action of the Bcl-2 family members. To date, Bcl-2 has been shown to interact with at least 13 different proteins including Bax, Bak, Bid, Bad, Raf-1, calcineurin, Ced-4, Bag-1, R-Ras, H-Ras, p53-BP2, survival motor neuron protein (SMN), and the prion protein, pr-1 *(24)*. These extensive interactions with proteins of different functions complicate the understanding of the mode of action of Bcl-2, although it is important to keep in mind that the physiological relevance of these interactions has not always been demonstrated.

THREE-DIMENSIONAL STRUCTURE

A hint into the function of the Bcl-2 family proteins came from the recent solved three-dimensional (3-D) structure of Bcl-X$_L$ *(25••)*. The protein showed structural homology to the pore-forming domains of certain bacterial toxins, in particular, diphtheria toxin and the colicins, A and E1. These bacteria toxins are pore-forming proteins that function as channels for ions or small proteins. The structure of Bcl-X$_L$, lacking the hydrophobic C-terminal domain, consists of two central hydrophobic helices (α5 and α6) surrounded by five amphipathic helices, as well as a 60-residue flexible loop. The BH1, BH2, and BH3 domains are all located in close proximity on the surface of the protein, and form an elongated hydrophobic cleft that has been shown to represent the binding site of the BH3 domain of Bak. The flexible loop, which is located N-terminal of the BH3 domain, contains several phosphorylation sites that have been shown to undergo extensive phosphorylation in the immature B-cell line WEHI-231, when exposed to anti-IgM *(26•)*. In these cells, Bcl-2 was ineffective at suppressing anti-IgM-mediated apoptosis. However, a Bcl-2-deletion mutant, lacking the predicted flexible loop, was not phosphorylated, and it was able to block apoptosis in the same cells. Bcl-2 has also been shown to be phosphorylated in many tumor cell lines, following exposure to paclitaxel (Taxol) *(27,28)*. Although the kinase responsible for Bcl-2 phosphorylation has not been clearly identified, a good candidate is the c-Jun N-terminal kinase/stress-activated protein kinase, which is activated by several stimuli known to lead to cell death *(29•)*.

CHANNEL-FORMING ACTIVITY

As predicted from the 3-D structure, Bcl-X$_L$ was subsequently shown to form channels in synthetic lipid membranes *(30•)*. Like Bcl-X$_L$, both Bcl-2 and Bax display channel-forming activity in synthetic lipid membranes *(31•,32•)*. The channels formed by the Bcl-2 family of proteins can be described as multiconductance (large conductances above 1 nS have been reported for both Bax and Bcl-2), pH-sensitive, voltage-sensitive channels with rather poor ion selectivity. Although both the pro- and antiapoptotic proteins can form channels, their channel-forming properties and requirements are not identical. Bcl-2 and Bcl-X$_L$ channels are rather cation-selective, but Bax is anion selective at physiological pH.

Differences in the amino-acid composition of the α5 and α6 helices might explain the differing electrophysiological properties of the channels formed by the pro- and antiapoptotic proteins. However, the channel-forming requirements also differ between the pro- and antiapoptotic proteins. Bax can form pores at neutral pH, whereas Bcl-2 requires low pH. Bax channel formation is dependent on the lipid composition of the membrane. In membranes composed of noncharged lipids, phosphatidylcholine, and cholesterol, no channel-forming activity was detected (unpublished observations). However, including a negatively charged phospholipid, phosphatidylserine, in the membrane enabled the protein to form channels. Thus, one possible regulatory mechanism of the channel-forming activity of Bax could be the lipid composition in the microenvironment of the mitochondrial membrane.

LOCALIZATION

Bcl-2 family members have been shown to be membrane-associated or cytosolic, depending on whether they contain a C-terminal hydrophobic domain *(13)*. Bcl-2 has been reported to be associated with endoplasmic reticulum, mitochondrial membranes (possibly at the contact sites), and the nuclear envelope *(33)*. In contrast, Bax seems to be exclusively associated with mitochondrial membranes. Recent studies indicate that protein localization changes during apoptosis. This is reviewed here in Chapter 8. For example, Bax was found to move from the cytosol to the mitochondria, when apoptosis was induced *(34•)*. The mechanisms responsible for this translocation are unknown, but it appears that the C-terminal domain is required. It is possible that, under normal conditions, the hydrophobic domain is masked, allowing the protein to remain cytosolic; its unmasking during apoptosis allows the protein to translocate to the mitochondria. Protein–protein interaction with Bcl-2 family members or other proteins, and/or conformational changes in the Bax protein, may protect or expose the hydrophobic domain, which could in turn lead to a change in the subcellular localization. However, specific association with the mitochondrial membrane suggests some form of specific interaction, rather than unspecific hydrophobic interactions induced simply by exposure of the hydrophobic domain. A proposed model for activation and translocation of Bax from the cytosol to the mitochondrial membrane is shown in Fig. 1.

MITOCHONDRIA DURING APOPTOSIS

Using a cell-free system, Newmeyer et al. *(35••)* first reported the requirement of mitochondria in an apoptosis system. Later on, Li

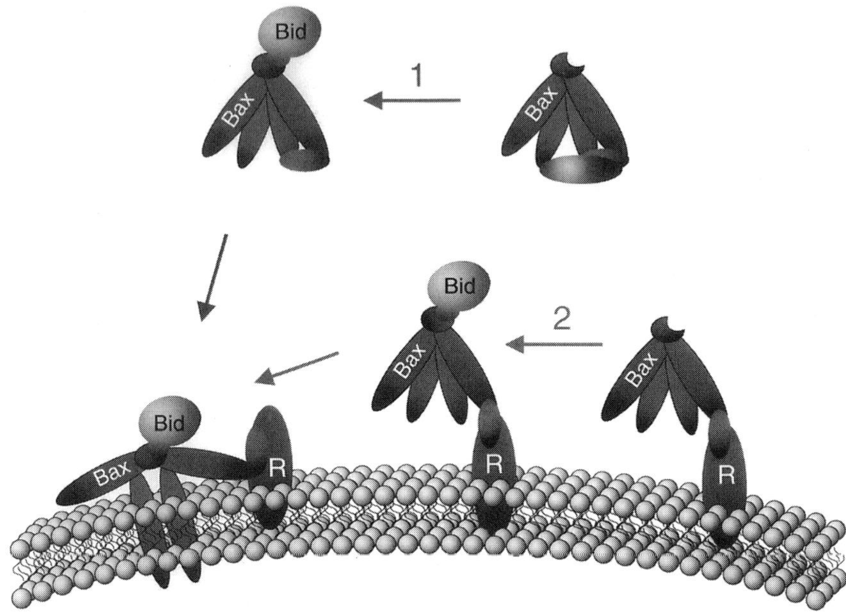

Fig. 1. Models for activation and translocation of Bax from the cytosol to the mitochondrial membrane. Bax is kept either in the cytosol, with the C-terminal hydrophobic domain masked through interactions with other proteins and/or conformational changes in the protein itself, or bound to its mitochondrial receptor in an inactive conformation. Pathway 1. Cytosolic Bax is activated through interaction with an activating protein (for example, Bid), which induces exposure of the C-terminal domain, resulting in translocation and binding to the mitochondrial receptor (R) and channel formation in the membrane. Pathway 2. Bax is bound to its mitochondrial receptor (R) in an inactive conformation. Binding of an activating protein (Bid) during apoptosis induces a conformational change in Bax that activates the protein, and leads to channel formation in the membrane.

et al. *(36••)* discovered that cytochrome-*c*, together with dATP and Apaf-1, a ced-4 homolog, were sufficient to induce proteolytic processing and activation of caspase-9 in vitro. They showed that cytochrome-*c* forms a complex with Apaf-1 and caspase-9 in the cytosol. The binding of caspase-9 to Apaf-1, cytochrome-*c*, and adenosine triphosphate (ATP), induces its activation. Using different cell types, several laboratories have now clearly demonstrated that cytochrome-*c* is released from mitochondria into the cytosol during apoptosis *(37,38)*. In addition to cytochrome-*c*, apoptosis-inducing factor (AIF), another protein present in the intermembrane space of the mitochondria, has been reported to be released during apoptosis *(39).* However, the molecular identity of AIF is still unknown. Cytochrome-*c* release is accompanied by a drop in the mitochondrial membrane potential. Whether the release of cytochrome-*c* precedes or follows the loss of membrane potential is still debatable. The idea that the membrane potential of mitochondria could play a major role in the apoptotic process was first announced by Hennet et al. *(40),* who reported that the membrane potential of L929 cells overexpressing Bcl-2 was elevated, compared to wild-type cells. Moreover, these authors showed that the resistance of L929 cells to tumor necrosis factor (TNF) was linked with the membrane potential level. Subsequently, several laboratories have reported a loss of mitochondrial membrane potential occurring early in the apoptotic process in several cell types. The loss of membrane potential, together with the leakage of mitochondrial proteins, has been attributed to opening of the so-called permeability transition pore (PTP). The PTP is an entity whose exact composition is unknown, though it is thought to be composed of several proteins, including hexokinase, mitochondrial creatine kinase, a voltage-dependent anion channel present on the outer membrane, the inner membrane adenine nucleotide translocator (ANT), and the matrix cyclophilin D. Finally, Bax (but not Bcl-2 or Bcl-X$_L$) was recently found to be part of this complex *(41).* The PTP has been proposed to be located at the contact sites between the inner and outer mitochondrial membranes. Bcl-2 was not found to co-purify with the PTP complex, although electron microscopy studies have shown localization of the protein to mitochondrial membrane contact sites *(33).* The PTP functions as a calcium-, voltage-, pH-, and redox-gated channel with several levels of conductance and poor ion selectivity. It can be inhibited by cyclosporin A, which binds to cyclophilin D, bongkrekic acid, a ligand of the ANT, and with Bcl-2 *(24,42,43).* Bcl-2 seems to act both upstream and downstream of the PTP. It prevents the release of cytochrome-*c* during apoptosis in several cell types; it also inhibits opening of the PTP induced by H$_2$O$_2$, calcium, and *ter*-butylhydroperoxide. This effect was observed in liver mitochondria isolated from transgenic mice, in which the transgene was specifically overexpressed in hepatocytes, which contained high levels of Bcl-2 *(44••).* It was also observed in liposomes, in which the PTP had been artificially reconstituted *(41).* According to Shimizu et al. *(44••),* Bcl-2 prevents PTP opening by increasing proton efflux from the matrix, thereby inhibiting mitochondrial respiration (only when mitochondria are stressed). Diminution of mitochondrial respiration could prevent the production of free radicals, which often occurs when mitochondria are decoupled. This would explain the antioxidant properties that had been previously attributed to Bcl-2 *(45,46)*. In addition, Bcl-2 was recently shown to inhibit apoptosis downstream of PTP opening, i.e., following cytochrome-*c* efflux from mitochondria *(37)*.

HOW IS CYTOCHROME-c RELEASED FROM MITOCHONDRIA DURING APOPTOSIS?

Because of its localization on mitochondria, and because of its channel activity, Bax was an obvious candidate to test as an inducer of cytochrome-*c* release. As predicted, it was found that Bax can directly induce the release of cytochrome-*c* from isolated mitochondria *(47,48)*. Bcl-2 can prevent the release of cytochrome-*c*, not only in cells undergoing apoptosis, but also when co-added with Bax to isolated mitochondria. Bcl-2 could act by inhibiting Bax insertion in the mitochondria, or by inhibiting, directly or indirectly, Bax-channel activity. The mechanisms by which Bax triggers cytochrome-*c* efflux from mitochondria are hitherto unknown. Cytochrome-*c* could be released directly through the channel formed by Bax, or could be the result of an indirect mechanism consecutive to Bax-channel formation. Whether or not the PTP is involved in this process is still controversial. It has indeed been shown that opening of the PTP results in a leakage of cytochrome-*c* and other mitochondrial proteins into the cytosol *(49)*. In support of this idea, mitochondrial swelling and rupture of the outer mitochondrial membrane have been reported to occur during Fas-mediated apoptosis of Jurkat T-cells, and in FL5.12 cells following IL-3 deprivation *(50)*. Mitochondrial swelling could be responsible for the rupture of the outer mitochondrial membrane, because the surface area of the inner membrane, with its cristae, is much larger than that of the surrounding outer membrane. If this model is correct, many mitochondrial proteins, in addition to cytochrome-*c*, are expected to be released from mitochondria into the cytosol. Description of the rupture of the outer mitochondrial membrane during apoptosis was surprising because such a morphological change had never been described before, despite intensive electron microscopy studies. Instead, the opposite, i.e., a mitochondrial shrinkage described as mitochondrial pyknosis, has been reported: Mitochondria are condensed and their matrix becomes hyperdense *(51•)*. Mitochondrial condensation would be caused by loss of water and ions from the mitochondria, and would represent a point of no return for the cell. At this stage, however, no alteration of the nucleus morphology was detected. Changes in membrane permeability leading to mitochondrial swelling appeared, only as a late event, in a cell displaying a necrotic, rather than an apoptotic, appearance. Mitochondrial swelling or mitochondrial shrinkage: The question remains open. The answer to this question may be difficult to obtain, because not all mitochondria may be affected during the apoptosis process. Indeed, it is possible that just a subpopulation of mitochondria, (only those proximal to the cell death stimulus) might be affected, and, subsequently, may be responsible for the downstream activation of caspases, leading to cell death. Such a mechanism can be envisaged, particularly in neurons, in which mitochondria are present not only in the cell body, but also in the most distal part of the neurites. How attrition of mitochondria in neurites, with a release of cytochrome-c, could lead to the demise of the cell body is an important question that needs to be clarified if one wants to understand how neurons degenerate following lesions of their axons. It would also be interesting to determine what is the minimal amount of cytochrome-c (i.e., the minimal level of altered mitochondria) required to activate caspases.

Cytochrome-c release from isolated mitochondria, following addition of Bax, was also found to occur in the absence of mitochondrial swelling, suggesting that the PTP was not involved in this process. However, cyclosporin A, a strong inhibitor of the PTP, was able to inhibit the effect of Bax, suggesting that some components of the PTP (at least cyclophilin D) might participate in the release process *(47)*. This result contrasts with the authors' findings that cyclosporin A, in the absence of Ca^{2+}, failed to inhibit the Bax-induced cytochrome-c release from isolated mitochondria *(48)*. It has also been reported that the PTP, reconstituted in liposomes filled with cytochrome-c, failed to release cytochrome upon opening of the PTP with atractyloside *(41)*. Recently, it was found that, in both yeast and mammalian cells, Bax required a functional mitochondrial ATP synthase in order to be pro-apoptotic. Thus, the killing effects of Bax were abrogated by oligomycin. The relationship between Bax and the mitochondrial proton pump is not yet clear. Also, it is unknown whether Bax-induced release of cytochrome-c requires a functional ATP synthase.

CONCLUSION

Although a substantial amount of new information has emerged during the past year concerning the structure and function of the Bcl-2 family of proteins, the molecular mechanisms by which these proteins exert their pro- and antiapototic functions still remain unclear. It appears that the proteins might act at different levels in the apoptotic signaling cascades, and that they have several functional mechanisms; for example, channel-forming activity and receptor–ligand activity mediated through the BH domains. These two mechanisms may have an additive or cumulative effect, or work independently, dependent on the signaling pathways.

ACKNOWLEDGMENTS

The authors thank Tim Wells for encouraging support, Christopher Hebert for preparing the figures, and Charles Bradshaw for critical reading of the manuscript.

REFERENCES

• of special interest
•• of outstanding interest

1. Saunders JW. Death in embryonic systems. *Science* 1966; 154: 604–612.
2. Kerr JF, Wyllie AH, Currie AR. Apoptosis: a basic biological phenomenon with wide-ranging implications in tissue kinetics. *Br J Cancer* 1972; 26: 239–257.
3. McDonnell TJ. Cell division versus cell death: a functional model of multistep neoplasia. *Mol Carcinog* 1993; 8: 209–213.
4. Camilleri-Broet S, Davi F, Feuillard J, Bourgeois C, Seilhean D, Hauw JJ, Raphael M. High expression of latent membrane protein 1 of Epstein-Barr virus and BCL-2 oncoprotein in acquired immunodeficiency syndrome-related primary brain lymphomas. *Blood* 1995; 86: 432–435.
5. Migheli A, Cavalla P, Piva R, Giordana MT, Schiffer D. Bcl-2 protein expression in aged brain and neurodegenerative diseases. Neuroreport 1994; 5: 1906–1908.
6. Sukharev SA, Pleshakova OV, Sadovnikov V. Role of proteases in activation of apoptosis. *Cell Death Differ* 1997; 4: 457–462.
7. Möröy T, Zörnig M. Regulators of life and death: the Bcl-2 gene family. *Cell Physiol Biochem* 1996; 6: 312–336.
8. Kroemer G. Mitochondrial implication in apoptosis. Towards an endosymbiot hypothesis of apoptosis evolution. *Cell Death Differ* 1997; 4: 443–456.
9. Reed JC. Cytochrome c: Can't live with it—Can't live without it. *Cell* 1997; 91: 559–562.
10. Borner C, Olivier R, Martinou I, Mattmann C, Tschopp J, Martinou J-C. Dissection of functional domains in Bcl-2 alpha by site-directed mutagenesis. *Biochem Cell Biol* 1994; 72: 463–469.
11. Yin X-M, Olivai ZN, Korsmeyer SJ. BH1 and BH2 domains of Bcl-2 are required for inhibition of apoptosis and heterodimerization with Bax. *Nature* 1994; 369: 321–323.
12. Farrow SN, Brown R. New members of the Bcl-2 family and their protein partners. *Curr Opin Gen Dev* 1996; 6: 45–49.
13. Borner C, Martinou I, Mattmann C, Irmler M, Schaerer E, Martinou J-C, Tschopp J. Protein Bcl-2α does not require membrane attachment, but two conserved domains to suppress apoptosis. *J Cell Biol* 1994; 126: 1059–1068.
14. Cheng EH-Y, Levine B, Boise LH, Thompson CG, Hardwick JM. Bax-independent inhibition of apoptosis by Bcl-X$_L$. *Nature* 1996; 379: 554–556.

•15. Chittenden T, Flemington C, Houghton AB, Ebb RG, Gallo GJ, Elangovan B, Chinnadurai G, Lutz RJ. Conserved domain in Bak, distinct from BH1 and BH2, mediates cell death and protein binding functions. *EMBO J* 1995; 14: 733–736.

16. Hunter JJ, Parslow TG. Peptide sequence from Bax that converts Bcl-2 into an activator of apoptosis. *J Biol Chem* 1996; 271: 8521–8524.

•17. Wang K, Yin XM, Chao DT, Milliman CL, Korsmeyer SJ. BID: a novel BH3 domain-only death agonist. *Genes Dev* 1996; 10: 2859–2869.

•18. Zha J, Harada H, Osipov K, Jockel J, Waksman G, Korsmeyer SJ. BH3 domain of BAD is required for heterodimerization with BCL-XL and pro-apoptotic activity. *J Biol Chem* 1997; 272: 24,101–24,104.

19. Wang HG, Miyashita T, Takayama S, Sato T, Torigoe T, Krajewski S, et al. Apoptosis regulation by interaction of Bcl-2 protein and Raf-1 kinase. *Oncogene* 1994; 9: 2751–2756.

20. Shibasaki F, Kondo E, Akagi T, McKeon F. Suppression of signalling through transcription factor NF-AT by interactions between calcineurin and Bcl-2. *Nature* 1997; 386: 728–731.

21. Huang DC, Adams JM, Cory S. Conserved N-terminal BH4 domain of Bcl-2 homologues is essential for inhibition of apoptosis and interaction with CED-4. *EMBO J* 1998; 17: 1029–1039.

•22. Cheng EH, Kirsch DG, Clem RJ, Ravi R, Kastan MB, Bedi A, Ueno K, Hardwick JM. Conversion of Bcl-2 to a Bax-like death effector by caspases. *Science* 1997; 278: 1966–1968.

23. Grandgirard D, Studer E, Monney L, Belser T, Fellay I, Borner C, Michel MR. Alphaviruses induce apoptosis in Bcl-2-overexpressing cells: evidence for a caspase-mediated, proteolytic inactivation of Bcl-2. *EMBO J* 1998; 17: 1268–1278.

24. Reed JC. Double identity for proteins of the Bcl-2 family. *Nature* 1997; 387: 773–776.

••25. Muchmore SW, Sattler M, Liang H, Meadows RP, Harlan JE, Yoon HS, et al. X-ray and NMR structure of human Bcl-x$_L$, an inhibitor of programmed cell death. *Nature* 1996; 381: 335–341.
The 3-D structure of one of the Bcl-2 family proteins (Bcl-X$_L$) provided indications that these proteins could possess channel forming activity, which was also shown later (see refs. 30–32).

•26. Chang BS, Minn AJ, Muchmore SW, Fesik SW, Thompson CB. Identification of a novel regulatory domain in Bcl-X(L) and Bcl-2. *EMBO J* 1997; 16: 968–977.

27. Haldar S, Chintapalli J, Croce CM. Taxol induces bcl-2 phosphorylation and death of prostate cancer cells. *Cancer Res* 1996; 56: 1253–1255.

28. Roth W, Wagenknecht B, Grimmel C, Dichgans J, Weller M. Taxol-mediated augmentation of CD95 ligand-induced apoptosis of human malignant glioma cells: association with bcl-2 phosphorylation but neither activation of p53 nor G2/M cell cycle arrest. *Br J Cancer* 1998; 77: 404–411.

•29. Maundrell K, Antonsson B, Magnenat E, Camps M, Muda M, Chabert C, et al. Bcl-2 undergoes phosphorylation by c-Jun N-terminal kinase/stress-activated protein kinases in the presence of the constitutively active GTP-binding protein Rac1. *J Biol Chem* 1997; 272: 25,238–25,242.

•30. Minn AJ, Velez P, Schendal SL, Liang H, Muchmore SW, Fesik SW, Fill M, Thompson CB. Bcl-x$_L$ forms an ion channel in synthetic lipid membranes. *Nature* 1997; 385: 353–357.

•31. Schendel SL, Xie Z, Montal MO, Matsuyama S, Montal M, Reed J. Channel formation by antiapoptotic protein Bcl-2. *Proc Natl Acad Sci* 1997; 94: 5113–5118.

•32. Antonsson B, Conti F, Ciavatta AM, Montesuit S, Lewis S, Martinou I, et al. Inhibition of Bax channel forming activity by Bcl-2. *Science* 1997; 277: 370–372.

33. Krajewski S, Tanaka S, Takayama S, Schibler MJ, Fenton W, Reed JC. Investigation of the subcellular distribution of the bcl-2 oncoprotein: residence in the nuclear envelope, endoplasmic reticulum and outer mitochondrial membranes. *Cancer Res* 1993; 53: 4701–4714.

•34. Hsu Y-T, Wolter KG, Youle RJ. Cytosol-to-membrane redistribution of Bax and Bcl-x_L during apoptosis. *Proc Natl Acad Sci* 1997; 94: 3668–3672.

••35. Newmeyer DD, Farschon DM, Reed JC. Cell-free apoptosis in Xenopus egg extracts: inhibition by Bcl-2 and requirement for an organelle fraction enriched in mitochondria. *Cell* 1994; 79: 353–364.
First paper suggesting that mitochondria may play a key role in apoptosis.

••36. Li, P Nijhawan D, Budihardjo I, Srinivasula SM, Ahmad M, Alnemri ES, Wang X. Cytochrome c and dATP-dependent formation of Apaf-1/caspase-9 complex initiates an apoptotic protease cascade. *Cell* 1997; 91: 479–489.
Provides evidence for a direct involvement of cytochrome-c in the downstream apoptotic signaling cascade because it participates in the activation of a caspase.

37. Rosse T, Olivier R, Monney L, Rager M, Conus S, Fellay I, Jansen B, Borner C. Bcl-2 prolongs cell survival after Bax-induced release of cytochrome c. *Nature* 1998; 391: 496–499.

38. Bossy-Wetzel E, Newmeyer DD, Green DR. Mitochondrial cytochrome c release in apoptosis occurs upstream of DEVD-specific caspase activation and independently of mitochondrial transmembrane depolarization. *EMBO J* 1998; 17: 37–49.

39. Susin SA, Zamzami N, Castedo M, Hirsch T, Marchetti P, Macho A, et al. Bcl-2 inhibits the mitochondrial release of an apoptogenic protease. *J Exp Med* 1996; 184: 1331–1341.

40. Hennet T, Bertoni G, Richter C, Peterhans E. Expression of BCL-2 protein enhances the survival of mouse fibrosarcoid cells in tumor necrosis factor-mediated cytotoxicity. *Cancer Res* 1993; 53: 1456–1460.

41. Marzo I, Brenner C, Zamzami N, Sisin SA, Beutner G, Brdiczka D, et al. The permeability transition pore complex: a target for apoptosis regulation by caspases and Bcl-2-related proteins. *J Exp Med* 1998; 187: 1261–1271.

42. Halestrap AP, Davidson AM. Inhibition of Ca2(+)-induced large-amplitude swelling of liver and heart mitochondria by cyclosporin is probably caused by the inhibitor binding to mitochondrial-matrix peptidyl-prolyl *cis-trans* isomerase and preventing it interacting with the adenine nucleotide translocase. *Biochem J* 1990; 268: 153–160.

43. Petronilli V, Nicolli A, Costantini P, Colonna R, Bernardi P. Regulation of the permeability transition pore, a voltage-dependent mitochondrial channel inhibited by cyclosporin A. *Biochim Biophys Acta* 1994; 1187: 255–259.

••44. Shimizu S, Eguchi Y, Kamiike W, Funahashi Y, Mignon A, Lacronique V, Matsuda H, Tsujimoto Y. Bcl-2 prevents apoptotic mitochondrial dysfunction by regulating proton flux. *Proc Natl Acad Sci* 1998; 95: 1455–1459.
Bcl-2 is shown to function as a proton channel, and, through this activity, protects mitochondria during apoptosis.

45. Kane DJ, Sarafian TA, Anton R, Hahn H, Gralla EB, Valentine JS, Ord T, Bredesen DE. Bcl-2 inhibition of neural death: decreased generation of reactive oxygen species. *Science* 1993; 262: 1274–1277.

45. Hockenbery DM, Oltvai ZN, Yin XM, Milliman CL, Korsmeyer SJ. Bcl-2 functions in an antioxidant pathway to prevent apoptosis. *Cell* 1993; 75: 241–251.

47. Jurgensmeier JM, Xie Z, Deveraux Q, Ellerby L, Bredesen D, Reed JC. Bax directly induces release of cytochrome c from isolated mitochondria. *Proc Natl Acad Sci* 1998; 95: 4997–5002.

48. Eskes R, Antensson B, Osen-Sand A, Montessuit S, Richter C, Sadoul R, Mazzi G, Nichols A, Martinen J-C. Bax-induced cytochrome C release from mitochondria is independent of the permeability transition pore but highly dependent on Mg^{2+} ions. *J Cell Biol* 1998; 143: 217–224.

49. Kantrow SP, Piantadosi CA. Release of cytochrome c from liver mitochondria during permeability transition. *Biochem Biophys Res Commun* 1997; 232: 669–671.

50. Vander Heiden MG, Chandel NS, Williamson EK, Schumacker PT, Thompson CB. Bcl-X_L regulates the membrane potential and volume homeostasis of mitochondria. *Cell* 1997; 91: 627–637.

•51. Mancini M, Anderson BO, Caldwell E, Sedghinasab M, Paty PB, Hockenbery DM. Mitochondrial proliferation and paradoxical membrane depolarization during terminal differentiation and apoptosis in a human colon carcinoma cell line. *J Cell Biol* 1997; 138: 449–469.

52. Hsu SY, Kaipia A, McGee E, Lomeli M, Hsueh AJ. Bok is a pro-apoptotic Bcl-2 protein with restricted expression in reproductive tissues and heterodimerizes with selective anti-apoptotic Bcl-2 family members. *Proc Natl Acad Sci USA* 1997; 94: 12,401–12,406.

53. O'Connor L, Strasser A, O'Reilly LA, Hausmann G, Adams JM, Cory S, Huang DC. Bim: a novel member of the Bcl-2 family that promotes apoptosis. *EMBO J* 1998; 17: 384–395.

54. Inohara N, Ekhterae D, Garcia I, Carrio R, Merino J, Merry A, Chen S, Nunez G. Mtd, a novel Bcl-2 family member activates apoptosis in the absence of heterodimerization with Bcl-2 and Bcl-XL. *J Biol Chem* 1998; 273: 8705–8710.

10 Mechanism of Action of the Proapoptotic Gene *Bak*

Robin Brown, PhD

CONTENTS

ABSTRACT

The mechanism of action of the proapoptotic gene bak *has so far proved to be elusive. Early molecular clues suggested that* bak *functioned as a competitive inhibitor of protective members of the* bcl-2 *family. Although this might still be an explanation for the apoptosis induced by overexpression of* bak *in cells, more recent data suggests that* bak *may have a role distinct from the other proapoptotic gene,* bax. *Important questions remain to be answered for* bak. *In particular, different subcellular compartments are occupied in different cells, suggesting that a general* bak *function might disrupt the integrity of a number of key organelles.*

From: *Apoptosis and Cancer Chemotherapy*
Edited by: J. A. Hickman and C. Dive © Humana Press Inc., Totowa, NJ

INTRODUCTION

With the general availability of sequence databases, investigators have become used to identifying the unknown function of a protein by searching for regions of homology shared with proteins of known function. For the Bcl-2 family of proteins, shared domains have been identified within the family, but have not been ascribed a function by homology with other proteins in the databases. Expression experiments in cells in culture have established that the proteins fall essentially into two classes, proapoptotic or antiapoptotic. It is these experiments, coupled with genetic experiments in the nematode *Caenorhabditis elegans (1•)* and in *Drosophila (2)*, that have defined the cell-death pathway. Despite this, it is still not certain what the mechanism of action of these proteins is.

HOW DOES bak COMMIT CELLS TO APOPTOSIS?

Expression studies have provided few surprises and remarkably little insight into *bak*'s mechanism of action. In almost all cells, overexpression of bak from a strong viral or cellular promoter results in the induction of cell death. A single exception exists in the lymphoblastoid cell line WI-L2 *(3)*. In this background, cells expressing *bak* are actually protected from apoptosis induced by serum withdrawal or by menadione treatment. Clearly, some different set of molecular interactions occurs in these cells that may be related to the presence of Epstein-Barr virus (EBV) proteins (the line is EBV-transformed), some of which have been implicated in the control of apoptosis.

Bak AND Bcl-2 STRUCTURE

Bak, together with the five mammalian proteins most closely related to Bcl-2, contains the three signature domains of the family, the BH1, BH2, and BH3 domains *(4–6)*. It is these domains that allow bak to interact with a subset of protective members of the family. Data from the solution of the crystal structure of Bcl-X_L underscores the structural significance of the three domains, which appear to be in close proximity and form an extended hydrophobic cleft. Together with the arrangement of the α helices, the Bcl-X_L structure is reminiscent of the structure of the transmembrane domain of the diphtheria toxin *(7••)*. The bacterial toxins, by insertion into membranes, form pH-dependent membrane pores. This attractive hypothesis may account for the ability of the Bcl-2 proteins to regulate the ionic balance of the mitochondria or the endoplasmic reticulum. It should be noted, however, that the bacterial toxins form channels in the plasma membrane of cells, and not in intracellular membranes. This is important because the plasma membrane is exposed to much greater pH fluctuation. Experiments in vitro that have examined the ability of Bcl-2 family proteins to form channels in isolated lipid membranes have proved inconclusive *(8,9)*. Although the channel formed by Bcl-X$_L$ appears to be cation selective, it is only mildly selective at pH 4.0. These experiments have also established that *bax* channel-forming activity can be blocked in vitro by Bcl-2 *(10)*. This pH is very unlikely to be sustained in mammalian cells, but it is not possible to rule out transient pH changes regulating the channel. Acidic pH does appear to induce dimerization of Bcl-2 *(11)*.

A fourth domain, shared only by the protective members of the family Bcl-2, Bcl-X$_L$ and Bcl-w, is not present in either proapoptotic *bak* or *bax*. Recent experiments have suggested that the function of this domain is to allow the protective members of the family to interact with Apaf-1, the mammalian homolog of ced-4 *(12)*. Apaf-1 was purified as an adenosine triphosphate (ATP)-dependent activity required for caspase activation in cytosolic extracts *(13••)*. In addition to Apaf-1, Apaf-2 is also capable of activating the caspases in vitro. Apaf-2 was identified as cytochrome-*c;* Apaf-1 shares homology with the *C. elegans* genes ced-3 and ced-4. This remarkable protein, therefore, contains domains that allow it to function like these two central cell-death regulators of programmed cell death in *C. elegans*. The ced-3 homology domain is involved in protein–protein interactions with caspases, and the ced-4 homology domain may be responsible for nucleotide (ATP) binding.

In the nematode *C. elegans,* activation of the cell death caspase ced-3 requires ced-4 function. The two proteins form a complex with the worm Bcl-2 homolog ced-9 *(14–16)*. By analogy, the role of the protective members of the family in mammalian cells is to prevent the interaction of Apaf-1 with one of the caspases, blocking cell death. Because neither bak nor bax can interact with Apaf-1, the death-promoting potential of these proteins resides in their ability to compete with Apaf-1 for binding to Bcl-2, resulting in the interaction of free Apaf-1 with a caspase. In mammalian cells, the caspase present in the complex may vary, depending on the localization of the complex in the cell.

ORGANELLE-SPECIFIC DEATH COMPLEXES

Access to the death pathway from the plasma membrane receptor Fas requires the recruitment of a signal-transduction complex, termed the death-inducing signaling complex (DISC) *(17•)*. This complex results in the direct activation of procaspase-8 at the plasma membrane, and the initiation of an apoptotic cascade *(18)*. The protein–protein interactions required to assemble the complex depend on the presence of a number of domains on the proteins, including death domains, death-effector domains, and card domains. The complex includes both activators and inhibitors of the pathway, and the exact composition of proteins in the complex determines whether the pathway is activated or not *(19)*. It appears from recent experiments that the plasma membrane may not be the only site within the cell at which such a complex is assembled. In the endoplasmic reticulum, recent results suggest the presence of such a complex.

The endoplasmic reticulum-resident protein Bap31 is an integral membrane protein that binds to Bcl-X$_L$ *(20)*. By analogy with the Fas receptor disk complex, Bap31 contains a weakly homologous death-effector domain. Other components of the complex include procaspase-8 and ced-4 (Apaf-1) *(21)*. The chain of events following an endoplasmic reticulum-resident apoptotic signal appears to involve the cleavage of Bap31 by activated caspase-8. The resulting Bap31 fragment (p20) is then capable of initiating apoptosis. Bax, and presumably Bak, then prevent Bcl-X$_L$ from becoming a part of this complex, and so probably allow procaspase-8 activation and cleavage to take place *(20)*. Therefore, the endoplasmic reticulum has direct access to the apoptotic machinery via the protein Bap31.

In HeLa cells, the subcellular distribution of Bak suggests that the protein is predominantly localized to the endoplasmic reticulum, with only a small fraction of the protein in mitochondria. This pattern does appear to vary with cell type, but strongly suggests that, at least in HeLa cells, Bak initiates apoptosis from the endoplasmic reticulum. Bak was originally cloned as a binding partner of the adenovirus E1B19k protein, which is localized to the endoplasmic reticulum/ nuclear envelope *(22)*. In a wide variety of cells, E1B19k is a potent blocker of apoptosis, and its correct subcellular distribution is required for its antiapoptotic function *(22)*. Therefore, E1B19K must be blocking apoptosis by preventing activation of a program resident in the endoplasmic reticulum.

Recent genetic analysis of the lethality of bak in the yeast *Saccharomyces pombe* and bax in *Saccharomyces cerevisiae* has implicated two endoplasmic reticulum-resident proteins, calnexin and BI-1, respectively. In *S. pombe,* calnexin is required for bak-dependent lethality *(23)*. Because calnexin functions as part of the quality-control machinery involved in the folding of *N*-glycosylated proteins, this is another example of the link between endoplasmic reticulum function and apoptosis. By contrast, bax lethality in yeast is blocked by the overexpression of the human endoplasmic reticulum-resident protein BI-1, suggesting that a critical component of bax lethality in yeast is present in the endoplasmic reticulum *(24)*.

The presence of these interacting proteins in the endoplasmic reticulum suggests that a number of signaling pathways exist in this organelle, some of which are coupled to apoptosis, because it appears that the endoplasmic reticulum's response to insult is graded (Fig. 1). Initially, in an attempt to cope with increasing accumulation of denatured or unfolded proteins, the cell upregulates a number of proteins, including the endoplasmic reticulum chaperone, glucose response protein 78 (GRP78 [Bip]) *(25)*. These proteins attempt to restrict the damage by functioning as molecular chaperones to the lumenal proteins, and, in the case of Bip, may help to restore integrity to the endoplasmic reticulum membrane *(26)*. Renal epithelial cells treated with iodoacetamide, an alkylating agent, upregulate both HSP70 and GRP78 *(25)*. Antisense experiments that target the downregulation of GRP78 lower the tolerance of the cell to idoacetamide. In a similar model, using the mouse lymphoma cell line WEHI7.2, cells that undergo apoptosis in response to thapsigargin fail to generate a calcium-mediated GRP78/GRP94 stress response *(26)*. The GRP78/GRP94 stress response is operative in these cells when induced by tunicamycin, which suggests that there are at least two signaling pathways to GRP/74/GRP98 in most cells, but, in WEHI7.2 cells, one of these has been lost.

The pathway that signals from the endoplasmic reticulum to the nucleus is poorly understood, but it is possible to partially reconstruct a framework for the response. The pathway is probably initiated by different forms of endoplasmic reticulum stress, including Ca^{2+} disturbances, defects in N-linked glycosylation, or trafficking. The secondary response to some of these insults will be the stress response mediated by GRP78/GRP94. Finally, a transcriptional response initiated through the transcription factor C/EBP homologous protein (CHOP) induces apoptosis *(27)*. CHOP is a member of the HLH family of transcription factors that is induced by endoplasmic reticulum stress *(28)*. Expression of CHOP alone in cells does not commit them to apoptosis in the absence of a stress response, suggesting that CHOP targets in the nucleus, or that an essential cofactor is not present unless the cell has received a previous insult. Finally, embryo-derived fibroblasts from CHOP−/− mice are significantly less susceptible to apoptosis induced by endoplasmic reticulum stress *(28)*.

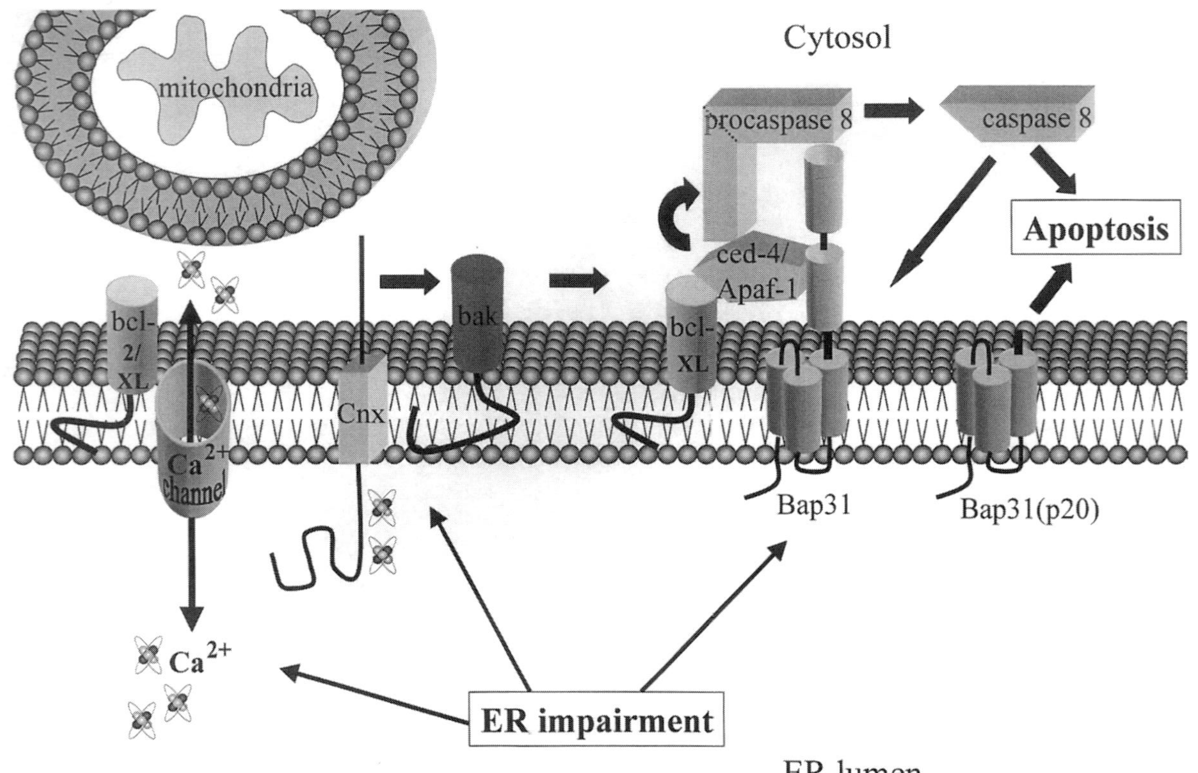

Fig. 1. Endoplasmic reticulum membrane is the site for regulation of an apoptosis program (*see* text for explanation).

But how do bak and Bcl-2 interact with this pathway? One possibility is that critical pores on the endoplasmic reticulum membrane are under bak/Bcl-2 control. This may be related to the function of GRP78 (Bip) in the endoplasmic reticulum, which, during translocation of the nascent polypeptide chain from the ribosome to the lumen of the endoplasmic reticulum, seals the translocon pore, and maintains the integrity of the endoplasmic reticulum at a time when it would be vulnerable to the collapse of its permeability barrier with the cytosol *(26)*. If some of the antiapoptotic effects of bip are related to this activity, Bcl-2-family protectors and killers might play a role in modifying the opening of endoplasmic reticulum membrane pores during normal and pathological conditions.

SUBCELLULAR DISTRIBUTION

A substantial body of evidence now exists suggesting that the Bcl-2 family of proteins initiate or block apoptosis through effects on the mitochondria. Apoptosis appears to be initiated at the mitochondria by the release of cytochrome-*c*, resulting in the activation of cytoplasmic caspases. This is blocked in cells overexpressing bcl-2 *(30,31)*. In addition, Bcl-X_L appears to play a role in the regulation of mitochondrial volume and the maintenance of membrane potential *(32)*.

In the case of CD95 (APO-1/Fas) signaling, an intriguing set of results suggest that there may be considerable complexity to a cell's response, which depends once more on the particular cell background *(33)*. Cells in culture can be classified based on their response to Fas-induced apoptosis. Type 1 cells are characterized by the rapid activation of caspase-3 and caspase-8 by the receptor-associated DISC complex. Type 2 cells show limited recruitment of a DISC complex and much slower activation of both proteases. In both these cell types all mitochondrial apoptogenic activities are blocked by Bcl-2 or Bcl-X_L over expression, but only in type 2 cells does this result in blocking CD95-mediated apoptosis. Consistent with these data is the observation that only in type 2 cells does Bcl-2/Bcl-X_L overexpression inhibit caspase activation. The model for the interaction of the cell-death proteins present in the nematode *C. elegans* suggests that they are most similar to the type 2 cells. Type 1 cells, which bypass the mitochondria in favor of direct activation of the caspase(s), would then appear to use a more recent evolutionary program. In this sense, one must return to the concept that what is true for the DISC complex at the plasma membrane is also true for the distribution of other caspase complexes in the cell. So mitochondria may be involved both mechanistically and incidentally in cell death.

Although the timing of these events relative to each other remains controversial, both events appear to be blocked by overexpression of Bcl-2 or Bcl-X$_L$, strongly suggesting that the mitochondria are one of the key organelles involved in the initiation of apoptosis.

Although the Bcl-2 family proteins are normally resident in the mitochondrial outer membrane, they are also present on the endoplasmic reticulum/nuclear membrane *(34,35)*. Bcl-2, targeted to the mitochondria or to the endoplasmic reticulum, protects the cell from apoptosis to a degree that is dependent on the cell background. In MDCK cells, Bcl-2, localized to the mitochondria, is effective in blocking apoptosis induced by serum starvation, but is inactive if localized to the endoplasmic reticulum *(36)*. Conversely, endoplasmic reticulum-localized Bcl-2, expressed in Rat1/myc cells, is much more effective if targeted to the endoplasmic reticulum. These results suggest that Bcl-2 does not have access to the apoptotic complex in Rat1/Myc cells, if located on the mitochondria. Because the proapoptotic genes antagonize this function, each organelle may have its own complex with which it can initiate apoptosis. In addition, cells may respond to specific apoptotic stimuli by initiating apoptosis in specific organelles.

The Bcl-2 family of proteins, therefore, have a role in the function and policing of the endoplasmic reticulum, which is both a specialized compartment for folding and assembly of secretory proteins, and an important sensor for some types of cell stress that can lead to apoptosis. Indeed, the normal endoplasmic reticulum responds to a variety of signals from cell-surface receptor, which maintain the expression of the endoplasmic reticulum-resident chaperones GRP78 (Bip) and GRP94 under nonstressed conditions *(37)*. In addition, there is also a well-described endoplasmic reticulum signaling pathway that responds to the accumulation of unfolded proteins by transcriptionally upregulating a set of endoplasmic reticulum-resident proteins *(38,39)*. This endoplasmic reticulum lumen-to-nucleus signaling pathway is referred to as the unfolded protein response, and is activated by the accumulation of misfolded proteins in the endoplasmic reticulum. Compounds that affect endoplasmic reticulum function, such as thapsigargin and tunicamycin, are capable of inducing GRP78 and GRP94 and apoptosis *(40)*. But how is this sensing capability of the endoplasmic reticulum linked to apoptosis?

DEATH IS A CALCIUM-MEDIATED EVENT

The involvement in apoptosis of both the mitochondria and the endoplasmic reticulum raises the question of how these organelles activate the effector caspases. Because it is most likely that bak and bax perform the same function in both these organelles, there should be some common effector. The best candidate for this is probably calcium. The cells' intracellular stores of calcium reside principally in two organelles, the endoplasmic reticulum and mitochondria. The calcium concentration in cells is extremely low in the cytosol ($<10^{-7}M$), but in the endoplasmic reticulum it is orders of magnitude higher. This gradient between the endoplasmic reticulum and cytosol is maintained by a Ca^{2+}-ATPase that pumps calcium back into the endoplasmic reticulum against the concentration gradient. Opening of endoplasmic reticulum-resident calcium channels in response to signals permits a rapid and transient rise in the cytosolic calcium concentration and the activation of calcium-sensitive enzymes. Should the calcium concentration rise to dangerous levels in the cytosol, a low-affinity but high-capacity calcium pump in the mitochondria removes calcium from the cytosol to the mitochondria. Agents that disrupt the function of the endoplasmic reticulum calcium pump induce apoptosis. In particular, the plant-derived compound, thapsigargin, selectively inhibits the endoplasmic reticulum pump, and induces apoptosis in a wide variety of cells. Apoptosis induced by thapsigargin is blocked by overexpression of bcl-2. The antiapoptotic effect appears to be caused by the ability of bcl-2 to maintain calcium uptake into the endoplasmic reticulum *(41)*. Low extracellular calcium reduces the ability of Bcl-2 to protect thapsigargin (TG)-treated cells. This capacitative re-entry of calcium may be required to fully restore the endoplasmic reticulum calcium pool, which is something Bcl-2 cannot completely accomplish, because of its endoplasmic reticulum membrane-bound location. The Bcl-2 block of TG-induced apoptosis also appears to restore function to the endoplasmic reticulum. TG induces a delay in the endoplasmic reticulum processing of cathepsin D *(42)*. In Bcl-2-overexpressing cells, this delay is prevented, suggesting some restoration of endoplasmic reticulum function in the presence of calcium disturbances. Because the proapoptotic proteins also reside on this membrane, and can bind to Bcl-2, an attractive model based on competition would be that Bcl-2 is, or directly affects a calcium-dependent pore or transporter in the endoplasmic reticulum, and this is blocked by bak or bax. Overexpression of bak in cells results in apoptosis, because of the inability of the cell to maintain calcium homeostasis in the endoplasmic reticulum, or to employ the emergency calcium response of the mitochondria. This leads to the idea that the stress response is the sensor, and that while calcium disregulation is the effector. The link between the stress response and the disregulation of calcium is not clear, principally because we do not know which calcium-response proteins induce apoptosis.

ORDER ON THE PATHWAY

Understanding the chronology of molecular events related to bak and the Bcl-2 family proteins is an important issue. In the nematode *C. elegans,* the apoptotic pathway has been ordered genetically, but, crucially, this gives only limited insight into the biochemical events. Epistatic analysis places ced-9 upstream of the caspase ced-3, but does not establish whether ced-9 inhibits the active or inactive form of ced-3. Mammalian experiments aimed at elucidating this aspect of the pathway have established that both Bik (a BH3-containing Bcl-2 family member) and Bak appear to act downstream of the block imposed by Crm-A *(43).* Because Crm-A preferentially inhibits ced-3/ICE and caspase-8, these experiments suggest that Bak induces apoptosis by activation of downstream caspases. This does indeed seem to be the case. Cell death induced by Bik and Bak results in the activation of caspase-7 and the cleavage of the caspase substrate poly-ADP ribose polymerase (PARP). In contrast to Crm-A, apoptosis induced by Bak and Bik is inhibitable by the caspase inhibitor *iap.* Although these experiments provide a more complete picture of the order of events related to Bak killing and caspase activation, they still do not establish whether Bak and Bik act directly on the caspase enzyme itself, or on an inhibitor of the enzyme. The most compelling evidence that Bcl-X_L interacts directly in a complex comes from experiments in which epitope-tagged Apaf-1 was immunoprecipitated with caspase-9 from 293 cells. Apaf-1 appears to co-immunoprecipitate only with caspase-9 *(44,45),* unlike the *C. elegans* homolog of Apaf-1, ced-4, which associates with most prodomain caspases. In addition to its association with caspase-9, Apaf-1 also interacts with Bcl-X_L, but this binding is mediated by a region on Apaf-1 distinct from that which binds caspase-9. Thus, Apaf-1, caspase-9, and Bcl-X_L form a complex in 293 cells, under conditions in which one of the components is overexpressed. Co-expression of either Bak or Bax attenuated the interaction of Bcl-X_L with Apaf-1. Although this is a predictable result, because both Bak and Bax bind directly to Bcl-X_L, it once again suggests that pro- and antiapoptotic protein of the Bcl-2 family alter the composition of a caspase–Apaf-1 complex, resulting in the activation or repression of caspase activity.

CONCLUSIONS

The distribution of the Bcl-2 family of proteins, including Bak, suggests two things. First, the mechanism of action of the proteins will be concerned with membrane-mediated events. Because the proteins are present in both the mitochondrial and endoplasmic reticulum/nuclear membrane, models of apoptosis must explain the mechanism of action, so that, despite the downstream consequences of Bak or Bcl-2 activity, the primary effect at the membrane is the same. Second, a great deal of the evidence that suggests the mechanism of action of the Bcl-2 family proteins is based on overexpression experiments. These data, coupled with immunoprecipitation in overexpressing cells, has a great potential to mislead. These experiments are almost never quantitative, and it is often impossible to assess the relative affinities for the protein partners. If one molecule in 1000 of protein X binds to a Bcl-2 protein, is this significant? It is possible to do these experiments using quantitative methodology, or even to use specific crosslinking agents in nonoverexpressing cells.

Finally, it must be acknowledged that these interactions take place within the architecture of the cell, and if one speculates on the interaction of Bcl-2 with other proteins, the subcellular localization must justify this. The Bcl-2 proteins can now be put, with some confidence, at their correct position on the apoptotic pathway, but their mechanism of action is still elusive.

Apoptosis research has been in a phase that has greatly helped in understanding the complexity of the protein–protein interactions that control the pathway. What is needed for the future are much more careful experiments in normal cells that will reveal the specifics.

REFERENCES

• of special interest
•• of outstanding interest

•1. Hengartner MO. Death cycle and Swiss army knives. *Nature* 1998; 391: 441,442. *Up-to-date review, despite the somewhat obscure analogy.*

2. McCall K, Steller H. Facing death in the fly: genetic analysis of apoptosis in Drosophila. *Trends Genet* 1997; 13: 222–226.

3. Keifer MC, Brauer MJ, Powers VC, Wu JJ, Umansky SR, Tomei LD, Barr PJ. Modulation of apoptosis by the widely distributed bcl-2 homologue *Bak. Nature* 1995; 374: 736–739.

4. Yin X-M, Oltvai ZN, Korsmeyer SJ. BH1 and BH2 domains of bcl-2 are required for inhibition of apoptosis and heterodimerisation with bax. *Nature* 1994; 369: 321–323.

5. Zha H, Aime-Sempe C, Sato T, Reed JC. Proapoptotic protein Bax heterodimerizes with Bcl-2 and homodimerizes with Bax via novel domain (BH3) distinct from BH1 and BH2. *J Biol Chem* 1996; 271: 7440–7444.

6. Chittenden T, Flemington C, Houghton AB, Ebb RG, Gallo GJ, Elangovan B, Chinnadurai G, Lutz RJ. Conserved domain in bak, distinct from BH1 and BH2, mediates cell death and protein binding functions. *EMBO J* 1995; 14: 5589–5596.

••7. Muchmore SW, Sattler M, Liang H, Meadows RP, Harlan JE, Yoon HS, et al. X-ray and NMR structure of human bcl-X_L, an inhibitor of programmed cell death. *Nature* 1996; 381: 335–341.
   Is this the most significant paper since the cloning of Bcl-2? Maybe.

8. Minn AJ, Velez P, Schendel SL, Liang H, Muchmore SW, Fesik SW, Fill M, Thompson CB. Bcl-x(L) forms an ion channel in synthetic lipid membranes. *Nature* 1997; 385: 353–357.

9. Schendel SL, Xie Z, Montal MO, Matsuyama S, Montal M, Reed JC. Channel formation by antiapoptotic protein Bcl-2. *Proc Natl Acad Sci USA* 1997; 94: 5113–5118.

10. Antonsson B, Conti F, Ciavatta A, Montessuit S, Lewis S, Martinou I, et al. Inhibition of Bax channel-forming activity by Bcl-2. *Science* 1997; 277: 370–372.

11. Xie Z, Schendel S, Matsuyama S, Reed JC. Acidic pH promotes dimerization of Bcl-2 family proteins. *Biochemistry* 1998; 37: 6410–6418.

12. Huang DCS, Adams JM, Cory S. Conserved N-terminal BH4 domain of bcl-2 is essential fo inhibition of apoptosis and interaction with ced-4. *EMBO J* 1998; 17: 1029–1039.

••13. Zou H, Henzel WJ, Liu X, Lutschg A, Wang X. Apaf-1, a human protein homologous to *C. elegans* CED-4 participates in cytochrome c-dependent activation of caspase-3. *Cell* 1997; 90: 405–413.
   For a long time, the missing piece in the puzzle to connect the nematode cell-death pathway to its mammalian equivalent was ced-4. This paper described the mammalian homolog, and one would not have guessed how interesting it turned out to be.

14. Wu D, Wallen HD, Nunez G. Interaction and regulation of subcellular localization of CED-4 by CED-9. *Science* 1997; 275: 1126–1129.

15. Chinnaiyan AM, O'Rourke K, Lane BR, Dixit VM. Interaction of CED-4 with CED-3 and CED-9: a molecular framework for cell death. *Science* 1997; 275: 1122–1126.

16. Seshagiri S, Miller LK. *Caenorhabditis elegans* CED-4 stimulates CED-3 processing and CED-3-induced apoptosis. *Curr Biol* 1997; 7: 455–460.

•17. Kischkel FC, Hellbardt S, Behrmann I, Germer M, Pawlita M, Krammer PH, Peter ME. Cytotoxicity-dependent APO-1 (Fas/CD95)-associated proteins form a death-inducing signaling complex (DISC) with the receptor. *EMBO J* 1995; 14: 5579–5588.
   Demonstrated that one should think much more about complexes in apoptosis.

18. Medema JP, Scaffidi C, Kischkel FC, Shevchenko A, Mann M, Krammer PH, Peter ME. FLICE is activated by association with the CD95 death-inducing signaling complex (DISC). *EMBO J* 1997; 16: 2794–2804.

19. Yuan J. Transducing signals of life and death. *Curr Opin Cell Biol* 1997; 9: 247–251.

•20. Ng FWH, Nguyen M, Kwan T, Branton PE, Nicholson DW, Cromlish JA, Shore GC. p28 Bap31, a bcl-2/bcl-XL-and procaspase-8-associated protein in the endoplasmic reticulum. *J Cell Biol* 1997; 139: 327–338.
Endoplasmic reticulum comes in from the cold, although it has been known for a long time that apoptotic machinery exist in organelles other than mitochondria.

21. Ng FWH, Shore GC. Bcl-XL cooperatively associates with the Bap31 complex in the endoplasmic reticulum dependent on Procaspase-8 and ced-4 adaptor. *J Biol Chem* 1998; 273: 3140–3143.

22. Rao L, Modha D, White E: The E1B 19K protein associates with lamins *in vivo* and its proper localization is required for inhibition of apoptosis. *Oncogene* 1997; 15: 1587–1597.

23. Torgler CN, deTiani M, Raven MTE, Aubry J-P, Brown R, Meldrum E. Expression of bak in *S. pombe* results in a lethality mediated through interaction with the calnexin homologue cnx1. *Cell Death Differ* 1997; 4: 263–271.

24. Xu Q, Reed JC. *Mol Cell* 1: 337–346.

25. Liu H, Bowes RC, van der Walter B, Sillence C, Nagelkerke JF, Stevens JL. Endoplasmic reticulum chaperones GRP78 and calreticulin prevent oxidative stress, Ca^{2+} disturbances, and cell death in renal epithelial cells. *J Biol Chem* 1997; 272: 21,751–21,759.

26. Hamman BD, Hendershot LM, Johnson AE. BiP maintains the permeability barrier of the endoplasmic reticulum membrane by sealing the lumenal end of the translocon pore before and early in translocation. *Cell* 1998; 92: 747–758.

27. McCormick TS, McColl KS, Distelhorst CW. Mouse lymphoma cells destined to undergo apoptosis in response to thapsigargin treatment fail to generate a calcium-mediated grp78/grp94 stress response. *J Biol Chem* 1997; 272: 6087–6092.

28. Ron D, Habener JF. CHOP, a novel developmentally regulated nuclear protein that dimerizes with transcription factors C/EBP and LAP and functions as a dominant negative inhibitor of gene transcription. *Genes Dev* 1992; 6: 439–453.

29. Wang X-Z, Lawson B, Brewer JW, Zinszer H, Sanjay A, Mi L-J, et al. Signals from the stressed endoplasmic reticulum induce C/EBP-Homologous protein (CHOP/GADD153). *Mol Cell Biol* 1996; 16: 4273–4280.

30. Yang J, Liu X, Bhalla K, Kim CN, Ibrado AM, Cai J, et al. Prevention of apoptosis by Bcl-2: release of cytochrome c from mitochondria blocked. *Science* 1997; 275: 1129–1132.

31. Kluck RM, Bossy-Wetzel E, Green DR, Newmeyer DD. The release of cytochrome c from mitochondria: a primary site for Bcl-2 regulation of apoptosis. *Science* 1997; 275: 1132–1136.

32. Vander Heiden MG, Chandel NS, Williamson EK, Schumacker PT, Thompson CB. Bcl-xL regulates the membrane potential and volume homeostasis of mitochondria. *Cell* 1997; 91: 627–637.

33. Scaffidi C, Fulda S, Srinivasan A, Friesen C, Li F, Tomaselli KJ, et al. Two CD95 (APO-1/Fas) signaling pathways. *EMBO J* 1998; 17: 1675–1687.

34. Lithgow T, van Driel R, Bertram JF, Strasser A. The protein product of the oncogene bcl-2 is a component of the nuclear envelope, the endoplasmic reticulum, and the outer mitochondrial membrane. *Cell Growth Differ* 1994; 5: 411–417.

35. Akao Y, Otsuki Y, Kataoka S, Ito Y, Tsujimoto Y. Multiple subcellular localization of bcl-2: detection in nuclear outer membrane, endoplasmic reticulum membrane, and mitochondrial membranes. *Cancer Res* 1994; 54: 2468–2471.

36. Zhu W, Cowie A, Wasfy GW, Penn LZ, Leber B, Andrews DW. Bcl-2 mutants with restricted subcellular location reveal spatially distinct pathways for apoptosis in different cell types. *EMBO J* 1996; 15: 4130–4141.
37. Brewer JW, Cleveland JK, Hendershot LM. A pathway distinct from the mammalian unfolded protein response regulates expression of endoplasmic reticulum chaperones in non-stressed cells. *EMBO J* 1997; 16: 7207–7216.
38. Welihinda AA, Tirasophon W, Green SR, Kaufman RJ. Gene induction in response to unfolded protein in the endoplasmic reticulum is mediated through Ire1p kinase interaction with a transcriptional coactivator complex containing Ada5p. *Proc Natl Acad Sci USA* 1997; 94: 4289–4294.
39. Sidrauski C, Walter P. Transmembrane kinase Ire1p is a site-specific endonuclease that initiates mRNA splicing in the unfolded protein response. *Cell* 1997; 90: 1031–1039.
40. Liu H, Bowes RC, van der Walter B, Sillence C, Nagelkerke JF, Stevens JL. Endoplasmic reticulum chaperones GRP78 and calreticulin prevent oxidative stress, Ca^{2+} disturbances, and cell death in renal epithelial cells. *J Biol Chem* 1997; 272: 21,751–21,759.
41. Lam M, Dubyak G, Chien L, Nunez G, Miesfield RL, Distelhorst CW. Evidence that bcl-2 represses apoptosis by regulating endoplasmic reticulum-associated Ca^{2+} fluxes. *Proc Nat Acad Sci USA* 1994; 91: 6569–6573.
42. He H, Lam M, McCormick TS, Distelhorst CW. Maintenance of calcium homeostasis in the endoplasmic reticulum by bcl-2. *J Cell Biol* 1997; 138: 1219–1228.
43. Orth K, Dixit VM. Bik and bak induce apoptosis downstream of CrmA but upstream of inhibitor of apoptosis. *J Bio Chem* 1997; 272: 8841–8844.
44. Hu Y, Benedict MA, Wu D, Inohara N, Nunez G. Bcl-XL interacts with Apaf-1 and inhibits Apaf-1-dependent caspase-9 activation. *Proc Natl Acad Sci USA* 1998; 95: 4386–4391.
45. Pan G, O'Rourke K, Dixit VM. Caspase-9, Bcl-XL, and Apaf-1 form a ternary complex. *J Biol Chem* 1998; 273: 5841–5845.

11 Killers or Clean-Up Crew
How Central Are the Central Mechanisms of Apoptosis?

Douglas R. Green, PhD
and Helen M. Beere, PhD

CONTENTS

ABSTRACT

Recent characterization of a number of specific mitochondrial events, which may or may not precede the downstream activation of apoptotic caspases, has focused much attention on the absolute requirement for mitochondrial involvement in the irreversible commitment to cell death. Although the mitochondrial release of cytochrome-c appears essential for at least some caspase-mediated events, the mechanism of its release is not clear, particularly regarding a specific role for changes in mitochondrial membrane potential and permeability transition. Mitochondrial-independent activation of caspases via activation of the Fas–FasL pathway is certainly well defined, although a role for mitochondrial-mediated amplification of this cascade supports the concept that mitochondria may ultimately authorize cellular

From: *Apoptosis and Cancer Chemotherapy*
Edited by: J. A. Hickman and C. Dive © Humana Press Inc., Totowa, NJ

demise. Evidence that mitochondria represent a convergence point for a number of pro- and antiapoptotic signals is presented in the context of their dependence on caspase-mediated events.

"We may be lost, but we're making great time."—**Yogi Berra**

INTRODUCTION

The field of apoptosis appears to be undergoing something akin to the process itself. For decades, the phenomenon was recognized, but, for a variety of regulatory reasons, this failed to trigger an explosion of activity until the late 1980s, when research into active cell death suddenly attracted significant attention. A process of initiation began in which the molecular events accompanying apoptosis were identified, and this eventually coalesced, in the mid 1990s, into a view of how apoptosis occurs. As we enter the final executioner phase the remarkable success in delineating the process is already prompting pundits to say that the field is dead, and that its researchers are poised to be eaten up by the voracious scientific community.

The goal in this brief discussion is to outline progress toward the elucidation of the central mechanisms of apoptosis, and to then take stock of what these may tell about those cell death processes we actually care about. We will argue that, although an understanding of the process is indeed being reached, there is very little real knowledge about how the mechanisms of apoptosis relate to those events in which it plays a role, in terms of the molecular control of cell life and death. One may feel confident that apoptosis is of major importance in development, oncogenesis, and inflammation, but there is only a glimmer of how the mechanisms of apoptosis are engaged in any of these situations (or even which mechanisms are the most critical).

We will propose several pathways to account for the different ways in which the central mechanisms of apoptosis function, and begin to outline models for how these mechanisms interact. By the time this discussion sees print, many more insights will have been gained from new developments in the field, and these too will need to be incorporated, if these models are to have any value.

TOWARD THE CENTRAL
MECHANISMS OF APOPTOSIS

The finding that the *Caenorhabditis elegans* ced-3 gene is a cysteine protease, with homology to interleukin-1β converting enzyme (ICE) (caspase-1) *(1••)*, and when overexpressed induced apoptosis in mam-

Table 1
Some Substrates Whose Cleavage Contributes to Cell Death
and/or Apoptotic Phenotype

Substrate	Caspase	Function	Ref.
Inhibitor of caspase-activated DNase/DNA Fragmentation Factor 45	3	Associates with and inhibits endonuclease; cleavage releases endonuclease function	7••
Lamin	6	Cleavage dissolves nuclear envelope	8
p21-activated kinase 2	3 and ?	Cleavage activates kinase, which mediates blebbing	9••
Gelsolin	3 and ?	Cleavage activates actin binding, which contributes to blebbing	10••
MEKK	3 ?	Accelerate apoptosis	11
Retinoblastoma	3 ?	Accelerate TNF-induced apoptosis	12
Focal adhesion kinase (pp125^{FAK})	3	Accelerates apoptosis by disruption of matrix signaling complex	13

malian cells *(2),* led to the discovery and characterization of the caspase family. Based on knockout studies, caspases-1 *(3)* and -2 *(4)* probably do not play major roles in apoptosis, but other caspases, such as caspase-3 *(5••),* certainly do. Most available evidence suggests that orchestration of the final death throes of a cell is mediated by caspases-3, -6, and -7 via cleavage of key protein substrates. Although many caspase substrates have been identified, only a few have been clearly demonstrated to play defined roles in the apoptotic process. These are summarized in Table 1. In addition to these are the caspases themselves, which are activated by caspase cleavage *(6).* Many other caspase substrates have been identified, but their roles in the apoptotic process have generally not been established (i.e., their cleavage occurs, but may be a bystander phenomenon).

It is clear that other key substrates remain to be identified. The apoptosis-associated phosphatidylserine externalization (FLOP) *(14),* which helps to target apoptotic cells for phagocytosis *(15),* depends on caspase activation *(16),* but proceeds independent of blebbing or nuclear changes *(17).* Adherent cells lose adhesion during apoptosis, although the caspase-dependent events leading to this phenotype have not been completely elucidated. Finally, the plasma membrane eventually loses integrity, and, although this is dramatically accelerated by caspase activation *(18)*, it is not obvious which substrates are intimately involved in the final dissolution of the cell.

How are the initial caspases activated? When cytotoxic T cells induce apoptosis in their targets, they do this through introduction of a serine protease, granzyme B, into the cytosol, which in turn directly activates caspase-3 *(19)*. Although serine proteases have been implicated in some forms of apoptosis, other examples of caspase activation by serine proteases during apoptosis have not been described. Therefore, whether or not lysosomal proteases (for example) may participate in triggering apoptosis via caspases (or other means?) remains an open question.

A second way in which caspases are activated is via their aggregation and autocleavage. Crosslinking of so-called death receptors, such as Fas and TNFR, recruits adaptor proteins, such as Fas-associated death domain (FADD) *(20)* and RIP (receptor-interacting protein)-associated ICH1/Nedd 2 homologous protein with a death domain (RAIDD) *(21)*, which in turn bind the prodomains of some caspases. FADD interacts via the binding of its death-effector domain (DED) to another DED in the prodomain of caspase-8 *(22,23)*. RAIDD interacts with caspase-2 via the binding of its caspase activation recruitment domain (CARD) with a CARD in the prodomain of caspase-2 *(21)*. By bringing the caspase molecules together, the weak protease activity of the procaspases cleaves each partner, causing two caspases to process and assemble into one active caspase complex (with two independent active proteolytic sites) *(24)*. Once activated, the caspase can now cleave and activate key substrates, as well as other, downstream, caspases.

In *C. elegans*, the ced-3 caspase is activated via its interaction with ced-4 *(25,26)*. In the vertebrates, a ced-4-like protein apoptotic protease-activating factor (Apaf)-1, is similarly capable of activating one of the caspases, caspase-9 *(27)*. However, this effect of Apaf-1 is dependent on at least one co-factor, cytochrome-*c* *(28)*. In the current version of the vertebrate apoptosome model, cytochrome-*c* binds to a WD repeat region at the C-terminus of Apaf-1, and this allows the Apaf-1 CARD domains located toward the N-terminus to bind the CARD in the prodomain of caspase-9. The effect is to allow the caspase to auto-process, and, in turn, to activate downstream caspases *(29)*.

This role of cytochrome-*c* in caspase activation via Apaf-1 has focused attention on the significance of the mitochondria in apoptosis. Several years ago, any role for mitochondria in apoptosis was questioned, when cells lacking mitochondrial DNA were shown to nevertheless undergo apoptosis *(30)*. Subsequently, a correlation between onset of apoptosis and a loss of mitochondrial inner transmembrane potential ($\Delta\Psi_m$) was noted, and attributed to a mitochondrial permeability transition (PT) *(31)*. Pharmacologic inhibitors of the PT were reported to inhibit apoptosis, thus supporting a role for this mitochondrial event. A simple model suggested that triggering the PT (which could occur as a consequence of reactive oxygen species [ROS], ceramide, Ca^{2+}, or caspases) would lead to cytochrome-*c* release, and this would thus link upstream signals to downstream caspase activation via the apoptosome.

Although it is likely that this represents one pathway to apoptosis, some evidence suggests that mitochondria can release apoptosis without undergoing a PT *(32•)*. However, the alternative mechanisms for cytochrome-*c* release without loss of $\Delta\Psi_m$ have been elusive (which, of course, doesn't mean they don't exist).

Significantly, Bcl-2 acts to inhibit the release of cytochrome-*c* from mitochondria, in vitro and in vivo *(33•,34••)*. In one cell-free system, cytochrome-*c* completely bypassed the Bcl-2 block, suggesting that the site of action of Bcl-2 was at the level of the mitochondria and not downstream of it *(34••)*. However, some recent studies have suggested that Bcl-X can bind to Apaf-1 *(35)*, and that Bcl-X and Bcl-2 can function downstream of cytochrome-*c* release to prevent caspase activation *(36,37)*. Interpretation of these results is, however, confounded by the fact that these inhibitors often delay, rather than fully inhibit, apoptosis. Thus, it is possible that cytochrome-*c* release may appear in cells that had not yet died, whereas control cells without Bcl-2 had already passed the point of detectable apoptosis. Future studies, particularly using in vitro systems, will help to resolve this issue.

EYES ON THE PRIZE:
BIOLOGICAL IMPACT OF THE EXECUTIONER

With so much effort devoted to identifying the central mechanisms of apoptosis, it is easy to lose sight of some of the principal reasons for caring about these mechanisms. One is the reasonable assumption that a detailed understanding of how cells actually die, when apoptosis is induced, will provide insights into those biological processes that depend on (or are influenced by) apoptosis. An overview of such processes might therefore be useful. We will then ask whether understanding of the nature of the executioner has helped to shed light on these processes.

Although there are many biological processes that are influenced by the process of apoptosis, they can be grouped into four general types: cell death caused by damage or stress, cell death in developmental processes, the regulation of cell death in neoplastic transformation, and the impact of different modes of cell death on whether or not an inflammatory response will occur.

Cellular damage or stress, leading to apoptosis or other forms of cell death, has obvious ramifications for health and disease. For example, the initial focal necrotic lesions seen in ischemic injury of the heart or brain subsequently extend through apoptosis, and pharmacologic control of this death is therefore desirable. Because caspases are central to the apoptotic process, inhibition of caspase activity holds promise as a way to manage this intervention. Indeed, some studies have suggested that cell death in neurons, induced by withdrawal of survival factors, can be effectively blocked by caspase inhibitors *(38)*. Similarly, caspase inhibitors have been shown to block cell death induced by ligation of Fas/CD95 *(39)*.

However, it is by no means clear that inhibition of caspases is an effective way to block cell death induced by a variety of agents in a variety of cells. Caspase inhibitors block the appearance of the apoptotic phenotype in virtually all cases, but several studies have shown that cell death proceeds nevertheless. This has been seen for cell death induced by c-Myc, withdrawal of growth factors, glucocorticoids, and transient treatment with DNA-damaging agents, staurosporine, and ultraviolet (UV) radiation *(40)*. Instead of a relatively rapid apoptotic death, a slower nonapoptotic death was often observed *(41)*. These studies suggested that the commitment to cell death occurs prior to the activation of caspases in these cases. The nature of this commitment remains hypothetical (*see* Understudy for the Executioner: When Are Mitochondria Central?).

In development, apoptosis contributes to form by removing cells from the embryo *(42)*. Apoptosis also contributes in a more subtle way to shape repertoires of functional cells, such as in the selection of lymphocytes to respond to foreign, but not self, peptides *(43),* and in the death of neurons that form complex connections in the brain. Here, the role of caspases is supported by mutations that alter or destroy caspase activity. In *C. elegans,* mutation of the ced-3 caspase effectively eliminates cell death from the developing animal *(44)*. Significant effects of caspase mutations have also been seen in other organisms. In *Drosophila,* mutation of one caspase, *Drosophila* cell-death protease (DCP)-1, results in a defect in cell death in the larvae *(45)*. In the mouse, targeted disruption of the caspase-3 gene results in severe defects in developmental apoptosis of neurons, so that the brain is grossly enlarged, although apoptosis in developing lymphocytes remains intact *(5••)*. If caspases are not essential for cell death, as suggested above, then why do these mutations have such effects? However, if caspases are essential for cell death, then why are the effects in mice and flies only partial? Presumably, redundancy plays a role in answering the last question, but more information is needed before this can be stated with confidence.

A third area in which the process of apoptosis appears to play a central role is that of oncogenesis. The realization that transformation occurs not only as the result of an increase in cell proliferation, but also as a consequence of a decrease in cell death, has provided a new framework for thinking about cancer. But it has also been realized for many years that cancer also presents a paradox: If the central mechanisms of apoptosis antagonize cellular oncogenesis, why have most tumors not simply lost the ability to undergo apoptotic cell death? Although many tumors can resist death through expression of antiapoptotic mechanisms (e.g., Bcl-2) or mutations in the initiation pathways (e.g., p53), when the cells do undergo death, it is via apoptosis. This suggests (at least at first approximation) that the central mechanisms of execution are intact in most (if not all) tumors. Why? Is there something wrong with our assumption that the central mechanisms of apoptosis offset oncogenesis? Or is there a factor we are not taking into consideration? There are at least three possibilities, and they are all interesting.

The first is that the molecular mediators of execution also participate in cellular events that are essential to the survival (as well as the death) of the cell. This idea, that death is the mirror of life *(46)*, predicts that loss of the executioner mechanism would fail to give cells a growth advantage. For example, cytochrome-*c* is critical both for activation of the vertebrate apoptosome (*see* Toward the Central Mechanisms of Apoptosis) and for oxidative phosphorylation in the cell. Simply losing cytochrome-*c* does not provide a growth advantage. In contrast, it is not as clear that the other members of the apoptosome complex, Apaf-1 and caspase-9 *(27)*, might not be completely dispensable for cell survival. Yet, inhibition of caspase activation does not appear to give a growth advantage to cells *(40)*, and, although absence of caspase-3 produces dramatic developmental defects in mice *(5••)*, there is no obvious oncogenic effect of this knockout. This explanation does not provide a satisfactory solution to the cancer/apoptosis paradox.

The second possible explanation is that the real executioner has not been properly identified. Focus has been directed at the activation and function of the caspases. Evidence that these are required for the orchestration of apoptosis is convincing, but inhibition of caspases does not necessarily block cell death *per se,* even if it blocks the manifestation of apoptosis. A tumor that loses the ability to activate caspases may still be fully susceptible to cell death that is initiated by the triggers of apoptosis, even though this death will not show the apoptotic phenotype. A more careful consideration of the role of the mitochondria and the point of no return in the process of active cell death may help to support this explanation. However, this position is incompatible with the observations, mentioned above, that disruption of caspase-3 expression does effectively prevent developmental cell death in the brain, and thus caspase activation does appear to be necessary and sufficient for at least some forms of death.

This leads to a third possibility: There are multiple pathways of cell death, in which caspases (and other players) have fundamentally different roles in determining whether a cell lives or dies. If so, then it may be that simply not enough tumors and lines have been examined for defects in specific elements of the apoptotic executioner, in order to make a definitive statement. That is, a tumor may have gained a growth advantage through disruption of part of the apoptosome, but the pathway leading from a death receptor to caspase activation may still be intact. If only the latter is examined, one might conclude that the central mechanisms of apoptosis in this cell are fully functional. Further studies will help to confirm or refute this idea.

We have considered the perspective that the main issue of biological importance is cell life vs cell death, placing essentially no emphasis on the mode of cell death. This would clearly be a mistake. If a cell dies in the body, how it dies is of tremendous importance, because this will determine how it will be cleared. Apoptotic cells are cleared silently by phagocytosis, and recent evidence suggests that apoptotic bodies actually inhibit the expression of proinflammatory cytokines from some phagocytes *(47)*. This is a function of apoptosis-associated plasma membrane changes that include an externalization of phosphatidylserine, which is dependent on caspase activation (*see* Toward the Central Mechanisms of Apoptosis). In contrast, cells that die via necrosis recruit a potent inflammatory response. How this occurs, i.e., what intracellular molecules that are released during necrosis induce inflammation, is almost completely unexplored. Further, despite predictions of the role of necrosis vs apoptosis in initiating inflammation, and the dramatically different outcomes of these modes of cell death, little information exists regarding how the elements and regulators of the central executioner of apoptosis (say, in tissue cells) influence the nature of an inflammatory response upon damage to that tissue.

Thus, the activation and function of caspases during the executioner phase of apoptosis clearly has an impact on the inflammatory consequences of the death of the cell. The extent to which it influences cell death, in response to damage, during development, or as a safeguard against oncogenic transformation, is not universal, and might vary with the specific cell type and the pathway of apoptosis that has been initiated.

UNDERSTUDY FOR THE EXECUTIONER: WHEN ARE MITOCHONDRIA CENTRAL?

As discussed above, mitochondrial release of cytochrome-c is an important step in caspase activation under some circumstances. Is this the point at which a cell becomes irrevocably committed to die? If mitochondria have irreversibly lost the ability to generate adenosine triphosphate, and if, instead, the oxidative phosphorylation machinery generated ROS, this might indeed commit a cell to die. If so, then caspase-dependent and -independent mechanisms leading to cytochrome-c release might account for caspase-dependent and -independent commitment to death. Following cytochrome-c release, the activation of the apoptosome and subsequent function of caspases would orchestrate the apoptotic events that normally follow this commitment, but the fate of the cell would be sealed when the mitochondria became involved.

A simple answer to the question, "When are mitochondria central to the apoptotic process?" might be this: when Bcl-2 effectively blocks apoptosis. Bcl-2 binds to the mitochondrial outer membrane, and effectively inhibits the release of cytochrome-c during the apoptotic process, thereby preventing apoptosome activation and caspase function. When Bcl-2 fails to protect cells, as is the case with Fas-ligation in T lymphocytes *(48),* a reasonable explanation is that, in such cells, the activated caspase-8 acts directly to process and activate caspase-3, with no requirement for mitochondrial amplification of this effect, and thus no opportunity for Bcl-2 to block. As attractive as this answer is, however, it is probably only an approximation. In addition to targeting mitochondria, Bcl-2 also resides on the nuclear membrane and the endoplasmic reticulum. Enforced targeting to the endoplasmic reticulum (vs mitochondria) results in a Bcl-2 that is still capable of blocking apoptosis under limited conditions *(49).* In addition, Bcl-2 can interfere with the

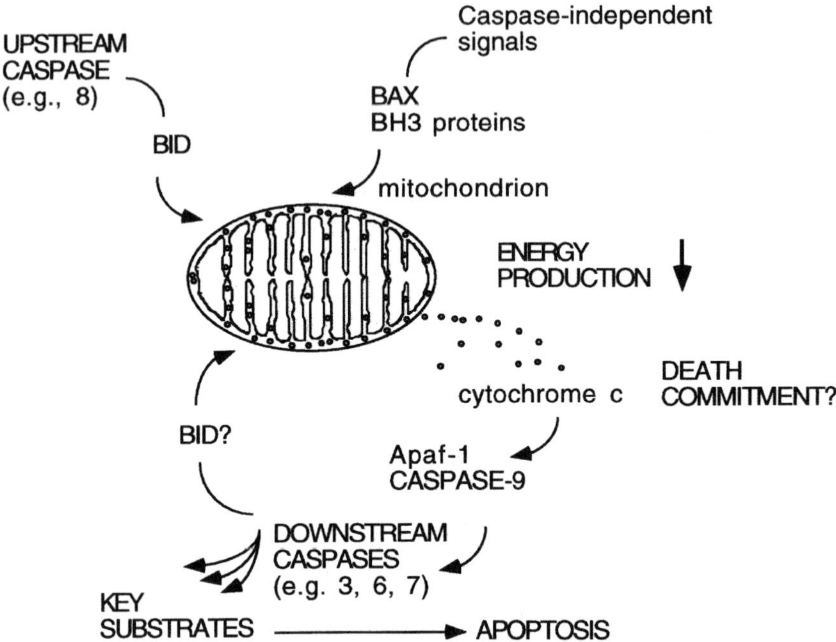

Fig. 1. A few of the pathways through which extracellular signals may be linked to downstream caspase activation via the intermediate mitochondrial release of cytochrome-*c:* one in which stimuli, such as DNA damage, mediate the caspase-independent release of cytochrome-*c,* potentially via translocation of the proapoptotic protein Bax; alternatively, FasL ligates to its receptor, Fas, resulting in upstream activation of caspase-8, which may in turn mediate cytochrome-*c* release via its interaction with proapoptotic Bcl-2 family members, including Bid. Potentially, the mitochondrial release of cytochrome-*c* may also function as an amplification loop to potentiate the effects of caspase-3.

function or activation of some transcription factors *(50•),* and this may also contribute to its antiapoptotic effect. Thus, the ability of Bcl-2 to block an apoptotic pathway is at best only an approximation of the importance of mitochondria in that pathway.

Essentially three pathways connect cellular signals to the mitochondrial release of cytochrome-*c* (Fig. 1). One pathway is triggered by stressors (e.g., UV, staurosporine) via an unknown mechanism, which results in cytochrome-*c* release without a requirement for caspase function. Potentially, this could involve the translocation of Bax from the cytosol to the mitochondria, an event that has been observed under such circumstances *(51,52).* Furthermore, Bax has also been shown to be capable of inducing cytochrome-*c* release in mammalian cells *(37,53),* and even in yeast mitochondria *(54),* where there appear to be no other Bcl-2 family members. A second pathway is triggered by ligation of death receptors, such as Fas (CD95), resulting in activation of upstream caspases such as caspase-8. Caspase-8 then is capable of triggering cytochrome-*c* release, probably through interaction with a subset of proapoptotic Bcl-2 family members, especially Bid *(55).* A third pathway, also caspase-dependent, may involve obligate activation of downstream caspases, such as caspase-3 (this is to account for those cases in which caspase-3 is required for cell death, as discussed above). It may simply be that, when levels of active procaspase exceed a threshold, which is set by the relative levels of such inhibitors as X-linked inhibitors of apoptosis (XIAP) *(56),* the caspase autoprocesses and triggers death. This threshold may also depend on the presence of a pathway that amplifies the downstream caspase signals through mitochondrial release of cytochrome-*c*.

Figure 1 makes it clear that mitochondria are not obligatory for all forms of apoptosis. When the apoptotic signal involves the ligation of a death receptor, this can lead directly to caspase activation, such as occurs when Fas-ligation activates caspase-8. The upstream caspase can, in turn, directly cleave and activate downstream caspases, such as caspase-3. In this setting, Bcl-2 family members probably do not act to inhibit apoptosis, as mentioned above. However, in some cells, ligation of a death receptor leads to apoptosis that can be inhibited by Bcl-2 family members. This is most easily explained by the idea that, when the upstream caspase is limiting, an amplification signal is needed. For example, in a cell-free system of caspase-3 activation, addition of mitochondria greatly enhanced the effects of low doses of caspase-8 *(57),* and this effect was inhibited by Bcl-2 in vitro. In intact cells, therefore, mitochondrial amplification (via cytochrome-*c* release) of a small caspase-8 signal might be needed if the effect is going to proceed through caspase-3 to apoptosis.

Basically, there are two major pathways: one in which caspases are directly activated through ligation of death receptors, and one in which caspases are activated via the function of Apaf-1. These are not as disparate as they might appear, because both appear to act via the aggregation of caspases containing large prodomains. The aggregation of even two caspase molecules is sufficient to cleave and activate the protease, via the relatively weak enzymatic activity of the zymogens *(6).* The fundamental difference between these two modes of activation is the role of cytochrome-*c* in the vertebrate apoptosome, engaging Apaf-1, and allowing it to aggregate procaspase-9 *(29)*. This, of course, brings the mitochondria into the picture, and the release of cytochrome-*c* by the mitochondria is an important focus for regulatory influences.

WHEN MITOCHONDRIA KILL

To return, then, to the problem raised in the chapter title: Apoptotic stimuli can cause a commitment for cell death in the absence of caspases, and this may be coincident with mitochondrial changes, such as the release of cytochrome-*c*. But what is this commitment? The release of cytochrome-*c* has been reported to correlate with the opening of the mitochondrial permeability transition pore in the inner membrane, and this has been suggested to be responsible for caspase-independent death *(41)*. One way this might occur is through the generation of ROS *(58,59)*, which may then kill the cell. Evidence that caspase-independent apoptosis is via ROS is lacking, however.

Cytochrome-*c* release can occur without the loss of mitochondrial transmembrane potential that accompanies the permeability transition, especially in the absence of caspase function *(32•)*. Therefore, other mechanisms for mitochondria-mediated cell death might be necessary to explain caspase-independent death. One that quickly comes to mind is the loss of electron transport function that occurs when cytochrome-*c* is not associated with its partners in the electron transport chain. Such loss of electron transport during apoptosis has been described *(60)*. Is this responsible for caspase-independent cell death? One can certainly envision cells starving to death as a major mechanism if energy metabolism is disrupted. One prediction, then, might be that cells that are adapted to grow in the absence of mitochondrial function (e.g., rho0 cells) will not display mitochondrial-mediated, caspase-independent cell death upon induction of apoptosis in the presence of caspase inhibitors.

Another possible mechanism for the caspase-independent death may be the release of apoptosis-inducing factor (AIF) *(61)*. This partially characterized factor is reported to activate caspases, and also to act directly on nuclei to induce DNA fragmentation *(62)*. It is released from mitochondria following a permeability transition, but it may be released under other conditions as well. Further characterization of this factor will be enlightening.

In any case, it is important to note that, whatever the mechanism of caspase-independent death following apoptosis induction and inhibition of caspases, the effects are inhibitable by other antiapoptotic molecules. These include Bcl-2, Bcl-X, and Bcr-Abl *(63)*. The ability of Bcl-2-family proteins to interfere with this mode of cell death further implicates the mitochondria as a likely source of the execution, whatever it is.

It is not always the case, however, that inhibition of caspases fails to protect cells from caspase-independent death. Caspase-3 and -9 knockout mice accumulate neurons to an astonishing extent *(5••,64,65),* and these cells appear by all accounts to be fully alive. Earlier reports also showed that caspase inhibitors block cell death in neurons, induced by withdrawal of growth factors, and again, these cells appeared to remain fully functional (i.e., there was no evidence of a slow death that followed the apparent survival in the presence of the caspase inhibitors) *(38).* Assuming that these forms of death do not go through a death receptor with an upstream caspase (and it is not clear that this can be assumed), then, in some cells, apoptosis-inducing signals might proceed to the mitochondria, but not result in death, unless caspases are activated. It is interesting to speculate that some primary cells can perhaps withstand some or all of their mitochondrial function being disrupted, and simply repair those functions (and/or make new mitochondria). The induction of mitochondrial biogenesis, following an apoptotic signal, has been described *(66).*

This raises a potentially exciting possibility. If primary cells can survive apoptosis-inducing signals, provided that caspases are blocked, and if proliferating or transformed cells cannot (because of the mitochondrial death mechanism), then new approaches to cancer chemotherapy are suggested in which the death of the primary cells is prevented by caspase inhibitors. The transformed cells will not only die, but may die in a manner that promotes inflammation, perhaps adding to the anticancer effect. Although highly speculative, the failure to detect any obvious defects in the executioner mechanism of tumors (e.g., defects in Apaf-1/caspase-9 function) might be a consequence of the failure of such defects to provide a growth advantage to a transformed cell.

Is the field of apoptosis dead? If the death of this field is as fundamentally interesting as the death of a cell, then we will happily acknowledge it. One might even say "Death is dead, long live death," but no one would get it. Except, of course, the reader who has stayed with us this far.

REFERENCES

• of special interest
•• of outstanding interest

••1. Yuan J, Shaham S, Ledoux S, Ellis H, Horvitz HR. The C. elegans cell death gene ced-3 encodes a protein similar to mammalian interleukin-1β-converting enzyme. *Cell* 1993; 75: 641–652.

With the finding that the C. elegans death gene ced-3 encoded a protein with homology to ICE (caspase-1), which was able to induce apoptosis in mammalian cells, created the link between caspase activation and cell death.

2. Miura M, Zhu H, Rotello R, Hartwieg EA, Yuan J. Induction of apoptosis in fibroblasts by IL-1β-converting enzyme, a mammalian homolog of the *C. elegans* cell death gene ced-3. *Cell* 1993; 75: 653–680.

3. Kuida K, Lippke JA, Ku G, Harding MW, Livingston DJ, Su MSS, Flavell RA. Altered cytokine export and apoptosis in mice deficient in interleukin-1β converting enzyme. *Science* 1995; 267: 2000–2003.

4. Bergeron L, Perez GI, Macdonald G, Shi L, Sun Y, Jurisicova A, et al. Defects in regulation of apoptosis in caspase-2-deficient mice. *Genes Dev* 1998; 12: 1304–1314.

••5. Kuida K, Zheng TS, Na S, Kuan C, Yang D, Karasuyama H, Rakic P, Flavell RA. Decreased apoptosis in the brain and premature lethality in CPP32-deficient mice. *Nature* 1996; 384: 368–372.
First knockout with a dramatic effect on development and apoptosis, underscoring the importance of caspase-3. The continued detection of apoptosis also indicated the presence of redundancy, as suggested by the other caspase knockouts. Caspase-3 knockouts, described by other groups, have contributed equally important findings.

6. Cryns V, Yuan J. Proteases to die for. *Genes Dev* 1998; 12: 1551–1570.

••7. Enari M, Sakahira H, Yokoyama H, Okawa K, Iwamatsu A, Nagata S. A caspase-activated DNase that degrades DNA during apoptosis and its inhibitor ICAD. *Nature* 1998; 391: 43–50.
One of a number of papers describing the nuclease responsible for apoptosis-associated DNA fragmentation. Caspases cleave the inhibitor to release the active nuclease.

8. Takahashi A, Alnemri ES, Lazebnik YA, Fernandes-Alnemri T, Litwack G, Moir RD, et al. Cleavage of lamin A by Mch2 alpha but not CPP32: multiple interleukin 1 beta-converting enzyme-related proteases with distinct substrate recognition properties are active in apoptosis. *Proc Natl Acad Sci USA* 1996; 93: 8395–8400.

••9. Rudel T, Bokoch GM. Membrane and morphological changes in apoptotic cells regulated by caspase-mediated activation of PAK2. *Science* 1997; 276: 1571–1574.
First paper to describe the role of caspase-mediated activation of p21 rac-activated kinase in apoptosis-associated blebbing. Some of the experiments used PAK1 (referred to only as PAK in the paper), and the conclusions had to be confirmed by other groups.

••10. Kothakita S, Azuma T, Reinhard C, Klippel A, Tang J, Chu K, et al. Caspase-3-generated fragment of gelsolin: effector of morphological change in apoptosis. *Science* 1997; 278: 294–298.
Caspase cleavage of gelsolin activates its actin-binding activity and mediates membrane blebbing during apoptosis.

11. Cardone MH, Salvesen GS, Widmann C, Johnson G, Frisch SM. The regulation of anoikis: MEKK-1 activation requires cleavage by caspases. *Cell* 1997; 90: 315–323.

12. Tan X, Martin SJ, Green DR, Wang JYJ. Degradation of retinoblastoma protein in tumor necrosis factor- and CD95-induced cell death. *J Biol Chem* 1997; 15: 9613–9616.

13. Levkau B, Herren B, Koyama H, Ross R, Raines EW. Caspase-mediated cleavage of focal adhesion kinase pp125^{FAK} and disassembly of focal adhesions in human endothelial cell apoptosis. *J Exp Med* 1998; 187: 579–586.

14. Martin SJ, CP, McGahon AJ, Rader JA, Van-Schie RC, LaFace DM, Green DR. Early redistribution of plamsa membrane phosphatidylserine is a general feature of apoptosis regardless of the initiating stimulus: inhibition by overexpression of Bcl-2 and Abl. *J Exp Med* 1995; 182: 1545–1556.

15. Fadok VA, Voelker DR, Campbell PA, Cohen JJ, Bratton DL, Henson PM. Exposure of phosphatidylserine on the surface of apoptotic lymphocytes triggers specific recognition and removal by macrophages. *J Immunol* 1992; 148: 2207–2216.

16. Martin SJ, Finucane DM, Amarante-Mendes GP, O'Brian GA, Green DR. Phosphatidylserine externalization during CD-95-induced apoptosis of cells and cytoplasts requires ICE-CED-3 protease activity. *J Biol Chem* 1996; 271: 28,753–28,756.

17. Bratton D, Fadok VA, Richter DA, Kailey JM, Guthrie LA, Henson PM. Appearance of phosphatidylserine on apoptotic cells requires calcium-mediated nonspecific flip-flop and is enhanced by loss of the aminophospholipid translocase. *J Biol Chem* 1997; 272: 26,159–26,165.

18. Xiang J, Chao DT, Korsmeyer SJ. BAX-induced cell death may not require interleukin 1 beta-converting enzyme-like proteases. *Proc Natl Acad Sci USA* 1996; 93: 14,559–14,563.

19. Darmon AJ, Nicholson DW, Bleackley RC. Activation of the apoptotic protease CPP32 by cytotoxic T-cell-derived granzyme B. *Nature* 1995; 377: 446–448.

20. Chinnaiyan AM, O'Rourke K, Tewari M, Dixit VM. FADD, a novel death domain-containing protein, interacts with the death domain of Fas and initiates apoptosis. *Cell* 1995; 81: 505–512.

21. Duan H, Dixit VM. RAIDD is a new 'death' adaptor molecule. *Nature* 1997; 385: 86–89.

22. Muzio M, Chinnaiyan AM, Kischkel FC, O'Rourke K, Shevchenko A, Ni J, et al. FLICE, a novel FADD-homologous ICE-CED-3-like protease, is recruited to the CD95 (Fas/APO-1) death inducing signaling complex. *Cell* 1996; 85: 817–827.

23. Boldin MP, Goncharov TM, Goltsev YV, Wallach D. Involvement of MACH, a novel MORT1/FADD-interacting protease, in Fas/APO-1 and TNF receptor-induced cell death. *Cell* 1996; 85: 803–815.

24. Martin D, Siegel R, Zheng L, Lenardo M. Membrane oligomerization and cleavage activates the caspase-8 (FLICE/MACH-1) death signal. *J Biol Chem* 1998; 273: 4345–4349.

25. Wu D, Wallen HD, Inohara N, Nunez G. Interaction and regulation of the Caenorhabditis elegans death protease CED-3 by CED-4 and CED-9. *J Biol Chem* 1997; 272: 21,449–21,454.

26. Chinnaiyan AM, Chaudhary D, O'Rourke K, Koonin EV, Dixit VM. Role of CED-4 in the activation of CED-3. *Nature* 1997; 388: 728,729.

27. Li P, Nijhawan D, Budihardjo I, Srinivasula SM, Ahmad M, Alnemri ES, Wang X. Cytochrome c and dATP-dependent formation of Apaf-1/caspase-9 complex initiates an apoptotic protease cascade. *Cell* 1997; 91: 479–489.

28. Liu X, Kim CN, Yang J, Jemmerson R, Wang X. Induction of apoptotic program in cell-free extracts: requirement for dATP and cytochrome c. *Cell* 1996; 86: 147–157.

29. Srinivasula SM, Ahmad M, Fernandes-Alnemri T, Alnemri E. Autoactivation of procaspase-9 by apaf-1 mediated oligomerization. *Mol Cell* 1998; 1: 949–957.

30. Jacobson MD, Burne JF, King MP, Miyashita T, Reed JC, Raff MC. Apoptosis and bcl-2 protein in cells without mitochondrial DNA. *Nature* 1993; 361: 365–368.

31. Marchetti P, Castedo M, Susin SA, Zamzami N, Hirsch T, Macho A, et al. Permeability transition is a central coordinating event of apoptosis. *J Exp Med* 1996; 184: 1155–1160.

•32. Bossy-Wetzel E, Newmeyer DD, Green DR. Mitochondrial cytochrome c release in apoptosis occurs upstream of DEVD-specific caspase activation and independently of mitochondrial transmembrane depolarization. *EMBO J* 1998; 17: 37–49. *One of the first papers to challenge the role of the mitochondrial permeability transition in cytochrome-c release.*

•33. Yang J, Liu X, Bhalla K, Kim CN, Ibrado AM, Cai J, et al. Prevention of apoptosis by Bcl-2: release of cytochrome c from mitochondria blocked. *Science* 1997; 275: 1129–1132.

••34. Kluck RM, Bossy-Wetzel E, Green DR, Newmeyer DD. The release of cytochrome c from mitochondria: a primary side for Bcl-2 regulation of apoptosis. *Science* 1997; 275: 1132–1136.
As the title says, Bcl-2 blocks cytochrome-c release. Unlike ref. 33, this paper shows effects in vitro and in vivo, and uses a system in which Bcl-2 can inhibit cytochrome-c release, although cytochrome-c can bypass the effects of Bcl-2.

35. Pan G, O'Rourke K, Dixit VM. Caspase-9, bcl-X$_L$ and apaf-1 form a ternary complex. *J Biol Chem* 1998; 273: 5841–5845.

36. Hu Y, Benedict MA, Wu D, Inohara N, Nunex G. Bcl-X$_L$ interacts with Apaf-1 and inhibits Apaf-1-dependent caspase-9 activation. *Proc Natl Acad Sci USA* 1998; 95: 4386–4391.

37. Rosse T, Olivier R, Monney L, Rager M, Conus S, Fellay I, Jansen B, Borner C. Bcl-2 prolongs cell survival after Bax-induced release of cytochrome c. *Nature* 1998; 391: 496–499.

38. McCarthy MJ, Rubin LL, Philpott KL. Involvement of caspases in sympathetic neuron apoptosis. *J Cell Sci* 1997; 110: 2165–2173.

39. Longthorne VL, Williams GT. Caspase activity is required for commitment to Fas-mediated apoptosis. *EMBO J* 1997: 16: 3805–3812.

40. McCarthy NJ, Whyte MK, Gilbert CS, Evan GI. Inhibition of Ced-3/ICE-related proteases does not prevent cell death induced by oncogenes, DNA damage, or the Bcl-2 homologue Bak. *J Cell Biol* 1997; 136: 215–227.

41. Hirsch T, Marchetti P, Susin SA, Dallaporta B, Zamzami N, Marzo I, Geuskens M, Kroemer G. The apoptosis-necrosis paradox. Apoptogenic proteases activated after mitochondrial permeability transition determine the mode of cell death. *Oncogene* 1997; 15: 1573–1581.

42. Glucksman A. Cell deaths in normal vertebrate ontogeny. *Biol Rev* 1951; 26: 59–86.

43. Green DR, Scott DW. Activation-induced apoptosis in lymphocytes. *Curr Opin Immunol* 1994; 6: 476–487.

44. Ellis HM, Horvitz HR. Genetic control of programmed cell death in the nematode C. elegans. *Cell* 1986; 44: 817–829.

45. Song Z, McCall K, Steller H. DCP-1, a *Drosophila* cell death protease essential for development. *Science* 1997; 275: 536–540.

46. Green DR, Martin SJ. Killer and the executioner: how apoptosis controls malignancy. *Curr Opin Immunol* 1995; 7: 694–703.

47. Fadok VA, Bratton DL, Konowal A, Freed PW, Westcott JY, Henson PM. Macrophages that have ingested apoptotic cells in vitro inhibit proinflammatory cytokine production through autocrine/paracrine mechanisms involving TGF-beta, PGE2, and PAF. *J Clin Invest* 1998; 101: 890–898.

48. Strasser A, Harris AW, Huang DC, Krammer PH, Cory S. Bcl-2 and Fas/APO-1 regulate distinct pathways to lymphocyte apoptosis. *EMBO J* 1995; 14: 6136–6147.

49. Zhu W, Cowie A, Wasfy GW, Penn LZ, Leber B, Andrews DW. Bcl-2 mutants with restricted subcellular localization reveal spatially distinct pathways for apoptosis in different cell types. *EMBO J* 1996; 15: 4130–4141.

•50. Shibasaki F, Kondo E, Akagi T, McKeon F. Suppression of signaling through transcription factor NF-AT by interactions between calcineurin and Bcl-2. *Nature* 1997; 386: 728–731.
The observation that Bcl-2 is able to modulate NF-AT-mediated transcription introduces a modulatory mechanism for the antiapoptotic effects of this protein.

51. Wolter KG, Hsu Y-T, Smith CL, Nechushtan A, Xi X-G, Youle RJ. Movement of Bax from the cytosol to mitochondria during apoptosis. *J Cell Biol* 1997; 139: 1281–1292.

52. Hsu Y-T, Wolter KG, Youle RJ. Cytosol-to-membrane redistribution of members of Bax and Bcl-X_L during apoptosis. *Proc Natl Acad Sci USA* 1997; 94: 3668–3672.

53. Jurgensmeier JM, Xie Z, Deveraux Q, Ellerby L, Bredesen D, Reed JC. Bax directly induces release of cytochrome c from isolated mitochondria. *Proc Natl Acad Sci. USA* 1998; 95: 4997–5002.

54. Manon S, Chaudhuri B, Guerin M. Release of cytochrome c and decrease of cytochrome c oxidase in Bax-expressing yeast cells, and prevention of these effects by coexpression of Bcl-X_L. *FEBS Lett* 1997; 415: 29–32.

55. Wang K, Yin W-M, Chao DT, Milliman CL, Korsmyer SJ. BID: a novel BH3 domain-only death agonist. *Genes Dev* 1996; 10: 2859–2869.

56. Deveraux Q, Takahashi R, Salvesen GS, Reed JC. X-linked IAP is a direct inhibitor of cell-death proteases. *Nature* 1998; 388: 300–304.

57. Kuwana T, Smith JJ, Muzio M, Dixit V, Newmeyer DD, Kornbluth S. Apoptosis induction by caspase-8 is amplified through the mitochondrial release of cytochrome c. *J Biol Chem* 1998; 26: 16,589–16,594.

58. Zamzami N, Marchetti P, Castedo M, Decaudin D, Macho A, Hirsch T, et al. Sequential reduction of mitochondrial transmembrane potential and generation of reactive oxygen species in early programmed cell death. *J Exp Med* 1995; 182: 367–377.

59. Tan S, Sagara Y, Liu Y, Maher P, Schubert D. The regulation of reactive oxygen species production during programmed cell death. *J Cell Biol* 1998; 141: 1423–1432.

60. Vayssiere JL, Petit PX, Risler Y, Mignotte B. Commitment to apoptosis is associated with changes in mitochondrial biogenesis and activity in cell lines conditionally immortalized with simian virus 40. *Proc Natl Acad Sci USA* 1994; 91: 11,752–11,756.

61. Susin SA, Zamzami N, Castedo M, Hirsch T, Marchetti P, Macho A, et al. Bcl-2 inhibits the mitochondrial release of an apoptogenic protease. *J Exp Med* 1996; 184: 1331–1341.

62. Susin SA, Zamzami N, Castedo M, Daugas E, Wang HG, Geley S, et al. The central executioner of apoptosis: multiple connections between protease activation and mitochondria in Fas/APO-1/CD95- and ceramide-induced apoptosis. *J Exp Med* 1997; 186: 25–37.

63. Amarante-Mendes GP, Naekyung-Kim C, Liu L, Huang Y, Perkins CL, Green DR, Bhalla K. Bcr-Abl exerts its antiapoptotic effect against diverse apoptotic stimuli through blockage of mitochondrial release of cytochrome c and activation of caspase-3. *Blood* 1998; 91: 1700–1705.

64. Kuida K, Haydar TF, Kuan C-Y, Gu Y, Taya C, Karasuyama H, et al. Reduced apoptosis and cytochrome c-mediated caspase activation in mice lacking caspase 9. *Cell* 1998; 94: 325–337.

65. Hakem R, Hakem A, Duncan GS, Henderson JT, Woo M, Soengas MS, et al. Differential requirement for caspase 9 in apoptotic pathways in vivo. *Cell* 1998; 94: 339–352.

66. Mancini M, Anderson BO, Caldwell E, Sedghinasab M, Paty PB, Hockenberry DM. Mitochondrial proliferation and paradoxical membrane depolarization during terminal differentiation and apoptosis in a human colon carcinoma cell line. *J Cell Biol* 1997; 138: 449–469.

12 Role of CD95 (APO-1/Fas) System in Chemotherapy

Klaus-Michael Debatin, MD, PhD

CONTENTS

ABSTRACT

The CD95 ligand-receptor pathway for the induction of apoptosis can be induced by chemotherapeutic drugs. In addition to its direct role in the induction of drug-mediated apoptosis, upregulation of CD95 (APO-1/Fas) may make tumor cells more susceptible to the activities of immune killers that express CD95 ligand. One mechanism for CD95 induction appears to involve p53; CD95 ligand may be induced via *ceramide synthesis and activation of stress-activated kinases. Stimulation of NF-κB by oxygen free radical generation may also upregulate ligand synthesis.*

From: *Apoptosis and Cancer Chemotherapy*
Edited by: J. A. Hickman and C. Dive © Humana Press Inc., Totowa, NJ

INTRODUCTION

The first antitumor drug (aminopterin) was introduced into the treatment of childhood leukemia by Farber almost 50 yr ago. In the past 20 yr combination therapy with several anticancer drugs has been used successfully, for example, to achieve long-term remission and cure in 70–80% of patients with acute lymphoblastic leukemia (ALL). Anticancer drugs have not been designed for a specific cellular or molecular target, but have been identified in assays based on their capacity to inhibit proliferation and clonogenicity of tumor cell lines. Cell death induced by anticancer agents has been considered to be a consequence of a block in proliferation, or simply of toxicity, but recent studies, many of which are reviewed in this book, have shown that most anticancer agents induce apoptosis *(1)*. In agreement with this, several key molecules in the cellular death program, such as p53 and proteins of the Bcl-2 family, are involved in the regulation or modulation of chemosensitivity of tumor cells *(2,3••,4,5•,6•,7)*, and they form the core of the texts contained herein. Although a controversial issue, recent data also suggest an active role for the CD95 (APO-1/Fas) system, one of the best-characterized ligand-driven apoptosis pathways, in the process of drug-induced cell death, or the cellular response, one of the best-characterized ligand-driven apoptosis pathways, to stress stimuli *(8,9••)*.

THE CD95 (APO-1/Fas) SYSTEM

CD95, a member of the tumor necrosis factor (TNF) nerve growth factor superfamily, is constitutively expressed in many cells and may be induced in many tissues by appropriate stimuli. Likewise, CD95 ligand (CD95L) is constitutively expressed in several tissues, and may be induced, for example, in T-cells following activation *(10)*. After crosslinking of the receptor by the multimeric ligand in the membrane-bound or soluble form, a death-inducing signaling complex (DISC) is formed that involves the death domains of the receptor: the adapter molecule Fas-associated death domain (FADD) and a chimeric adapter/caspase (caspase-8, or FADD-like ICE [FLICE]) *(9,11)*. Activation of caspase-8 leads to the cleavage of downstream caspases, such as caspase-3 and its substrates. In some cells, this pathway, making use of only a few signal transduction molecules, directly leads to cell death (type 1 cells); in other cells (type 2), mitochondrial function seems to be a critical checkpoint for execution of the death program initiated by CD95 triggering *(12••)*. CD95 is constitutively expressed in many tumor cells, including hematopoietic malignancies *(13,14)*. However, sensitivity for CD95-mediated apoptosis requires additional stimuli, as demonstrated for T-lymphocytes, in which prolonged activation is associated with increased susceptibility toward the CD95 death signal *(15).* In addition, studies on constitutive expression of CD95 in clinical tumor samples may be misleading, because exogenous stimuli, such as cytokines or cytotoxic drugs, may be able to upregulate receptor expression.

Failure to upregulate CD95 expression, or defects in the receptor, ligand, or signal molecules, may result in apoptosis defects in tumor cells *(13).* Mutations of CD95 in *lpr* mice and CD95L in *gld* mice constitute important descriptions of a pathology associated with an apoptosis gene defect *(8).* However, neither *lpr* nor *gld* mice develop overt malignancy, but rather suffer from a lymphoproliferative syndrome cause by the inability to delete long-term activated T-cells. The disease of *lpr* and *gld* mice is recapitulated in patients with an autoimmune lymphoproliferative syndrome, in which mutations of the CD95 receptor, primarily located in the death domain of the molecule, have been found *(16•,17•).* Patients develop extensive lymphadenopathy and hepatosplenomegaly with autoimmune disease such as immune thrombocytopenia. In contrast to this nonmalignant polyclonal disease, mutations of CD95, e.g., in lymphoblastic leukemia, are rare *(18•).* Thus, in contrast to p53 deficiency in knockout mice and in patients with Li-Fraumeni syndrome, there is no evidence for a direct tumor-suppressor function of CD95. However, in some affected families, an increased tumor incidence, in particular, Hodgkin's disease, has been described.

CD95 and CD95L may be present in the same cell after appropriate stimulation, leading to autocrine suicide or paracrine death by binding of the membrane-bound or soluble ligand to its cognate receptor. Activation-induced cell death (AICD) in T-cells appears to be an important mechanism for the termination of an immune response *(19•).* The pathology of *lpr* and *gld* mice and patients with CD95 deficiency may be explained by a deficient AICD. However, other death-inducing ligand (DIL)–receptor systems, such as TNF, may also play a role, and may contribute to the redundancy of apoptosis-inducing systems under conditions in which one system is blocked. Induction of CD95L–receptor interaction has been suggested to occur in different pathological conditions, such as HIV-infection, aplastic anemia, and liver failure *(13).* Various stimuli, including cytotoxic drugs, may exert cytotoxicity by increasing CD95 receptor, induction of CD95L and triggering of an autocrine or paracrine form of cell death: these findings will now be reviewed.

ACTIVATION OF THE CD95 SYSTEM DURING ANTICANCER THERAPY

Based on the concept of activation-induced death in T-cells, the cytotoxicity of anticancer treatment using cytotoxic drugs or γ-irradiation has been studied with respect to involvement of CD95 receptor–ligand interaction. In human T-cell lines, such as CEM cells or Jurkat cells derived from patients with ALL of T-cell type, but also in patients' cells, doxorubicin (DXR) and other cytotoxic drugs used in the chemotherapy of leukemias were found to induce CD95L expression *(20••)*. The cell lines used constitutively express CD95, and are sensitive for CD95-mediated apoptosis, that is, triggered via the agonistic antibody, anti-APO-1, or the natural ligand. By blocking CD95 receptor–ligand interaction, for example, by using antibodies such as an anti-APO-1 F(ab)$_2$ fragment, or by using cell lines in which CD95 was downregulated on prolonged exposure to an anti-CD95 antibody, drug-induced apoptosis was strongly diminished. The contribution of CD95L–receptor interaction to drug-induced cytotoxicity was most pronounced at the lower concentrations of DXR (up to 50–100 ng/mL), which may be achieved during therapy in vivo. In DXR cell lines derived from parental-sensitive cells, no induction of CD95L was found *(21)*. Induction of CD95L and activation of CD95L–CD95 interaction have also been found with bleomycin in hepatoblastoma cells, 5-fluorouracil (5-FU) in colon carcinoma cells, and with various chemotherapeutic drugs in medulloblastoma and neuroblastoma cells *(22,23•,24)*. Also, γ-irradiation has been observed to lead to increased transcription of CD95L mRNA *(25•,26••)*. In addition to CD95L, increased expression of CD95 is induced in cells with a low constitutive level of CD95 expression *(22,24)*. However, induction of CD95L is not found in all tumor cells *(27)*.

DRUG-INDUCED SENSITIZATION OF THE CD95 PATHWAY

Induction of increased CD95 expression in tumor cells by cytotoxic drugs provides an additional aspect, because cytotoxic T-cells use the CD95 system as one of the key mechanisms to kill their target cells *(28)*. Thus, tumor cells that express the CD95 receptor may be turned into highly susceptible targets for killer cells (T-cells, NK cells, LAK cells). Clinically relevant concentrations of diverse anticancer drugs, such as cisplatin, DXR, mitomycin, fluorouracil, or camptothecin, have been shown to sensitize colon carcinoma cell lines and leukemic cell lines for CD95-induced apoptosis by an agonistic antibody, CD95L, or activated killer cells *(29••,30)*. Sensitization may be caused by upregulation of CD95 expression and/or opening of downstream pathways. This finding may explain why low-dose chemotherapy is effective in certain tumors. For example, maintenance therapy is an indispensable element in the treatment of acute leukemias. The doses of methotrexate and 6-mercaptopurine used may be too low to mediate a direct cytotoxic effect, but may be sufficient to sensitize leukemia cells for physiological apoptosis signals by upregulating expression of death regulators, such as CD95. Likewise, 5-FU is successfully used in adjuvant therapy of colon carcinomas. These data suggest that chemotherapy not only has an immunosuppressive effect on the effector side, but may also be immunomodulating on the side of the target cell.

CD95 SYSTEM AND CELLULAR STRESS RESPONSE

The findings described so far imply that activation of DIL/receptor systems, such as CD95L and CD95, are part of the cellular response to cytotoxic treatments that damage DNA, disturb metabolism, or affect the mitotic apparatus, at least in some cells. How is the initial insult translated into a molecular response that leads to activation of apoptosis pathways? Genotoxic damage has been found to cause increased accumulation of p53, and consequently p53 has been implicated in the apoptosis response of a cell following DNA damage, as described elsewhere in this book. Indeed, increased transcriptional activity of p53 seems to be involved in upregulation of CD95 in drug-treated hepatoblastoma cells, and temperature-sensitive mutants of p53 have been found to directly upregulate CD95 *(22,31•)*. Thus, upregulation of the receptor may be a direct link to established apoptosis pathways, by which p53 may initiate the apoptosis response of the cell. The p53 protein may function by sensing damaged DNA and transcriptionally activating the expression of apoptosis-promoting molecules, such as CD95 and Bax. However, increased expression of CD95L appears to be independent of p53.

Recent data suggest a close connection between induction of CD95L, following treatment with anticancer drugs or γ-irradiation, and the activation of the cellular stress response, which includes the sphingolipid second messenger, ceramide, and stress-activated protein kinases (SAPK). This leads to phosphorylation and activation of the transcription factor c-jun (SAPK/[JNK]) *(25•26••,32•,34••)*. Ceramide, which accumulates in response to different types of cellular stress, such as chemo- and radiotherapy, induces expression of CD95L, cleavage of caspases, and apoptosis. Phorbol ester treatment, known to antagonize ceramide generation and JNK/SAPK activity, as well as γ-irradiation and CD95-mediated cytotoxicity, downregulates DXR-induced upregulation of CD95L, cleavage of CPP32, and cell death. Fibroblasts from type A Niemann-Pick patients (NPA), genetically deficient in ceramide synthesis, fail to upregulate CD95L expression, and to undergo apoptosis after γ-irradiation or DXR-treatment *(32•)*. CD95L expression and apoptosis in NPA fibroblasts can be restored by exogenously added ceramide. The critical role of the SAPK/JNK cascade is further demonstrated by recent findings showing that the CD95L promoter contains an (AP)-1 side *(25•)*. Thus, a transcriptionally active AP-1 complex may be formed following phosphorylation of c-jun by the JNK pathway, activated in response to cytotoxic drugs, γ-irradiation, and UV-irradiation *(25•,26••,32–33•)*. In addition to AP-1, nuclear factor-κB (NF-κB) has been identified as a transcription factor that also leads to increased expression of CD95L *(25•)*.

REACTIVE OXYGEN SPECIES AND MITOCHONDRIA

Various anticancer agents, including DXR, have been described as leading to increased production of reactive oxygen species (ROS). ROS production may be a consequence of physical–chemical interaction of drugs with intracellular molecules, or, alternatively, may be a consequence of disturbed mitochondrial function, leading to generation of oxygen radicals and formation of a permeability transition pore complex *(35••)*. Release of ROS from the mitochondria may then activate NF-κB. Topoisomerase-2 inhibitors, such as etoposide and teniposide, have been found to directly induce NF-κB-mediated transcription using a CD95L-promoter construct *(25•)*. Thus, stress-induced apoptosis involving increased expression of CD95L seems to be mediated by activation of NF-κB and AP-1. The finding that ROS are involved in driving activation of the CD95 system point to an important role for mitochondria, not only in the execution, but also in the control phase of cell death *(35••,36)*. However, mitochondria can also be targeted directly by chemotherapeutic drugs that do not lead to increased expression of CD95 or CD95L *(37•)*. Betulinic acid, a newly discovered anticancer compound with specific activity toward neuroectodermal tumors, leads to rapid activation of mitochondria in the absence of activation of CD95L or CD95-expression. CD95-resistant neuroblastoma cells still retain sensitivity for betulinic-acid-induced death; blockade of the apoptogenic mitochondrial function, for example, by enforced overexpression of Bcl-2 or Bcl-X$_L$ inhibits cell death.

COMPLEXITY CONTROVERSY
AND PARALLEL PATHWAYS

The signaling pathway initiated by formation of a death-inducing signaling complex, e.g., by crosslinking of CD95 death domains, may also be triggered independently of CD95L. This has been demonstrated for UV-induced apoptosis, which was found to involve signaling through CD95, without requirement for endogenously produced CD95L *(38,39)*. Likewise, induction of apoptosis in serum-starved fibroblasts by enforced expression of c-*myc* seems to involve signaling through the CD95 receptor in the absence of increased production of CD95L *(40)*.

The data from in vivo studies do not provide a definitive answer on the role of activation of the CD95 system in anticancer-therapy-induced apoptosis. Although *lpr* and *gld* mice do not exhibit an increased tumor rate, *lpr* mice do exhibit reduced radiosensitivity *(26••)*. In a series of 120 ALL patients, only two cases with mutations of the CD95 receptor have been identified *(18•)*. Both cases were refactory to chemotherapy. A clinical study in acute myelogenous leukemia (AML) patients has also found a favorable outcome for $CD95^+$ AML vs $CD95^-$ AML *(7)*. Thus, the relative importance of a cell-death pathway, such as CD95, may become evident under conditions of cellular stress and activation, for example, by drugs. In addition, the relative impact of a single death pathway, such as the CD95 pathway, may be overcome by the redundancy in the system of DILs, such as TNF or TNF-related apoptosis-inducing ligand (TRAIL) *(41••)*. Cellular stress leads to simultaneous expression of CD95L, TRAIL, and TNF (Herr and Debatin, unpublished data). Some of these pathways are reviewed in Chapter 3.

Irrespective of upstream triggers (DIL systems) or modulators (mitochondria), activation of the caspase cascade is involved in most forms of apoptosis as a downstream effector system. Chemoresistance can be mediated by inhibition of caspase activation, e.g., by zVAD-fmk or Crm-A, a poxvirus-derived serpin, or by antisense approaches targeting caspase-1 and -3, indicating that several members of this protease family are involved *(42)*. CD95-resistant cell lines that fail to activate caspases upon CD95 triggering, or have decreased expression of CD95, are cross-resistant to drug-mediated apoptosis *(42,43)*.

Caspase activation (particularly caspase-3) reflecting downstream activation of apoptosis effector molecules has been found in most studies on drug-induced apoptosis, but a requirement for activation of the CD95 system could not always be demonstrated. Using CD95-resistant cell lines and blockade of the receptor, several studies performed in leukemia cell lines have suggested that drug-induced apoptosis is CD95-independent *(44•,45••)*. However, induction of CD95L has only been investigated in one of these studies, and was found to be induced following drug treatment. In addition, experiments using FADD dominant-negative transgenic mice or FADD knockout mice, in which CD95-mediated death was abrogated, have found that cell death can still be induced by cytotoxic agents in fibroblasts, or by γ-irradiation in thymocytes *(46,47••)*. However, apoptosis signaling through death receptors may involve alternative pathways, such as the Daxx pathway *(48)*. So, the discrepancies may be explained by cell-type-specific differences, the different concentrations of the drugs used in the different studies, the effectiveness (affinity) of blocking agents, and/or redundancy in the system, such as participation of other DIL/receptor systems (TRAIL, TNF). Additionally, there are likely to be death-receptor-independent induction pathways of cell death, and these may work in parallel to the death-receptor-activated mechanisms.

Activation of mitochondria during drug-induced apoptosis seems to be a crucial event that coordinates and integrates several upstream and downstream pathways. This is reviewed in Chapter 11. Thus, activation of upstream (caspase-8) and downstream (caspase-3) effector caspases can be mediated by perturbance of mitochondria, leading to permeability transition (PT) and release of apoptogenic molecules, such as cyto-chrome-*c* and AIF *(35••)*. In so-called type 2 cells, perturbance of mito-chondrial function is an essential part of the CD95 pathway *(12••)*. Mitochondrial PT is induced by several anticancer drugs, and may be induced in a CD95-independent fashion, e.g., by betulinic acid, which leads to direct activation of mitochondria. Cytochrome-*c* and AIF re-leased from mitochondria are then able to directly activate caspase-3 and -8, respectively *(48)*. In neuroblastoma cells, activation of the receptor-proximal caspase-8 is upstream of mitochondrial PT in DXR-induced death, but downstream of mitochondria in betulinic-acid-induced apoptosis. In Bcl-2- or bcl-X_L-transfected cells, CD95-associated activa-tion of caspase-8 is still found following DXR, despite inhibition of both caspase-3 activation and apoptosis. In contrast, cleavage of caspase-8 and caspase-3, mitochondrial PT, and apoptosis are simultaneously blocked in Bcl-2-transfected cells treated with betulinic acid.

CONCLUSIONS

Most of the available data suggest that activation of caspase-3 or re-lated caspases is the crucial mediator of drug-induced apoptosis. In an integrated view, caspase-3 activation may result from multiple pathways that act as amplifiers of the death cascade (Fig. 1). These include signals

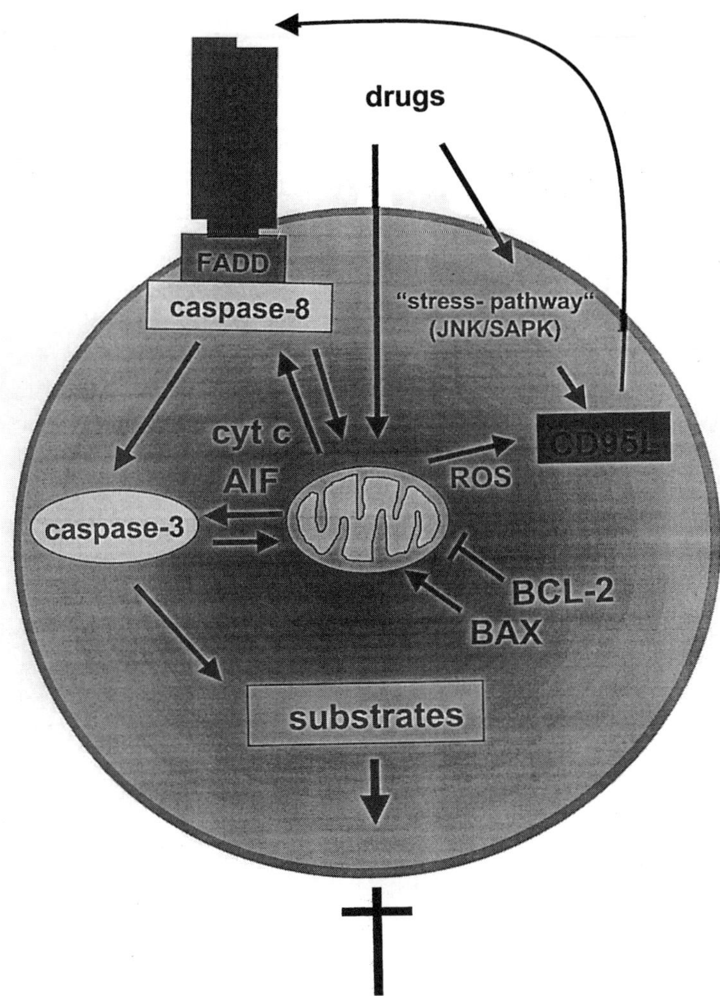

Fig. 1. The scenario integrates the various levels and compartments that have been shown to be triggered during cell death in response to treatment with cytotoxic drugs. Induction of death receptors, such as CD95 and DIL, such as CD95L, may result from activation of stress pathways (for the CD95L) and p53 (for the CD95 receptor). p53-dependent pathways are not included, for space reasons. As a critical event, activation of upstream (caspase-8 [FLICE]) or downstream (caspase-3 [CPP32]) caspases is shown, which act on substrates to exert the death program. Mitochondria appear to be a central coordinator or amplifier of various pathways, which include the generation of oxygen radicals (ROS) and the release of apoptogenic factors (cytochrome c and apoptosis-inducing factor [AIF]) to activate upstream and downstream caspases. The role of prototypic Bcl-2 family members that promote (Bax) or inhibit (Bcl-2) apoptosis is considered to be mediated through control of mitochondrial function.

by ligand-dependent (or -independent) crosslinking of death receptors, such as CD95, a mechanism which may be triggered by p53 and the ceramide-JNK-dependent stress pathway, and may depend on cell type and stimulus. In fact, in type I, but not type II, cells (for the CD95 pathway), doxorubicin induces the formation of a death-inducing signaling complex that contains CD95, FADD, and caspase-8 (S. Fulda and K.M. Debatin, unpublished). Signaling through this receptor-driven amplifier system seems to require additional amplification in most cells. Activation of mitochondrial PT by death-receptor-dependent signals, such as upstream caspases (caspase-8) or other mechanisms (e.g., ceramide), or by direct effects of the antitumor drug, appears to be the central amplifier in this process. This is suggested by the cytotoxicity of mitochondriotropic drugs, such as betulinic acid, and by the fact that drug-induced apoptosis is blocked by Bcl-2 in most cases. Activation of mitochondrial PT with sustained release of apoptogenic factors (cytochrome-*c*, AIF) is apparently able to mediate cleavage of upstream and downstream caspases, leading to final execution of the apoptosis program.

Current studies on a direct interplay between cytotoxicity of anticancer agents and apoptosis pathways may provide a new molecular understanding of sensitivity and resistance of tumor cells. The pathways critical for triggering of the cell-death program may serve as molecular targets for tailored therapy, using established anticancer agents, and may provide rational strategies for novel treatment approaches. The CD95 pathway appears to be an important component of cell death response in some forms of drug-induced apoptosis.

REFERENCES

• of special interest
•• of outstanding interest

1. Hannun YA. Apoptosis and the dilemma of cancer chemotherapy. *Blood* 1997; 89: 1845–1853.
2. Milner J. DNA damage, p53 and anticancer therapies. *Nature Med* 1995; 1: 879–880.
••3. Lowe SW, Bodis S, McClatchey A, Remington L, Ruley HE, Fisher DE, Housman DE, Jacks T. p53 Status and the efficacy of cancer therapy in vivo. *Science* 1994; 266: 807–810.
   Using p53-containing and p53-deficient tumors in an in vivo tumor model, the authors demonstrate that tumor regression and induction of apoptosis in response to γ-irradiation and Adriamycin therapy is linked to the presence of intact p53.
4. Campos L, Rouault J-P, Sabido O, Oriol P, Roubi N, Vasselon C, et al. High expression of Bcl-2 protein in acute myeloid leukemia cells is associated with poor response to chemotherapy. *Blood* 1993; 81: 3091–3096.
•5. Minn AJ, Rudin CM, Boise LH, Thompson CB. Expression of Bcl-x$_L$ can confer a multidrug resistance phenotype. *Blood* 1995; 86: 1903–1910.

•6. Krajewski S, Blomqvist C, Franssila K, Krajewska M, Wasenius V-M, Niskanen E, Nordling S, Reed JC. Reduced expression of pro-apoptotic gene bax is associated with poor response rates to combination chemotherapy and shorter survival in women with metastatic breast adenocarcinoma. *Cancer Res* 1995; 55: 4471–4478.

7. Min YH, Lee S, Lee JW, Chong SY, Hahn JS, Ko YW. Expression of Fas antigen in acute myeloid leukaemia is associated with therapeutic response to chemotherapy. *Br J Haematol* 1996; 93: 928–930.

8. Nagata S, Golstein P. Fas death factor. *Science* 1995; 267: 1449–1456.

••9. Nagata S. Apoptosis by death factor. *Cell* 1997; 88: 355–365.
Provides an in-depth summary and analysis of the known pathways initiated by triggering of death receptors, such as CD95 (APO-1/Fas).

10. Krammer PH, Dhein J, Walczak H, Behrmann I, Mariani S, Matiba B, et al. Role of APO-1 mediated apoptosis in the immune system. *Immunol Rev* 1994; 142: 175–191.

11. Peter ME, Kischkel FC, Hellbardt S, Chinnaiyan AE, Krammer PH, Dixit VM. CD95 (APO-1/Fas)-associating signalling proteins. *Cell Death Differ* 1996; 3: 161–170.

••12. Scaffidi C, Fulda S, Li F, Friesen C, Srinivasan A, Tomaselli KJ, et al. Two CD95 signaling pathways. *EMBO J* 1998; 17: 1675–1687.
The controversy of whether Bcl-2 overexpression blocks the CD95 death signal is clarified by the demonstration of two independent pathways that do or do not involve Bcl-2-inhibitable mitochondrial function in type 1 (mitochondria-independent) and type 2 (mitochondria-dependent) cells.

13. Debatin K-M. Disturbances of the CD95 (APO-1/Fas) system in disorders of lymphohematopoietic cells. *Cell Death Differ* 1996; 3: 185–189.

14. Debatin K-M, Goldman CK, Bamford R, Waldmann TA, Krammer PH. Monoclonal antibody-mediated apoptosis in adult T cell leukemia. *Lancet* 1990; 335: 497–500.

•15. Klas C, Debatin K-M, Jonker RR, Krammer PH. Activation interferes with the APO-1 pathway in mature human T cells. *Int Immunol* 1993; 5: 625–630.

•16. Rieux-Laucat F, Le Deist F, Hivroz C, Roberts IAG, Debatin K-M, Fischer A, De Villartay JP. Mutations in Fas associated with human lymphoproliferative syndrome and autoimmunity. *Science* 1995; 268: 1347–1349.

•17. Fisher GH, Rosenberg FJ, Straus SE, Dale JK, Middelton LA, Lin AY, et al. Dominant interfering Fas gene mutations impair apoptosis in a human autoimmune lymphoproliferative syndrome. *Cell* 1995; 81: 935–946.

•18. Beltinger CP, Kurz E, Böhler T, Schrappe M, Ludwig W-D, Debatin K-M. CD95(APO-1/Fas) mutations in childhood T-lineage acute lymphoblastic leukemia. *Blood* 1998; 91: 3943–3951.

•19. Dhein J, Walczak H, Bäumler C, Debatin K-M, Krammer PH. Autocrine T-cell suicide mediated by APO-1/Fas (CD95). *Nature* 1995; 373: 438–441.

••20. Friesen C, Herr I, Krammer PH, Debatin K-M. Involvement of the CD95 (APO-1/Fas) receptor/ligand system in drug induced apoptosis in leukemia cells. *Nature Med* 1996; 2: 574–577.
Using T-leukemia cell lines, induction of CD95L expression at the RNA and protein level, upon drug treatment and blocking of DXR apoptosis by interference with CD95 ligand–receptor interaction, is demonstrated (see ref. 45).

21. Friesen C, Fulda S, Debatin K-M. Deficient activation of the CD95 (APO-1/Fas) system in drug-resistant cells. *Leukemia* 1997; 11: 1833–1841.

22. Müller M, Strand S, Hug H, Heinemann EM, Walczak H, Hofmann WJ, et al. Drug-induced apoptosis in hepatoma cells is mediated by the CD95 (APO-1/Fas) receptor/ligand system and involves activation of wild-type p53. *J Clin Invest* 1997; 99: 403–413.

•23. Houghton JA, Harwood FG, Tillman DM. Thymineless death in colon carcinoma cells is mediated via Fas signaling. *Proc Natl Acad Sci USA* 1997; 94: 8144–8149. *Induction of apoptosis in colon cancer cell lines by 5-FU involves induction of CD95L, and is blocked by an anti-CD95L antibody.*

24. Fulda S, Sieverts H, Friesen C, Herr I, Debatin K-M. CD95 (APO-1/Fas) system mediates drug induced apoptosis in neuroblastoma cells. *Cancer Res* 1997; 57: 3823–3829.

•25. Kasibhatla S, Brunner T, Genestier L, Echeverri F, Mahboubi A, Green DR. DNA damaging agents induce expression of Fas ligand and subsequent apoptosis in T lymphocytes via the activation of NFκB and AP-1. *Mol Cell* 1998; 1: 543–551.

••26. Reap EA, Roof K, Maynor Kenrick, Borrero M, Booker J, Cohen PL. Radiation and stress-induced apoptosis: a role for Fas/Fas ligand interactions. *Proc Natl Acad Sci USA* 1997; 94: 5750–5755. *Demonstration that radiation-induced lymphocyte apoptosis is markly diminished in* lpr-*mice.*

27. Fulda S, Los M, Friesen C, Debatin K-M. Chemosensitivity of solid tumor cells *in vitro* is related to activation of the CD95 system. *Int J Cancer* 1998; 76: 105–114.

28. Debatin K-M. Cytotoxic drugs, programmed cell death, and the immune system: defining new roles in an old play. *J Natl Cancer Inst* 1997; 89: 750,751.

29. Micheau O, Solary E, Hammann A, Martin F, Dimanche-Boitrel MT. Sensitization of cancer cells treated with cytotoxic drugs to Fas-mediated cytotoxicity. *J Natl Cancer Inst* 1997; 89: 783–789. *Clinically relevant concentrations of cisplatin, DXR, mitomycin c, fluorouracil, or camptothecin enhance CD95 receptor expression on colon carcinoma cell lines, and induce increased sensitivity to anti-CD95 antibody, soluble CD95L, and T-cell-mediated toxicity.*

30. Yoshihiro K, Zhou YW, Zhang XL, Chen TX, Tanaka S, Azuma E, Sakurai M. Fas/APO-1 (CD95)-mediated cytotoxicity is responsible for the apoptotic cell death of leukaemic cells induced by interleukin-2-activated T cells. *Br J Haematol* 1997; 96: 147–157.

31. Owen-Schaub LB, Zhang W, Cusack JC, Angelo LS, Santee SM, Fujiwara T, et al. Wild-type human p53 and a temperature-sensitive mutant induce Fas/APO-1 expression. *Mol Cell Biol* 1995; 15: 3032–3040.

•32. Herr I, Böhler T, Wilhelm D, Angel P, Debatin K-M. Activation of CD95 (APO-1/Fas) signaling by ceramide mediates cancer therapy-induced apoptosis. *EMBO J* 1997; 16: 6200–6208.

•33. Faris M, Kokot N, Latinis K, Kasibhatla S, Green DR, Koretzky GA, Nel A. c-Jun N-terminal kinase cascade plays a role in stress-induced apoptosis in Jurkat cells by up-regulating Fas ligand expression. *J Immunol* 1998; 160: 134–144.

•34. Bose R, Verheij M, Haimovitz-Friedman A, Scotto K, Fuks Z, Kolesnick R. Ceramide synthase mediates daunorubicin-induced apoptosis: an alternative mechanism for generating death signals. *Cell* 1995; 82: 405–414.

••35. Kroemer G. Proto-oncogene bcl-2 and its role in regulating apoptosis. *Nature Med* 1997; 3: 614–620. *In-depth review on the role of mitochondria and Bcl-2 family proteins in the activation of caspases and regulation of apoptosis.*

36. Bauer MKA, Vogt M, Los M, Siegel J, Wesselborg S, Schulze-Osthoff K. Role of reactive oxygen intermediates in activation-induced CD95 (APO-1/Fas) ligand expression. *J Biol Chem* 1998; 273: 8048–8055.

•37. Fulda S, Friesen C, Los M, Scaffidi CA, Mier W, Benedict M, et al. Betulinic acid triggers CD95 (APO-1/Fas)- and p53-independent apoptosis via activation of caspases in neuroectodermal tumors. *Cancer Res* 1997; 57: 4956–4964.

38. Rehemtulla A, Hamilton CA, Chinnaiyan AM, Dixit VM. Ultraviolet radiation-induced apoptosis is mediated by activation of CD95 (Fas/APO-1). *J Biol Chem* 1997; 272: 25,783–25,786.

39. Aragane Y, Kulms D, Metze D, Wilkes G, Pöppelmann B, Luger TA, Schwarz T. Ultraviolet light induces apoptosis via direct activation of CD95 (Fas/APO-1) independently of its ligand CD95L. *J Cell Biol* 1998; 140: 171–182.

•40. Hueber A-O, Zörnig M, Lyon D, Suda T, Nagata S, Evan GI. Requirement for the CD95 receptor-ligand pathway in c-Myc-induced apoptosis. *Science* 1997; 278: 1305–1309.

••41. Golstein P. Cell death: TRAIL and its receptors. *Curr Biol* 1997; 7: R750–R753. *Review on a new family of recently cloned TNF-related death receptors.*

42. Los M, Herr I, Friesen C, Fulda S, Schulze-Osthoff K, Debatin K-M. Crossresistance of CD95- and drug-induced apoptosis as a consequence of deficient activation of caspases (ICE/Ced-3 proteases). *Blood* 1997; 90: 3118–3129.

43. Landowski TH, Gleason-Guzman MC, Dalton WS. Selection for drug resistance results in resistance to Fas-mediated apoptosis. *Blood* 1997; 89: 1854–1861.

•44. Eischen CM, Kottke TJ, Martins LM, Basi GS, Tung JS, Earnshaw WC, Leibson PJ, Kaufmann SH. Comparison of apoptosis in wild-type and Fas-resistant cells: chemotherapy-induced apoptosis is not dependent on Fas/Fas ligand interactions. *Blood* 1997; 90: 935–943.

••45. Villunger A, Egle A, Kos M, Hartmann B, Geley S, Kofler R, Greil R. Drug-induced apoptosis is associated with enhanced Fas (APO-1/CD95) ligand expression but occurs independently of Fas (APO-1/CD95) signaling in human T-acute lymphatic leukemia cells. *Cancer Res* 1997; 57: 3331–3334. *Induction of CD95L is shown upon drug treatment of leukemia cells with several anticancer drugs; however, cell death could not be blocked by CD95-blocking antibodies, overexpression of Crm-A, and in some CD95-resistant cells.*

46. Newton K, Harris AW, Bath ML, Smith KGC, Strasser A. Dominant interfering mutant of FADD/MORT1 enhances deletion of autoreactive thymocytes and inhibits proliferation of mature T lymphocytes. *EMBO J* 1998; 17: 706–718.

••47. Yeh W-C, de la Pompa JL, McCurrach ME, Shu H-B, Elia AJ, Shahinian A, et al. FADD: Essential for embryo development and signaling from some, but not all, inducers of apoptosis. *Science* 1998; 279: 1954–1958. *Although the FADD-null mutation is embryonically lethal because of developmental defects (e.g., in the heart), CD95-, TNF- and DR3-induced apoptosis is abrogated in FADD-deficient embryonic fibroblasts, but cell death induced via DR4, B1A, c-myc, and adriomycin can still be induced.*

48. Yang X, Khosravi-Far R, Chang HY, Baltimore D. Daxx, a novel Fas-binding protein that activates JNK and apoptosis. *Cell* 1997; 89: 1067–1076.

49. Fulda S, Susin SA, Kroemer G, Debatin K-M. Molecular ordering of apoptosis induced by anticancer drugs. *Cancer Res* 1998; in press.

13 Insulin-Like Growth Factor-1 Receptor as a Target for Anticancer Therapy

Renato Baserga, MD
and Mariana Resnicoff, PhD

CONTENTS

ABSTRACT

The type 1 receptor for the insulin-like growth factors (IGF-1R) plays a major role in the control of cell proliferation, both in vitro and in vivo. Downregulation of the IGF-1R function by a variety of procedures causes large-scale apoptosis of cells, especially when

From: *Apoptosis and Cancer Chemotherapy*
Edited by: J. A. Hickman and C. Dive © Humana Press Inc., Totowa, NJ

the cells are in anchorage-independent conditions. The ability of the targeted IGF-1R to discriminate between contact-inhibited cells and cells in anchorage-independence has been exploited in animals to abrogate tumor growth.

INTRODUCTION

It is self-evident that a tissue or an organ will grow when the number of cells produced per unit time exceeds the number of cells that die in the same period. More precisely, an increase in cell number of any given cell population (in vivo or in vitro, normal or abnormal) depends on three parameters *(1)*:

1. The length of the cell cycle. A shortening in the length of the cell cycle will cause cells to divide more frequently, resulting in an increase in cell number.
2. The growth fraction, i.e., the fraction of cells in a population that are actively in the cell cycle. In many cell populations, a sizable fraction of cells is in G/0, in which the cells are in a sort of hibernation, from which they can be rescued by appropriate stimuli. An increase in the growth fraction, i.e., a decrease in the G/0 fraction, will also result in an increased production of cells.
3. A decrease in the rate of cell death. This third parameter was all but ignored until recently, when apoptosis suddenly became the fashionable way for cells to die. But it had been known for several years that an increased production of cells (parameters 1 and 2) was not sufficient to explain why tumors grow and normal tissues in the adult animal do not *(2)*. Some extent of reduced cell death had to be postulated *(3)*.

Given that the growth of a cell population depends on a balance between cell division and cell death, it follows that both processes can be targeted in any strategy aimed at inhibiting the growth of tumors. There are countless ways to inhibit cell division or increase cell death, but one reasonable approach is to interfere with the function of growth factors and their receptors, which certainly play an important role among the environmental signals that regulate cell proliferation. Among the growth factor receptors, considerable interest has recently centered around the type 1 insulin-like growth factor receptor (IGF-1R), activated by its ligands. There are two good reasons for this. First, the IGF-1R acts in at least five different ways in controlling the size of a cell population. It is known to be mitogenic in vivo and in vitro, it promotes growth in size of the cell, sends a powerful antiapoptotic signal, it is quasi-obligatory for the transformation of cells, and it also modulates cell differentiation. Second, in general, most cells require more than one growth factor for cell division, and each cell type has its preferences. Epithelial cells prefer epidermal growth factor (EGF), fibroblasts like platelet-derived growth factor (PDGF), neuronal cells respond to several specific growth factors, and each lineage of hemopoietic cells has its own constellation of primary growth factors (interleukins, granulocyte-colony stimulating factor [G-CSF] and so on). But, in the great majority of cell types, the second growth factor is IGF-1, or IGF-2 *(4)*, both of which use the IGF-1R for their mitogenic activities. This means that the IGF-1R plays a role in regulating the growth of a large number of cell populations.

ANATOMY OF THE IGF-1 RECEPTOR

This is not the place to discuss in detail the structure and function of the IGF-1R, but a few essential things should be mentioned *(4•)*: The fully functional receptor is a dimer that is activated by at least three different ligands, IGF-1, IGF-2, and insulin (at high concentrations); it has 70% homology to the insulin receptor (IR), with which it shares some of the signaling pathways; it consists (like the IR) of an extracellular ligand-binding subunit, the α-subunit, and of a transmembrane β-subunit, which is bound to the α-subunit by disulfide bonds; ligand binding causes autophosphorylation of the receptor. Tyrosine autophosphorylation is by far the most studied as the mechanism of receptor activation, but it is perhaps time that other alternatives to receptor activation (or inactivation) should be considered, as for instance, serine/threonine phosphorylation or conformational changes; and the C-terminus of the IGF-1R (roughly the last 100 amino acids) has the least homology with the IR, and is dispensable for mitogenesis and protection from apoptosis, but is required for the transformation of cells in cultures.

TARGETING OF THE IGF-1 RECEPTOR

When dealing with growth factors and their receptors, there are two principal ways to interfere with their function: either by targeting the ligand(s) or the receptor. In the case of the IGF system, targeting of the ligands is not the proper choice, at least in humans, who produce both IGF-1 and IGF-2 through adult life. Given the functions of the IGF-1R mentioned in the Introduction, if one thinks of anticancer strategies, downregulation of the IGF-1R function is what one would like to accomplish. By downregulation of IGF-1R function, what is meant here is either interference with its function, or, more simply, decreasing the number of receptors. Five different approaches have been described in IGF-1R targeting: use of antibodies against the α-subunit, pioneered by Arteaga et al. *(5);* stable transfection, with a plasmid expressing an antisense RNA to the IGF-1R RNA *(6••);* use of antisense oligodeoxynucleotides against the IGF-1R *(7);* use of dominant-negative mutants of the receptor *(8,9);* and targeted disruption of the IGF-1R genes by homologous recombination *(10,11••).* All of these approaches have been shown to be effective both in vitro and in vivo, and each of them has its advantages and disadvantages.

EFFECTS OF TARGETING THE IGF-1 RECEPTOR

Effects In Vivo

When both the IGF-1R genes and the IGF-2 genes are disrupted, homozygous mouse embryos grow to a size that is 30% the size of wild-type littermates *(10,11).* The newborn mice are nonviable. Disruption of the IGF-2R genes markedly increases the levels of IGF-2 in the plasma, and the embryos are 145% the size of wild-type littermates *(12).* The newborn mice suffer perinatal mortality. Disruption of both the IGF-1R and the IGF-2R produce mice that are normal in size and viable *(12).* It is now agreed that, under these conditions, IGF-2 (the ligand) signals through the IR *(13,14).* Thus, strictly speaking, the IGF-1R may be required for optimal growth of embryos, but it is not an absolute requirement.

Effects In Vitro

This laboratory has developed a fibroblast cell line from the mouse embryos null for the IGF-1R genes *(10,11••)* designated as R− cells, which has been used extensively in many laboratories, because of the absence of background noise. R− cells grow in 10% serum *(15•),* which means again that the IGF-1R is not an absolute requirement for growth in tissue culture. However, R− cells are refractory to transformation by a variety of viral and cellular oncogenes *(4•),* indicating that the receptor is more important for transformation than for normal growth. Indeed, when the IGF-1R function is downregulated *(see above),* growth of cells in monolayer is very little affected, but colony formation in soft agar is drastically reduced *(16).* This finding raises the possibility that targeting of the IGF-1R may distinguish between normal growth and anchorage-independent growth, and, by extrapolation, between normal and abnormal growth.

Effect on Tumor Cells In Vivo

Targeting of the IGF-1R by antisense strategies or by dominant-negative mutants causes massive apoptosis of tumor cells in vivo *(9,17).* The extent of apoptosis is so dramatic that it results in abrogation of tumorigenesis in syngeneic rats and nude mice, and in inhibition of metastases *(18)*. The dominant-negative mutant described by D'Ambrosio et al. *(9)*, designated as 486/STOP, is especially interesting, because it consists of a 486-amino-acid polypeptide (from the α-subunit), which is partially secreted in the environment. Therefore, not only does it induce apoptosis in the tumor cells expressing it, but it also has a bystander effect. For instance, when tumor cells expressing 486/STOP are coinjected with wild-type tumor cells into nude mice, both types of cells undergo apoptosis, and the result is abrogation of tumor growth by the wild-type cells *(19)*. The ratio of wild-type: 486/STOP-expressing cells determines semiquantitatively the extent of inhibition of growth of wild-type tumor cells. At a ratio of 1:4, the coinjection of cells expressing 486/STOP can actually completely abrogate tumorigenesis by wild-type human lung cancer cells.

IGF-1 RECEPTOR SIGNALING IN APOPTOSIS

The induction of apoptosis by downregulation of IGF-1R function is the logical corollary of experiments showing that IGF-1 protects cells from apoptosis induced by a variety of agents *(20)*. This protective effect is even more dramatic when the IGF-1R is overexpressed *(21,22)*. Therefore, the mechanism by which targeting the IGF-1R induces apoptosis can be investigated by turning the problem around and asking how the IGF-1R, activated by its ligands, protects cells from apoptotic injuries.

The first step is the identification of the domain(s) in the receptor that are required for protection from apoptosis. As mentioned above, the C-terminus of the IGF-1R is dispensable for mitogenesis (but not for transformation). O'Connor et al. *(21)* carried out a mutational analysis of IGF-1R protection in cells that undergo apoptosis when interleukin-3 (IL-3) is withdrawn. Their findings can be summarized as follows: A mutation at the ATP-binding site (lysine 1003) completely inactivates the IGF-1R protective effect (as well as other functions); mutations in the tyrosine kinase domain (tyrosines 1131, 1135, and 1136) severely impair, but do not abrogate, protection; and the deletion of the C-terminus does not affect the antiapoptotic effect of the IGF-1R.

These findings are compatible with several papers that have recently appeared, which show that the antiapoptotic effect of the IGF-1R is dependent on the activation of phosphatidylinositol-3 kinase (PI-3K), which activates Akt/PKB *(23•–25)*. PI-3K associates with IRS-1 *(26)*, which is one of the major substrates of the IGF-1R. Activation of the IGF-1R causes tyrosyl phosphorylation of IRS-1 *(27)*, which then signals through PI-3K and Akt/PKB. Indeed, a recent paper has shown that Akt then phosphorylates Bad *(28,29)*, thus inactivating it. The circle is seemingly closed. From the receptor, through IRS-1, PI 3-K, and Akt/PKB, one arrives at the inhibition of proapoptotic members of the Bcl-2 family. There is no question that this pathway is involved in the mechanism by which IGF-1 protects cells from apoptosis. In fact, this pathway is also involved in other antiapoptotic signaling: for instance, the antiapoptotic effect of ras *(30••)*. However, this question can be legitimately posed: Is this the only pathway used by the IGF-1R?

THE STRANGE CASE OF 32D CELLS

The IGF-1R has been sending signals, not to the cell, but to investigators, that it has pathways that it does not share with other receptors. The PI-3K/ras pathway is a well-known mitogenic pathway that is shared not only by the IR and the IGF-1R, but also by several other growth factor receptors. The first clue that the IGF-1R was also using a pathway that was not shared with other receptors, and was ras-independent, came from observations with R− cells, especially the finding that an activated ras could not transform R− cells, although it easily transformed other 3T3 cells with a physiological number of IGF-1R *(31)*. Several investigators *(24,25,32)*, although acknowledging the importance of the PI-3Kinase–Akt pathway, have suggested that the IGF-1R also uses another, unidentified pathway. The best demonstration, though, that the IGF-1R uses another pathway comes from experiments with a murine hemopoietic cell line, called 32D cells (Fig. 1). These cells, like many other hemopoietic cell lines, are IL-3-dependent: Withdrawal of IL-3 causes rapid apoptosis. The peculiarity of 32D cells, however, is that they have no IRS-1 or IRS-2 *(33)*, and very low levels of either the IR *(33)* or the IGF-1R *(22)*. When the IR or IRS-1 are overexpressed, 32D cells still fail to survive IL-3 withdrawal: A combination of the two actually results in the growth of cells in the absence of IL-3 *(33)*. Thus far, everything is compatible with the proper activation of the PI-3K–Akt–PKB pathway, which requires a proper expression of both the receptor and IRS-1.

The surprise came with the IGF-1R, which, when overexpressed in 32D cells, completely protects them from IL-3 withdrawal *(22,34)*. Clearly, the IGF-1R sends an antiapoptotic signal in the absence of the IRS proteins that are needed by the IR. True, PI-3K can also bind directly to the C-terminus of the IGF-1R, but this alternative is rapidly ruled out by the observation that a truncated IGF-1R (lacking a direct binding site for PI-3K) also protects 32D cells from apoptosis *(34)*.

THE STRANGE CASE OF 32D CELLS

1) IL-3 withdrawal causes massive apoptosis
2) the IGF-I receptor gives full protection
3) the Insulin receptor gives no protection
4) 32D cells have no IRS-1 nor IRS-2
5) Insulin receptor plus IRS-1 give protection
The IRS-1 pathway, therefore, does protect cells
from apoptosis, but the IGF-I receptor has
a 2nd pathway, not IRS-1 dependent
and not shared with the Insulin receptor

Fig. 1. The strange case of 32d cells.

Finally, the protective effect of the IGF-1R in 32D cells is insensitive to very high doses of wortmannin, as high as 1,000 nM, and of LY294002, both of which are inhibitors of PI-3K. We are not fond of using inhibitory drugs to test pathways, because many specific drugs invariably turn out to inhibit other reactions. However, if an event (in this case, protection from apoptosis) is insensitive to high concentrations of PI-3K inhibitors, one has to conclude that PI-3K is not a requirement under those conditions, because these inhibitors may have other functions, but one thing they certainly do is to inhibit PI-3K. Clearly, the IGF-1R can protect cells from apoptosis by a mechanism that is IRS-proteins- and PI-3K-independent. This should not be construed as contradicting other reports on the activation of the PI-3K pathway *(see above)*. The IGF-1R certainly uses that pathway, but, in its absence, it uses another one that it does not share with the IR. The identification of this pathway is one of the most important challenges to be faced in this field, because it would allow the inhibition of IGF-1R signaling without inhibiting the IR signaling.

THERE IS STRENGTH IN NUMBERS

We have mentioned above that the various functions of the IGF-1R can be mapped to distinct domains of the receptor. Another way in which a cell can modulate its response to IGF-1 is by modulating the IGF-1R number. The levels of expression of the IGF-1R vary, even under physiological conditions, although the variations are modest. There are a number of other growth factors and intracellular molecules that either increase or decrease the number of IGF-1R/cell *(4•)*. Among the former are PDGF, basic fibroblast growth factor, and estrogens; among the factors that decrease IGF-1R levels are p53, interferon, and the tumor-suppressor gene *WT1*. These moderate changes in receptor levels can have dramatic effect. For instance, p53 cannot induce apoptosis in 32D cells, unless it can decrease IGF-1R levels *(22)*, and certain breast cancer cells do not respond to IGF-1, unless they are previously primed by estrogen *(35)*.

The best evidence of the importance of receptor number in modulating responses to IGF-1 comes from the experiments of Rubini et al. *(36)*, which show that a small increment in IGF-1R number can make a huge difference in the response to IGF-1.

IGF-1 RECEPTOR AND APOPTOSIS: PART OF THE PROBLEM OR PART OF THE SOLUTION?

Evidence is overwhelming that the IGF-1R activated by its ligands can protect from a variety of proapoptotic agents. Conversely, a functional impairment of the IGF-1R can cause massive apoptosis. Yet, the activation of the IGF-1R is known to stimulate the differentiation of myoblasts *(37,38)*, osteoblasts *(39)*, adipocytes *(40)*, neurons *(41,42)*, and several other cell lines. The role of the IGF system in differentiation has been studied in greater detail in myoblasts. Myoblasts in cultures are undifferentiated cells, which can grow indefinitely in serum, but differentiate into myocytes, if the serum is removed or decreased. If, after serum removal, the cells are incubated with either IGF-1 or IGF-2, they are stimulated to proliferate, but the stimulation is short-lived, and is followed by differentiation *(43,44)*. The problem is that differentiation, especially in neurons, keratinocytes, and hemopoietic cells, is usually followed by cell death. Terminally differentiated granulocytes have a very short life-span, and keratinocytes of the stratum corneum are constantly lost from the epidermis. This means that the IGF-1R, on one side, promotes survival, and, on the other side, by inducing differentiation, gently invites the cells to their own funerals (kindly supervised by the caspases).

ABSCOPAL EFFECT OF IGF-1 RECEPTOR TARGETING

The previous section asked a philosophical question, and involved a certain degree of metabiology, which can be defined as a discipline that is to biology what metaphysics is to physics. This section is not metabiological. On the contrary, it is based on facts that have been repeated many times, but, because they do not have any sensible explanation, they do not even allow intelligent questions, let alone philosophical ones. The original data was given by Resnicoff et al. *(6••),* and Baserga et al. *(45,46).* The basic experiment is very simple: Inject a syngeneic transplantable tumor in a naïve rat (or mouse), and let it grow to a palpable size. Then, inject the same tumor cells in the controlateral side, in which, however, the IGF-1R has been targeted (antisense strategies or dominant-negative mutants). The tumor cells with a targeted IGF-1R, of course, undergo apoptosis, and therefore do not grow at all *(see above).* The surprise is that the original wild-type tumor regresses. The effect is so rapid that it cannot be attributed to immunological mechanisms, at least not to conventional ones. This abscopal effect (i.e., an effect observed at a distance from the site of treatment) does not depend on how the IGF-1R has been downregulated: It can be induced whether the cells of the second injection have been treated with an antisense oligodeoxynucleotide, or are expressing an antisense RNA or a dominant-negative mutant *(45).* It can also be obtained by placing the targeted cells in a diffusion chamber, which is subsequently removed. Indeed, what is most bizarre is that other, unrelated cells can be placed in the diffusion chamber, and will produce the same abscopal effect, provided the IGF-1R is downregulated. The animal eventually becomes resistant to the subsequent implantation of wild-type cells. Because this resistance to subsequent implantation of syngeneic tumor cells lasts at least 1 yr, it is probably a classic immune response. But the immediate effect, which occurs quickly in naïve animals, does not have the characteristics of an immune response. If the receptor is not downregulated, the cells fail to induce regression of the wild-type tumor.

ON LEAPING BEFORE LOOKING

We would have classified the abscopal early effect of IGF-1R targeting in animals as a peculiar artifact, except that it seems to occur also in brain tumors of humans, based on clinical trials that are presently being conducted at Thomas Jefferson University (D. Andrews, et al., in preparation). This cannot, therefore, be ignored, and we are trying here to offer, not an explanation, but a hypothesis. There are, however, two other items that have to be considered at this point: Glioma cells expressing an antisense RNA to the IGF-1R RNA *(6••)* secrete in the medium, an activity that causes apoptosis of tumor cells. This activity has been identified as a peptide of mol wt <1 kDa. Second, a physical

Table 1
Induction of Apoptosis by Synthetic Peptides

Peptides	% cell recovery (24 h)
YL**E**PGPVTA	3.0 ± 0.1
YL**R**PGPVTA	3.5 ± 0.2
YL**A**PGPVTA	7.4 ± 0.4
Y**A**EPGPVTA	186.0 ± 4.6
YLE**A**GPVTA	221.0 ± 3.9
ALEPGPVTA	224.0 ± 3.2
YLEPG**A**VTA	260.0 ± 9.0

C6 cells were pretreated with the indicated peptides at 10^{-5} M for 24 h in serum-free medium, and tested for apoptosis, as described in ref. 17. Results are given in percentage recovery of cells, over initial seeding. Amino acid substitutions in the original peptide (the first one in the list) are indicated in bold characters. Each value represents the mean of four independent determinations. Adapted from ref. *48*.

association between the IGF-1R and (MHC)-class 1 complexes has previously been described *(47)*. Because the purification of the peptide from the medium conditioned by targeted glioma cells is still following the exasperating time schedule of protein purifications, we decided to take a huge leap and try the effect on tumor cells of other, already published, peptides associated with MHC-class 1 complexes. Three peptides were tested, and all three induced apoptosis, even at concentrations of $10^{-12} M$, in naïve animals *(48••)*. The results with representative peptides, at a concentration of $10^{-5} M$, are summarized in Table 1. Certain amino-acid substitutions (for instance, E to R) have no effect on the ability of the original peptide to induce apoptosis of tumor cells, but other substitutions (last four peptides in Table 1) completely abrogate the proapoptotic activity of the synthetic peptides. It is not, therefore, aspecific peptide toxicity, because one single amino-acid substitution can render a peptide inactive at concentrations of $10^{-5} M$ (although the active peptides are still active at $10^{-12} M$). We tested the YLRPGPVTA peptide on various human tumor cell lines, and it induced apoptosis (in increasing order of efficacy) in glioblastomas, ovarian carcinomas, colorectal carcinoma, small-cell lung carcinoma, and prostatic carcinoma. Finally, the active peptide can inhibit tumor growth when injected in vivo in nude mice.

We have therefore formulated the following hypothesis, which has the advantage that it can be tested: Downregulation of the IGF-1R causes the release of toxic peptides (probably MHC-class 1-associated peptides), which induce apoptosis of tumor cells; these peptides can also induce apoptosis of tumor cells, whose receptor has not been

IF THE IGF-I RECEPTOR IS UBIQUITOUS, HOW CAN IT BE A GOOD TARGET FOR ANTI-CANCER THERAPY?

➢ **IT IS NOT AN ABSOLUTE REQUIREMENT FOR NORMAL GROWTH, BUT...**

➢ **IT IS REQUIRED FOR GROWTH IN ANCHORAGE-INDEPENDENT CONDITIONS (LOCAL RECURRENCES ? METASTASES ?)**

➢ **TARGETING OF THE IGF-I RECEPTOR INDUCES MASSIVE APOPTOSIS OF TUMOR CELLS IN VIVO, WITH NEGLIGIBLE TOXICITY TO NORMAL CELLS**

➢ **IT INDUCES A HOST RESPONSE THAT CAN KILL SURVIVING TUMOR CELLS EVEN AT A DISTANCE (ABSCOPAL EFFECT)**

Fig. 2.

targeted, thus generating an abscopal effect; and since these peptides are antigenic, they eventually produce an immune response that renders the animals refractory to the subsequent injection of wild-type tumor cells. The principal merit of this hypothesis is that it is compatible with all the results obtained thus far. But there are some indications that this hypothesis may not be, after all, so outlandish. Thus, Lafarge-Frayssinet et al. *(49)* found that suppression of IGF-1 production in a rat hepatoma cell line (by an antisense approach) caused a fourfold increase in the expression of MHC-class 1 antigens. Albert et al. *(50)* reported that dendritic cells acquire peptide antigens from apoptotic cells. Most important, Marchand et al. *(51)* had very similar results to ours, in humans: They used a peptide encoded by gene *MAGE-3,* an antigen of melanoma cells that was supposed to induce immunity. To their surprise, the injection of this peptide caused regression of melanomas before an immune response could be detected. They were as puzzled of their results as we are of ours, but perhaps it is time that we let the results talk to us, instead of us talking to them.

CONCLUSIONS

Our conclusions are presented in Fig. 2, which is self-explanatory. This is not the only way of treating cancer, nor is this the only growth factor receptor that should be targeted. But we do think that the IGF-1R is such a good target in animals that it deserves to be tested in depth in humans.

REFERENCES

• of special interest
•• of outstanding interest

1. Baserga R. *Biology of Cell Reproduction.* Harvard University Press: Cambridge, MA. 1985.
2. Baserga R, Kisieleski. Comparative study of the kinetics of cellular proliferation of normal and tumorous tissues with the use of tritiated thymidine. *J Natl Cancer Inst* 1962; 28: 331–339.
3. Steel GG, Adams K, Barrett JC. Analysis of the cell population kinetics of transplanted tumours of widely differing growth rates. *Br J Cancer* 1966; 20: 784–800.
•4. Baserga R, Hongo A, Rubini M, Prisco M, Valentinis B. The IGF-I receptor in cell growth, transformation and apoptosis. *Biochim Biophys Acta* 1997; 1332: 105–126.
Comprehensive review of the diverse functions of the IGF-1R.
5. Arteaga CL, Kitten LJ, Coronado EB, Jacobs S, Kull FC, Allred DC, Osborne CK. Blockade of the type 1 somatomedin receptor inhibits growth of human breast cancer cells in athymic mice. *J Clin Invest* 1989; 84: 1418–1423.
••6. Resnicoff M, Sell C, Rubini M, Coppola D, Ambrose D, Baserga R, Rubin R. Rat glioblastoma cells expressing an antisense RNA to the insulin-like growth factor I (IGF-I) receptor are non-tumorigenic and induce regression of wild-type tumors. *Cancer Res* 1994; 54: 2218–2222.
Reports for the first time the abscopal effect obtained by targeting the IGF-1R.
7. Resnicoff M, Coppola D, Sell C, Rubin R, Ferrone S, Baserga R. Growth inhibition of human melanoma cells in nude mice by antisensense strategies to the type 1 insulin-like growth factor receptor. *Cancer Res* 1994; 54: 4848–4850.
8. Prager D, Li HL, Asa S, Melmed S. Dominant negative inhibition of tumorigenesis in vivo by human insulin-like growth factor receptor mutant. *Proc Natl Acad Sci USA* 1994; 91: 2181–2185.
9. D'Ambrosio C, Ferber A, Resnicoff M, Baserga R. A soluble insulin-like growth factor I receptor that induces apoptosis of tumor cells in vivo and inhibits tumorigenesis. *Cancer Res* 1996; 56: 4013–4020.
10. Baker J, Liu J-P, Robertson EJ, Efstratiadis A. Role of insulin-like growth factors in embryonic and postnatal growth. *Cell* 1993; 75: 73–82.
••11. Liu J-P, Baker J, Perkins AS, Robertson EJ, Efstratiadis A. Mice carrying null mutations of the genes encoding insulin-like growth factor I (igf-1) and type 1 IGF receptor (Igf1r). *Cell* 1993; 75: 59–72.
This elegant paper establishes the role of the IGF system in murine embryo development.
12. Ludwig T, Eggenschwiler J, Fisher P, D'Ercole JP, Davenport ML, Efstratiadis A. Mouse mutants lacking the type 2 IGF receptor (IGF2R) are rescued from perinatal lethality in Igf2 and Igf1r null backgrounds. *Dev Biol* 1996; 177: 517–535.
13. Morrione A, Valentinis B, Xu SQ, Yumet G, Louvi A, Efstratiadis A, Baserga R. Insulin-like growth factor II stimulates cell proliferation through the insulin receptor. *Proc Natl Acad Sci USA* 1997; 94: 3777–3782.

14. Louvi A, Accili D, Efstratiadis A. Growth-promoting interaction of IGF-II with the insulin receptor during mouse embryonic development. *Dev Biol* 1997; 189: 33–48.

•15. Sell C, Rubini M, Rubin R, Liu J-P, Efstratiadis A, Baserga R. Simian virus 40 large tumor antigen is unable to transform mouse embryonic fibroblasts lacking type-1 insulin-like growth factor receptor. *Proc Natl Acad Sci USA* 1993; 90: 11,217–11,221.
First paper dealing with the establishment of R− cells; very useful for studying the IGF-1R in the absence of endogenous background.

16. Baserga R. The price of independence. *Exp Cell Res* 1997; 236: 1–3.

17. Resnicoff M, Abraham D, Yutanawiboonchai W, Rotman H, Kajstura J, Rubin R, Zoltick P, Baserga R. The insulin-like growth factor I receptor protects tumor cells from apoptosis in vivo. *Cancer Res* 1995; 55: 2463–2469.

18. Long L, Rubin R, Baserga R, Brodt P. Loss of the metastatic phenotype in murine carcinoma cells expressing an antisense RNA to the insulin-like growth factor I receptor. *Cancer Res* 1995; 55: 1006–1009.

19. Reiss K, D'Ambrosio L, Yu X, Baserga R. Inhibition of tumor growth by a dominant negative mutant of the insulin-like growth factor 1 receptor with a bystander effect. *Clin Cancer Res* 1998; 4: 2647–2655.

••20. Harrington EA, Bennett MR, Fanidi A, Evan GI. c-myc-induced apoptosis in fibroblasts is inhibited by specific cytokines. *EMBO J* 1994; 13: 3286–3295.
Although not the first paper on the subject, this paper was the one that popularized the role of IGF-1 in protection from apoptosis.

21. O'Connor R, Kauffmann-Zeh A, Liu Y, Lehar S, Evan GI, Baserga R, Blattler WA. The IGF-I receptor domains for protection from apoptosis are distinct from those required for proliferation and transformation. *Mol Cell Biol* 1997; 17: 427–435.

22. Prisco M, Hongo A, Rizzo MG, Sacchi A, Baserga R. The IGF-I receptor as a physiological relevant target of p53 in apoptosis caused by Interleukin-3 withdrawal. *Mol Cell Biol* 1997; 17: 1084–1092.

•23. Kennedy SG, Wagner AJ, Conzen SD, Jodan J, Bellacosa A, Tsichlis PN, Hay N. The PI 3-kinase/Akt signaling pathway delivers an anti-apoptotic signal. *Genes Dev* 1997; 11: 701–713.
This paper, and others listed below, have made a cogent case for the PI-3K/ Akt pathway as one of the pathways followed by the IGF-1R in protection from apoptosis.

24. Kulik G, Klippel A, Weber MJ. Antiapoptotic signaling by the insulin-like growth factor I receptor, phosphatidylinositol 3-kinase, and Akt. *Mol Cell Biol* 1997; 17: 1595–1606.

25. Parrizas M, Saltiel AR, LeRoith D. Insulin-like growth factor I inhibits apoptosis using the phosphatidylinositol 3′-kinase and mitogen-activated protein kinase pathways. *J Biol Chem* 1997; 272: 154–161.

26. Backer JM, Myers MG Jr, Sun XJ, Chin DJ, Shoelson SE, Miralpeix M, White MF. Association of IRS-1 with the insulin receptor and the phosphatidylinositol 3′-kinase. *J Biol Chem* 1993; 268: 8204–8212.

27. White MF, Kahn CR. The insulin signaling system. *J Biol Chem* 1994; 269: 1–4.

28. Datta SR, Dudek H, Tao X, Masters S, Fu H, Gotoh Y, Greenberg ME. Akt phosphorylation of BAD couples survival signals to the cell-intrinsic death machinery. *Cell* 1997; 91: 231–241.

29. Dudek H, Datta SR, Franke TF, Birnbaum MJ, Yao R, Cooper GM, et al. Regulation of neuronal survival by the serine-threonine protein kinase Akt. *Science* 1997; 275: 661–665.

••30. Khwaja A, Rodriguez-Viciana P, Wennstrom S, Warne PH, Downward J. Matrix adhesion and ras transformation both activate a phosphoinositide 3-OH kinase and protein kinase B/Akt cellular survival pathway. *EMBO J* 1997; 16: 2783–2793.
An important paper connecting the ras pathway with Akt/PI-3K.

31. Sell C, Dumenil G, Deveaud C, Miura M, Coppola D, DeAngelis T, et al. Effect of a null mutation of the insulin-like growth factor I receptor gene on growth and transformation of mouse embryo fibroblasts. *Mol Cell Biol* 1994; 14: 3604–3612.

32. Zamorano J, Wang HY, Wang L-M, Pierce JH, Keegan AD. IL-4 protects cells from apoptosis via the insulin receptor substrate pathway and a second independent signaling pathway. *J Immunol* 1996; 157: 4926–4934.

33. Wang LM, Myers MG Jr, Sun XJ, Aaronson SA, White M, Pierce JH. IRS-1: essential for insulin- and IL-4-stimulated mitogenesis in hemopoietic cells. *Science* 1993; 261: 1591–1594.

34. Dews M, Nishimoto I, Baserga R. IGF-I receptor protection from apoptosis in cells lacking IRS proteins. *Receptors Signal Transduc* 1997; 7: 231–240.

35. Stewart AJ, Johnson MD, May, FEB, Westley BR. Role of insulin-like growth factors and the type-1 insulin-like growth factor receptor in the estrogen-stimulated proliferation of human breast cancer cells. *J Biol Chem* 1990; 265: 21,172–21,178.

36. Rubini M, Hongo A, D'Ambrosio C, Baserga R. The IGF-I receptor in mitogenesis and transformation of mouse embryo cells: role of receptor number. *Exp Cell Res* 1997; 230: 284–292.

37. Schmid C, Steiner T, Froesch ER. Preferential enhancement of myoblast differentiation by insulin-like growth factors (IGF-I and IGF-II) in primary cultures of chicken embryonic cells. *FEBS Lett* 1983; 161: 117–121.

38. Florini JR, Ewton DZ, Falen SL, van Wyk JJ. Biphasic concentration dependency of stimulation of myoblast differentiation by somatomedins. *Am J Physiol* 1986; 250: C771–C778.

39. Schmid C, Steiner T, Froesch ER. Insulin-like growth factor I supports differentiation of cultured osteoblast-like cells. *FEBS Lett* 1984; 173: 48–52.

40. Smith PJ, Wise LS, Berkowitz R, Wan C, Rubin CS. Insulin-like growth factor 1 is an essential regulator of the differentiation of 3T3-L1 adipocytes. *J Biol Chem* 1988; 263: 9402–9408.

41. Mill JF, Chao MV, Ishii DN. Insulin, insulin-like growth factor II, and nerve growth factor effects on tubulin mRNA levels and neurite formation. *Proc Natl Acad Sci USA* 1985; 82: 7126–7130.

42. Recio-Pinto E, Lanf FF, Ishii DN. Insulin and insulin-like growth factor II permit nerve growth factor binding and the neurite formation response in cultured human neuroblastoma cells. *Proc Natl Acad Sci USA* 1984, 81: 2562–2566.

43. Bach LA, Salemi R, Leeding KS. Roles of insulin-like growth factor (IGF) receptors and IGF-binding proteins in IGF-II-induced proliferation and differentiation of L6A1 rat myoblasts. *Endocrinology* 1995; 136: 5061–5069.

44. Coolican SA, Samuel DS, Ewton DZ, McWade FJ, Florini JR. The mitogenic and myogenic actions of insulin-like growth factors utilize distinct signaling pathways. *J Biol Chem* 1997; 272: 6653–6662.

45. Baserga R, Resnicoff M, D'Ambrosio C, Valentinis B: The role of the IGF-I receptor in apoptosis. *Vitam Horm* 1997; 53: 65–98.

46. Baserga R, Resnicoff M, Dews M. The IGF-I receptor and cancer. *Endocrine* 1997; 7: 99–102.
47. Hsu D, Olefsky JM. Effect of a major histocompatibility complex class I peptide on insulin-like growth factor I receptor internalization and biological signaling. *Endocrinology* 1993; 133: 1247–1251.
••48. Resnicoff M, Huang Z, Herbert D, Abraham D, Baserga R. A novel class of peptides that induce apoptosis and abrogate tumorigenesis in vivo. *Biochem Biophys Res Comm* 1997; 240: 208–212.
Demonstrates unequivocally that MHC-class 1-associated peptides can induce apoptosis of tumor cells directly, independent of the immune system. The release of these peptides may be related to the downregulation of the IGF-1R function.
49. Lafarge-Frayssinet C, Duc HT, Frayssinet C, Sarasin A, Anthony D, Guo Y, Trojan J. Antisense insulin-like growth factor I transferred into a rat hepatoma cell line inhibits tumorigenesis by modulating major histocompatibility complex 1 cell surface expression. *Cancer Gene Ther* 1997; 4: 276–285.
50. Albert ML, Sauter B, Bhardway N. Dendritic cells acquire antigen from apoptotic cells and induce class 1-restricted CTLs. *Nature* 1998; 392: 86–89.
51. Marchand M, Weynants P, Rankin E, Arienti F, Belli F, Parmiani G, et al. Tumor regression responses in melanoma patients treated with a peptide encoded by gene MAGE-3. *Int J Cancer* 1998; 63: 883–885.

14 Drug Resistance and the Survival Niche

Survival Signals Combine to Suppress Drug-Induced Apoptosis in B-Lymphoma Cells

Sian T. Taylor, BsC, PhD
and Caroline Dive, BPharm, PhD

CONTENTS

ABSTRACT

Studies of tumor cell lines in vitro *suggest that the suppression of drug-induced apoptosis may be an important drug-resistance mechanism applicable to all types of anticancer agents that are used* in

From: *Apoptosis and Cancer Chemotherapy*
Edited by: J. A. Hickman and C. Dive © Humana Press Inc., Totowa, NJ

vivo. Drug-induced apoptosis may be suppressed by overexpression of antiapoptotic proteins such as Bcl-2, and/or by survival signals in the tumor microenvironment. These signals include the action of growth factors, cell–cell interactions, and interactions of cells with extracellular matrix. In vivo, tumor cells are likely to receive a combination of these signals from within their microenvironment. These survival signals may be heterogeneously distributed, and some cells may therefore exist in what could be termed a survival niche; others may be more vulnerable to apoptosis, including that induced by anticancer agents. The emergence of a viable and drug-resistant subpopulation of tumor cells following drug treatment may therefore depend on the signals derived from within a particular survival niche. This chapter focuses on the survival niche, *in which drug-resistant B-cell lymphomas may reside, and describes attempts to create this cellular environment* in vitro.

INTRODUCTION: IMPORTANCE
OF THE SURVIVAL NICHE

Apoptosis is a default process *(1••)*. Cells require survival signals to prevent the engagement of this death program. Cell survival is mediated via signaling through various surface receptors, which, depending on cell phenotype include those for insulin-like growth factor-1 (IGF-1) *(2;* and *see* Chapter 13), interleukin-3 (IL-3) *(3)*, nerve growth factor (NGF) *(4)* and CD40 ligand (CD40L) *(5)*. Numerous studies of cell fate in vitro suggest that cell–cell interactions and interactions of cells with extracellular matrix (ECM) make an equally important contribution to the survival of a cell *(6,7)*. In vivo, cells are likely to receive a combination of signals from soluble factors and stromal interactions within their microenvironment. These survival signals may be heterogeneously distributed, and some cells may therefore exist in what could be termed a survival niche; others may be more vulnerable to apoptosis, including that induced by anticancer agents. Bcl-2 family members regulate the threshold at which apoptosis may be engaged (*see* Chapters 7–10), and it seems likely that these may be downstream recipients of survival signaling pathways. Indeed, the phosphorylation of Bad by protein kinase B (PKB), downstream of IGF-1 receptor ligation, is one established example of such a survival signaling pathway (*see* Chapter 13). However, in many studies of the effects of the enforced hyperexpression of Bcl-2 or of Bcl-X$_L$ on cell fate after damage, apoptosis is only delayed, rather than suppressed, and often no clonogenic advantage is observed (*see* Chapter 1). It is therefore questionable whether overexpression of one of the antiapoptotic members of the Bcl-2 family alone can provide true drug resistance, allowing survival and subsequent proliferation of drug-treated cells. We would argue that Bcl-2, in concert with other family members and binding proteins, only functions to suppress drug-induced apoptosis when the cells in which these proteins are expressed are stimulated by the appropriate cocktail of survival signals within the survival niche. The emergence of a viable and drug-resistant subpopulation of cells following drug treatment may depend on the signals derived from within a particular survival niche. One example of coordinated survival signaling was observed in our recent study of cultured primary breast epithelia, in which cells were plated on different types of ECM, and provided with various combinations of soluble factors. Apoptosis was only avoided when the cells received at least two signals, one via ligation of the IGF-1 (or insulin) receptor, and the other through interaction with the appropriate laminin-rich ECM. What is really interesting about these data is that although the IGF-1 receptor was autophosphorylated upon stimulation by IGF-1 in the absence of the appropriate ECM-cell interaction, the survival signal was only propagated through insulin receptor substrate-1 (IRS-1) and phosphatidylinositol-3 kinase (PI-3K), if the cells interacted with this ECM (N. Farrelley, C. Dive, and C. H. Streuli, et al., *J. Cell. Biol.,* in press). In these studies, IGF-1 and a laminin-rich ECM constitute a minimal survival niche for these alveolar epithelial cells in vitro.

Here we reviewed the clinical problem of drug resistance in B-cell lymphoma. We then discuss some of the survival signals that constitute the B-cell survival niche within the germinal center (GC) of secondary lymphoid tissues. We speculate on the effects on drug resistance of B-lymphoma cells within the survival niche in vivo, and describe attempts to create this protective B-cell environment in vitro. Finally, new data are presented on the mechanism of GC-derived signal-mediated suppression of drug-induced apoptosis in B-lymphoma cells.

THE CLINICAL PROBLEM: B-LYMPHOMA AND DRUG RESISTANCE

Approximately 15,000 new cases of non-Hodgkin's lymphoma arise in the United States each year, 50% of which are follicular lymphomas (FL). The median age of patients at presentation with FL is 50–60 yr *(8)*. Lymphomas are clinically divided into indolent low-grade or aggressive high-grade forms. The low-grade lymphomas have a favorable prognosis, and these patients have a mean survival time of 6–10 yr *(9,10)*. High-grade tumors have a poor prognosis, with a mean patient survival time of 6 mo from first diagnosis. Low-grade lymphomas initially regress, but relapse with drug-resistant tumors, and the disease becomes incurable *(8)*. Relapsed low-grade lymphomas re-emerge with a morphology similar to the original tumor, or undergo a blastic transformation to high-grade, in which the lymphoma becomes more aggressive. In the past, treatment involved radiotherapy, but it was later discovered that low-grade lymphomas were sensitive to alkylating agents, presenting a chemotherapeutic approach to the treatment of this type of disease. Although there are advantages to using chemotherapeutic drugs rather than irradiation, such as an increased mean survival time of 10 yr instead of 5 yr, the pattern of remission followed by relapse still occurs, and remains a major problem. New types of therapy, such as the administration of interferon-α or antibodies against specific tumor cell markers, are currently in development, but as yet do not appear to be successful *(9,10)*. So, although the majority of FL cells engage apoptosis after chemotherapy, a small, clinically undetectable subpopulation of cells do not die. These cells may remain dormant for a period of years (acquiring further mutations) before rapid regrowth, or they may continue to proliferate, slowly becoming clinically detectable as a relapsed drug-resistant tumor years later. What is different about this subpopulation of cells that increases their ability to resist drug-induced apoptosis? One of many possibilities is that the subpopulation of drug-resistant cells is located within an microenvironment that provides optimal survival signaling to suppress apoptosis. Unfortunately, there are no tractable FL cell lines available for mechanistic studies of survival signaling and drug resistance; Burkitt lymphoma (BL) cell lines are frequently used for this purpose. BL is a far more chemosensitive disease, although there is also a problem with disease relapse in African children.

REGULATION OF B-CELL FATE
AND THE ROLE OF CD40

To address questions of how B-lymphoma cells might exhibit an increased threshold for the induction of drug-induced apoptosis, we first need to know how apoptosis is regulated in normal B-cells. During normal immune function, the development of the B-cell compartment is regulated by ECM and cell–cell interactions in the GCs of secondary lymphoid tissues *(11••,12)*. This is accompanied by changes in the expression of Bcl-2 and Bcl-X_L *(13•)*. The CD40 molecule plays a central role in the regulation of Bcl-2 and Bcl-X_L, and of B-cell fate. CD40 is expressed on the surface of early and mature B-lymphocytes,

and is activated by its ligand, CD40L, which is presented on the surface
of T-lymphocytes and of follicular dendritic cells within the GC *(11••)*.
CD40 is a type 1 integral-membrane protein *(14)*, and a member of
the tumor necrosis factor receptor (TNF-R) superfamily, which includes
CD95 (Fas/APO-1) *(15, see* Chapter 12) and NGF *(16)*. Ligation of
CD40 activates a pathway that is essential for the initiation of immuno-
globulin (Ig)-isotype switching and secretion during the B-cell matura-
tion process. An important stage in B-cell development is the process
in which both positive and negative selection occurs. B-cells, which
produce high-affinity Ig protein/antigen receptors, are positively selected by signaling via CD40 *(13•)*. The great majority of B-cells express
low-affinity Ig, and these are eliminated in the GC by apoptosis *(17)*,
mediated via tumor necrosis factor (TNF)-α and CD95 ligand (CD95L)
(15,18) crosslinking their respective receptors and initiating an apoptotic caspase cascade.

CD40 signaling has been shown to rescue isolated GC B-cells from
spontaneous apoptosis in vitro *(19•)* and, in combination with interleu-
kin-4 (IL-4), to promote the proliferation of both normal and malignant
B-cells *(20,21)*. Crosslinking of the CD40 receptor correlates with an
increase in cell number, and recent evidence suggests that the CD40
molecule must be trimerized to produce a survival signal *(22)*. Paradoxically, CD40 contains a sequence with limited homology to the death-
effector domains found in CD95 and TNF-1. However, CD40 can signal
for death in certain circumstances, e.g., in neurons *(23)*, and in epithelia
and mesenchymal cells *(24)*, emphasizing the importance of cellular
context, as well as environmental factors.

SUPPRESSION OF APOPTOSIS BY CD40
IN B-LYMPHOMA CELL LINES

CD40 signaling can also suppress apoptosis induced by various
stimuli in B-lymphoma cell lines. It was shown by Parry et al. *(25)*
and Merino et al. *(26)* that CD40 could suppress IgM and IgD receptor-
mediated apoptosis, both in mature human B-cells and in the murine
WEHI-231 B-lymphoma cell line. CD40 signaling also suppressed
apoptosis induced by calcium (Ca) ionophore, anti-IgM, or irradiation
in Mutu Group I BL-cells *(21,27,28)*. CHEP BL and L3055 BL cells
were also protected from radiation-induced apoptosis by CD40 signaling, but only in those cells expressing wt p53 *(29•)*. The authors have
recently shown that CD40 signaling suppressed chlorambucil- and
etoposide-induced apoptosis in the Mutu BL-cell lines *(30••,* Taylor,
S. T. et al., unpublished results). Taken together, these data imply that

CD40 plays an important role in B-lymphoma cells to suppress apoptosis induced by chemotherapeutic agents, as well that induced by more physiological stimuli.

B-lymphoma cells in vivo receive multiple survival signals from the GC environment, including CD40 ligation. Many B-lymphomas exhibit deregulated expression of Bcl-2, and it is unknown to what extent this and environmental factors, such as cell–cell interactions to present CD40L, cytokine signaling, and ECM–cell interactions, may combine to enhance the survival of drug-treated B-lymphoma cells. This was one question that the authors set out to address (*see* Mimicking the B-Cell Survival Niche in Vitro; *30••*; S. T. Taylor et al., submitted).

EFFECT OF CD40 SIGNALING ON EXPRESSION OF BCL-2 AND BCL-X$_L$ IN B-CELLS

Initial studies showed that CD40 signaling caused an increase in Bcl-2 expression, implying that this was the mechanism by which CD40 promoted B-cell survival *(31•)*. Holder et al. *(21)* confirmed that CD40 stimulation rescued GC B-cells from apoptosis, but pointed out that the initial survival signal appeared to be unrelated to increased levels of Bcl-2. A delay in Bcl-2 induction suggested that a mechanism other than Bcl-2 upregulation mediated the survival of the B-cells during the first 24 h after CD40 stimulation *(21)*. Bcl-X$_L$ was subsequently implicated in CD40-mediated B-cell survival. CD40 ligation correlated with a rapid induction of Bcl-X$_L$ mRNA, and protein expression was detectable 3 h after CD40 stimulation *(32•,33•)*, when Bcl-2 and Bax levels were unaffected *(33)*. Thus, it appears that downstream targets of CD40 survival signaling in normal B-cells include both Bcl-X$_L$ and Bcl-2. Upregulation of Bcl-X$_L$ occurs within a few hours of CD40 stimulation, providing short-term protection; Bcl-2 levels increase later, and are associated with long-term cell survival.

HOW DOES CD40 SIGNAL FOR B-CELL SURVIVAL?

Although the signaling pathway(s) elicited by stimulation of CD40 have been extensively studied (Fig. 1), and several CD40-binding proteins have been identified, it remains unclear how this signaling pathway(s) results in changes in the expression of Bcl-2 or Bcl-X$_L$, to suppress apoptosis. Ligation of CD40 has several consequences, including initiation of phosphatidylinositol (PI) signaling cascades, effects on the intracellular concentration of cyclic adenosine monophosphate (cAMP), activation of nonreceptor tyrosine kinases (TKs), such as Lyn

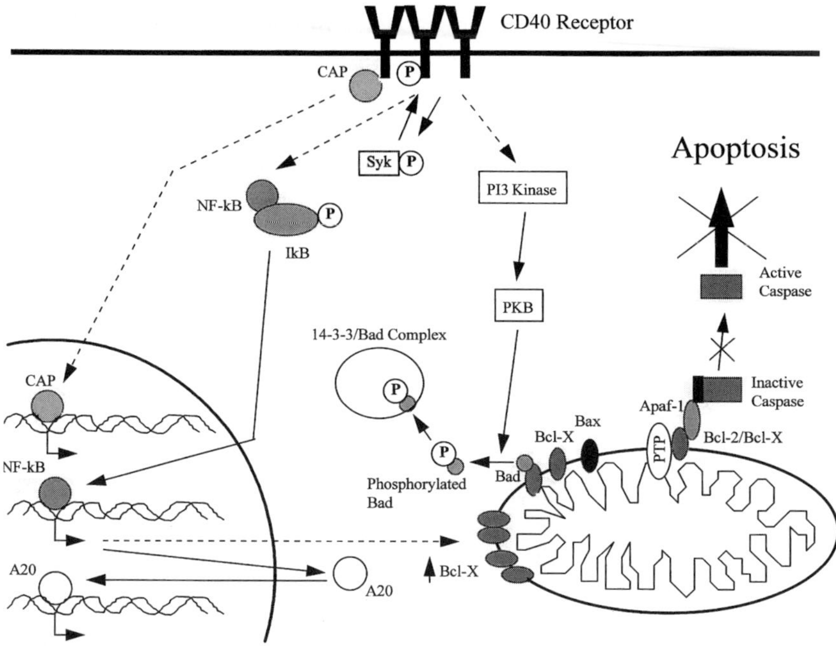

Fig. 1. Schematic of CD40 signaling pathways. *Dotted lines* represent putative pathways; *solid lines* represent pathways that have been demonstrated in B-cells.

and Syk, and activation of the transcription factors, nuclear factor-κB (NF-κB) and the zinc-finger protein, A20. These events are now briefly reviewed.

The CD40-binding partner, CD40-associated protein (CAP-1) *(34)* or CD40-receptor-associated factor-1 (CRAF-1) *(35)*, was identified using a yeast 2-hybrid screen. This protein appears to have homology with two putative signal-transduction molecules, the TNF-R-associated factors, TRAF-1 and TRAF-2 *(36)*. From its structure, it seems likely that CAP-1/CRAF-1 can interact with DNA, and possibly has the potential to regulate gene transcription, suggesting a direct link between CD40 ligation and an alteration in gene expression *(33,35)*. Phosphorylation of phospholipase Cγ1 (PLCγ1), phospholipase Cγ2 (PLCγ2), PI, PI-3K, and also GTPase-activating protein (GAP), are all sequelae of stimulation of CD40 *(37,38)*. In certain cell types, activation of PI-3K pathway leads to the phosphorylation of Bad, and thus a suppression of apoptosis, and, although it has yet to be demonstrated, the available evidence suggests that CD40 signaling may also result in Bad phosphorylation. Knox et al. *(39)* showed that tyrosine phosphorylation is important in CD40 signaling, especially in the rescue of GC B-cells from apoptosis. However, they disputed the link between CD40 and the PI signaling pathway, which involves phospholipase (PLC). These authors inferred that intracellular levels of cAMP increased in those GC B-cells that had been selected (i.e., those destined to survive), and suggested a link between CD40 signaling and the increase in intracellular cAMP concentration *(40)*. In 1991; Uckun et al. *(41)* linked CD40 signaling to a TK pathway. Stimulation of CD40 resulted in phosphorylation and activation of the nonreceptors TK, Lyn and Syk *(37,38)*. Association of CD40 with these TKs is indirect, and has been suggested to occur through adaptor molecules, such as a 28-kDa protein phosphorylated on CD40 signaling *(37)*. Crosslinking CD40 on B-cells activates NF-κB within 5–10 min *(42)*, and TK activity mediates this activation of NF-κB. Lyn, Fyn, and Syk nonreceptor TKs were phosphorylated by CD40 *(37)*; and Berberich et al. *(42)* investigated which of these PTKs was the most likely to be involved in activating NF-κB. Both herbimycin A and an antioxidant, ammonium pyrrolidine dithiocarbamate, blocked CD40-mediated NF-κB activation. This confirmed that TKs were involved, and that the likely candidate was Syk, because this TK is known to be activated by oxidants *(43)*. Recent evidence for the importance of NF-κB in apoptosis is linked primarily to the activation of NF-κB by the TNF-R1 *(44)*. In this scenario, NF-κB can be either pro- or antiapoptotic, depending on the cell type in which it becomes activated. However, when activated by CD40 in B-cells, it is associated with the suppression of apoptosis. A cytokine-inducible, zinc-finger protein, A20, has also been associated with CD40-dependent events, and is regulated by NF-κB *(45)*. However, the transcriptional targets of A20 are presently unknown, although their encoded proteins are involved in the suppression of apoptosis *(46)*. A great deal is known about the signaling events elicited by CD40 ligation, but how these events result in suppression of apoptosis remains to be deduced.

OTHER GC-DERIVED SIGNALS

What other survival signals would be found in a GC survival niche and could combine with CD40 signaling to enhance B-cell survival in vivo? Upon binding of its ligand, IL-4R is phosphorylated, and the protein 4PS binds to the phosphorylated receptor, either directly or through an adaptor protein. 4PS can then interact with a number of proteins, including PI-3K and Grb-2 *(47)*. It would be expected that association of Grb-2 and Sos to IL-4R would lead to Ras activation and signaling through the mitogen-activated protein kinase (MAPK) pathway. However, this does not occur, and it is suggested that, because of the absence of Shc phosphorylation by IL-4R, Ras cannot be not activated *(47)*. Therefore, although GrB-2 and Sos are activated, the downstream signaling events must still be elucidated. IL-4R also activates a Jak/Stat pathway involving Jaks 1 and 3 and Stat 6, which can mediate gene transcription *(47)*. IL-4 suppressed apoptosis in Ramos B-cells, but did not upregulate Bcl-X$_L$ *(48)*. Stimulation of the IL-4 receptor inhibited the expression of the early-response gene *Berg36,* which was induced by a lethal concentration of Ca ionophore. This inhibition of *Berg 36* was specific to IL-4, and could not be induced by CD40 ligation. Because *Berg36* was shown to mediate the proapoptotic stimulus, its inhibition by IL-4 contributed to a suppression of Ca ionophore-induced apoptosis. Thus, IL-4 signaling can suppress apoptosis in BL cells.

Cell–cell contacts are critically important for B-cells within the GC. Here, B-cells interact with T-cells (resulting in the presentation of CD40L), and with follicular dendritic cells, which express the $\alpha_4\beta_1$ integrin ligand, vascular cell adhesion molecule-1 (VCAM-1). Activation of this integrin has been shown to promote the survival of B-cells *(49)*. Other integrins in different cell types have also been shown to promote the suppression of apoptosis and survival *(50–52)*. Activation of the integrins leads to the phosphorylation of Shc and association with Grb and Sos, which in turn activates Ras and the MAPK pathway *(50–52)*. Integrins also activate focal adhension kinase (FAK) by phosphorylation, which leads to the activation of PI-3K, but whether this is a direct interaction is unclear. Thus, stimulation of B-cells via the VCAM-1–$\alpha_4\beta_1$ interaction and the T-cell interaction both activate the PI-3K, which, as previously discussed (*see* Chapter 13), can transmit a survival signal.

MIMICKING THE B-CELL SURVIVAL NICHE IN VITRO: EFFECTS OF THIS CELLULAR ENVIRONMENT ON DRUG-INDUCED APOPTOSIS

In 1991, Banchereau et al. *(19•)* published an elegant system in a successful attempt to promote tonsillar B-cell proliferation in vitro. B-cells were cocultured with confluent, irradiated murine fibroblasts transfected with the human CDw32 receptor (which binds an added agonistic anti-CD40 antibody), in the presence of recombinant human IL-4. We initially used this system to show that this combination of

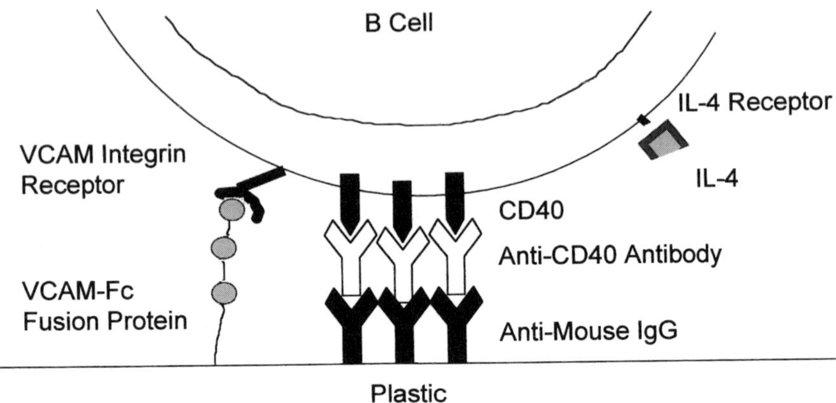

Fig. 2. Schematic of an in vitro survival niche for B-cells, incorporating VCAM–Fc fusion protein, antimouse IgG and anti-CD40, and IL-4.

survival signals delayed chlorambucil-induced apoptosis in Mutu BL cells *(30••)*. The system was then modified, that the feeder layer was replaced with a substrate coated with the VCAM–Fc fusion protein and antimouse IgG antibody to present anti-CD40 antibody (Fig. 2). This was used to show that the GC environment provided a clonogenic advantage to chlorambucil-treated BL cells. Parental BL cells upregulated Bcl-X_L, and clonogenicity was further enhanced in a BL clone (BL-Bcl-2), which overexpressed Bcl-2. Drug-treated BL Bcl-2 cells did not exhibit any clonogenic advantage in the absence of the GC-derived signals (Fig. 3).

Because of concerns about Epstein-Barr virus (EBV)-encoded proteins in Mutu BL and the unknown status of p53 in these cells, drug resistance in the GC environment was examined further, using JLP119 BL cells. This cell line is not so typical of BL; JLP119 cells express wt p53, and do not contain the EBV genome. Both of these features are important, because FL cells characteristically have wt p53, and are not associated with EBV infections. Therefore, the JLP119 cells allowed us to study the response of DNA-damaged malignant B-cells to extracellular signals similar to those that FL cells would encounter in the GC of a lymph node. JPL119 cells stabilize p53 after DNA damage, and upregulate p21[WAF-1]. This occurs in the presence or absence of GC-derived survival signals, and thus the survival niche does not prevent the activity of one sensor of drug-induced damage, but rather changes the cellular response to it downstream (*see* Fig. 4).

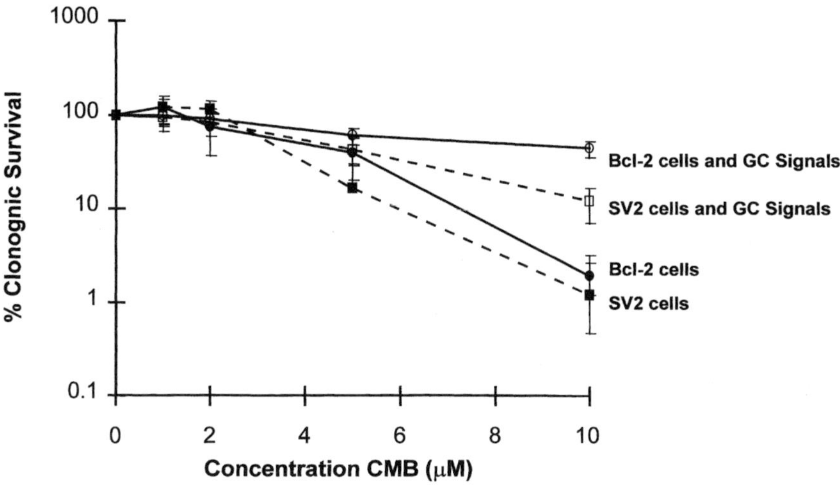

Fig. 3. Combined survival signals (IL-4, interaction with VCAM, and ligation of CD40) provide a clonogenic advantage to chlorambucil-treated BL cells, which is enhanced by the overexpression of Bcl-2.

The role of each individual survival signal in the suppression of etoposide-induced apoptosis in JLP119 cells was examined. The anti-CD40 antibody must be immobilized to stimulate the increase in Bcl-X_L protein, and this upregulation occurs faster when IL-4 is present (IL-4 does not upregulate Bcl-X_L itself). All three survival signals presented alone can delay etoposide-induced apoptosis, but combine to enhance cell survival. Given the finding that CD40-activated NF-κB in B-cells, its role in the upregulation of Bcl-X_L was investigated. In the presence of all three stimuli (CD40 activation, IL-4, and VCAM-Fc), NF-κB was translocated to the nucleus. This translocation of NF-κB to JLP119 nuclei by the GC signals could be blocked by (E)-capsaicin, which was previously shown to act by inhibiting the degradation of NF-κB inhibitor, IκB (53; S. T. Taylor et al., submitted). Bax immunoprecipitates with Bcl-X_L in JLP119 cells. Although the cellular content of Bax remained unaltered in the absence or presence of IL-4 and VCAM/Fc fusion and etoposide, the GC signals suppressed a pronounced etoposide-induced increase in immunofluorescence associated with the N-terminus of Bax, measured by flow cytometry. This change induced by etoposide suggests that Bax does not necessarily require an increase in overall protein levels to induce apoptosis, but Bax may undergo a conformational change and/or a change in binding partners to become an active promoter of apoptosis (S. T. Taylor et

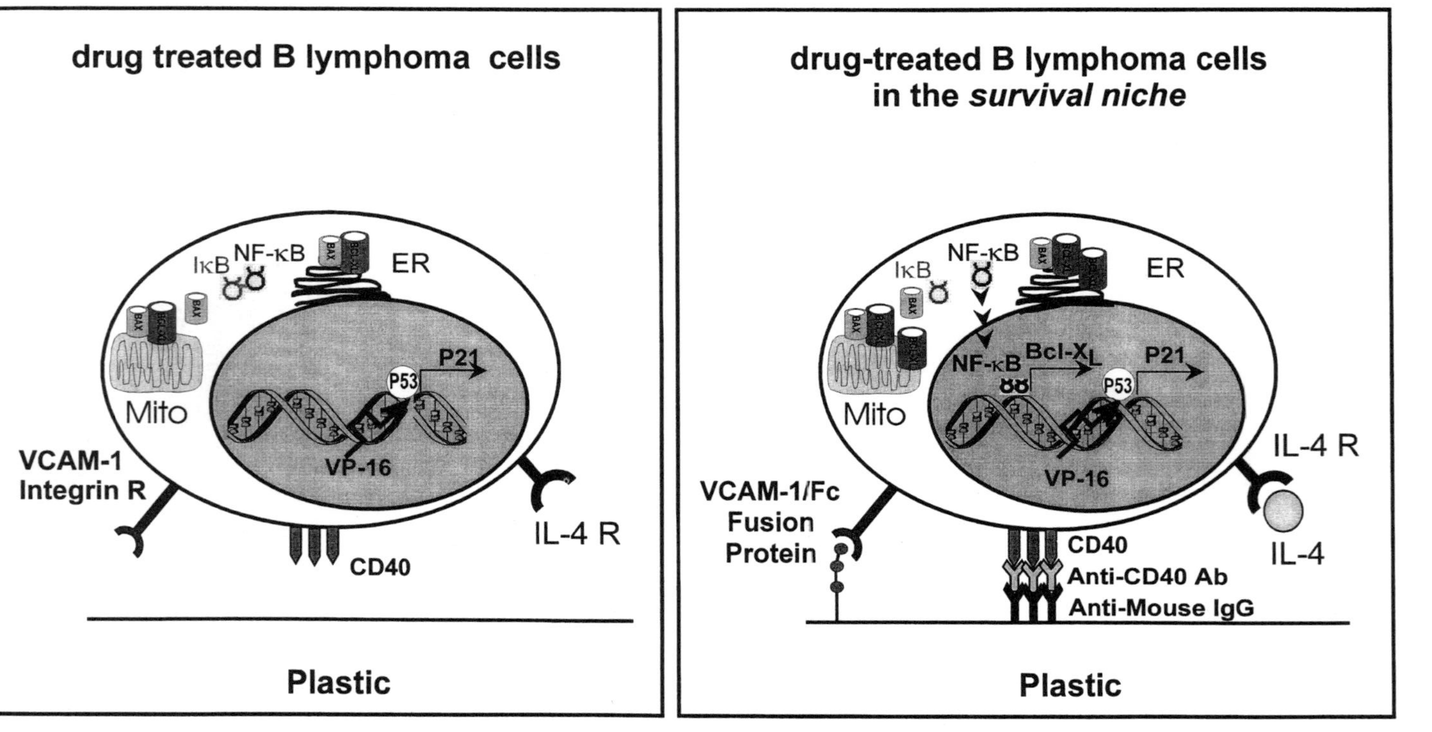

Fig. 4. Model for the suppression of drug-induced apoptosis in JLP119 BL cells within the GC environment, mimicked by the in vitro survival niche.

al., submitted), a change(s) suppressed by VCAM-Fc and IL-4. This observation shows that, although CD40 signaling is important in the suppression of apoptosis, other signals from the GC environment, such as IL-4 and VCAM-1, also play important roles (Fig. 4).

SUMMARY

This chapter puts forward the idea of a survival niche, in which combinations of environmental signals promote cell survival. Cells within such an environment may be particularly resistant to drug-induced apoptosis. Even in initially chemosensitive tumors, apparently successful drug treatment may select for that minor cell subpopulation within a survival niche, which subsequently repopulates the tumor. We predict that, only with fuller comprehension of the complexities of combinatorial survival signaling, will investigators be able to think rationally about design of novel agents that might overcome this type of drug resistance.

ACKNOWLEDGMENTS

The authors would like to acknowledge the patient efforts of Andrew Walker in setting up the B-cell survival niche in vitro, and John A. Hickman for his major input into the survival niche project, and for constructive criticism of the chapter.

REFERENCES

• of special interest
•• of outstanding interest

••1. Raff MC. Social controls on cell survival and cell death. *Nature (London)* 1992; 356: 397–400.
This is a landmark commentary. Raff may not have been the first to suggest that cells need one another to survive, but he certainly made everyone think hard about this, with respect to the control of apoptosis.

2. Harrington EA, Bennet MR, Fanidi A, Evan GI. c-Myc-induced apoptosis in fibroblasts is inhibited by specific cytokines. *EMBO J* 1994; 13: 3286–3295.

3. Collins MK, Marvel J, Malde P, Lopez-Rivas A. Interleukin 3 protects murine bone marrow cells from apoptosis induced by DNA damaging agents. *J Exp Med* 1992; 176: 1043–1051.

4. Edwards SN, Buckmaster AE, Tolkovsky AM. Death programme in cultured sympathetic neurones can be suppressed at the post-translational level by nerve growth factor, cyclic AMP, and depolarisation. *J Neurochem* 1991; 57: 2140–2143.

5. MacLennan ICM, Liu Y-J, Johnson GD. Germinal centres in the immune response. *Immunol Rev* 1992; 126: 143–161.

6. Frische SM, Francis H. Disruption of epithelial cell-matrix interactions induces apoptosis. *J Cell Biol* 1994; 124: 619–626.

7. Pullan S, Wilson J, Metcalfe A, Edwards GM, Goberdhan N, Tilly J, et al. Requirement of basement membrane for the suppression of programmed cell death in mammary epithelium. *J Cell Sci* 1996; 109: 631–642.

8. Stein RS. Clinical features and evaluation. In: *Cancer Treatment and Research,* vol. 4, Lymphomas 1, Bennett JM, ed. Martinus Nijhoff: The Hague/Boston/London. 1981; 150–157.

9. Rohatiner AZS, Lister TA. New approaches to treatment of follicular lymphoma. *Br J Haematol* 1991; 79: 349–354.

10. Horning SJ. Treatment approaches to the low-grade lymphomas. *Blood* 1994; 83: 881–884.

•• 11. Clark EA, Ledbetter JA. How B and T cells talk to each other. *Nature* (London) 1994; 367: 425–428.
    This review gives a comprehensive picture of the interactions between B- and T-cells in vivo. The authors also explain in detail some of the interactions that occur, including the CD40 receptor and its ligand. These interactions are described in the context of B cell maturation during an immune response.

12. Durie FH, Foy TM, Masters SR, Laman JD, Noelle RJ. Role of CD40 in regulation of humoral and cell-mediated immunity. *Immunol Today* 1994; 11: 101–105.

• 13. Liu Y-J, Joshua DE, Williams GT, Smith CA, Gordon J, MacLennan ICM. Mechanisms of antigen driven selection in germinal centres. *Nature* (London) 1989; 342: 929–931.
    This is one of the first papers to describe CD40 suppression of apoptosis. The authors were studying the mechanism by which some B-cells survive, although others undergo apoptosis during the B-cell maturation process in an immune response. They were able to show that engagement of the CD40 receptor was able to prevent centrocytes from undergoing apoptosis.

14. Paulie S, Rosen A, Ehlin-Henriksson B, Braesch-Andersen S, Jakoson E, Koho H, Perlmann P. Human B lymphocyte and carcinoma antigen CDw40 is a phosphoprotein involved in signal transduction. *J Immunol* 1989; 142: 590–595.

15. Krammer PH, Dhein J, Walczak H, Berhmann I, Mariani S, Matiba B, et al. Role of APO-1-mediated apoptosis in the immune system. *Immunol Rev* 1994; 142: 175–191.

16. Smith CA, Farrah T, Goodwin RG. TF superfamily of cellular and viral proteins: activation, co-stimulation, and death. *Cell* 1994; 76: 959–962.

17. Nunez G, Merino R, Grillot D, Gonzalez-Garcia M. Bcl-2 and Bcl-X: regulatory switches for lymphoid death and survival. *Immunol Today* 1994; 15: 582–587.

18. Lagresle C, Bella C, Daniel PT, Krammer PH, DeFrance T. Regulation of GC B cell differentiation: Role of the human APO-1/ Fas (CD95) molecule. *J Immunol* 1995; 154: 5746–5756.

• 19. Banchereau J, Depaoli P, Valle A, Garcia E, Rousset F. Long-term human B-Cell lines dependent on interleukin-4 and antibody to CD40. *Science* 1991; 251: 70–72.
    The feeder-layer GC system was developed by this group, who, in studying both normal B-lymphocytes and B-cell lines were able to show that CD40 and IL-4 signaling could lead to long-term proliferation. This observation showed that

normal B-cells, which undergo apoptosis if placed in culture, could not only survive, but also proliferate.

20. Johnson PWM, Watt SM, Betts DR, Davies D, Jordan S, Norton AJ, Lister TA. Isolated follicular lymphoma-cells are resistant to apoptosis and can be grown in-vitro in the CD40 stromal cell system. *Blood* 1993; 82: 1848–1857.

21. Holder MJ, Wang H, Milner AE, Casamayor M, Armitage R, Spriggs MK, et al. Suppression of apoptosis in normal and neoplastic human B-lymphocytes by CD40 ligand is independent of Bcl-2 induction. *Eur J Immunol* 1993; 23: 2368–2371.

22. Mazzei GJ, Edgerton MD, Losberger C, Lecoanethenchoz S, Graber P, Durandy A, et al. Recombinant soluble trimeric CD40 ligand is biologically-active. *J Biol Chem* 1995; 270: 7025–7028.

23. Ruan YL, Rabizadeh S, Camerini D, Bredesen DE. Expression of CD40 induces neural apoptosis. *J Neurosci Res* 1997; 50: 383–390.

24. Hess S, Engelmann H. Novel function of CD40—induction of cell-death in transformed cells. *J Exp Med* 1996; 183: 159–167.

25. Parry SL, Hasbold J, Holman M, Klaus GGB. Hypercross-linking surface IgM or IgD receptors on mature B-cells induces apoptosis that is reversed by costimulation with IL-4 and anti-CD40. *J Immunol* 1994; 152: 2821–2829.

26. Merino R, Grillot DAM, Simonian PL, Muthukkumar S, Fanslow WC, Bondada S, Nunez G. Modulation of anti-IgM-induced B-cell apoptosis by Bcl-X(L) and CD40 in WEHI-231 cells—dissociation from cell-cycle arrest and dependence on the avidity of the antibody-IgM receptor interaction. *J Immunol* 1995; 155: 3830–3838.

27. Gregory CD, Dive C, Henderson S, Smith, CA, Williams GT, Gordon J, Rickinson AB. Activation of Epstein-Barr virus latent genes protects human B cells from death by apoptosis. *Nature* (London) 1991; 349: 612–614.

28. An SK, Knox KA. Ligation of CD40 rescues Ramos-Burkitt lymphoma B-cells from calcium ionophore-triggered and antigen receptor-triggered apoptosis by inhibiting activation of the cysteine protease Cpp32/Yama and cleavage of its substrate Parp. *Febs Lett* 1996; 386: 115–122.

•29. Milner AE, Grand RJA, Vaughan ATM, Armitage RJ, Gregory CD. Differential effects of BCL-2 on survival and proliferation of human B-lymphoma cells following gamma-irradiation. *Oncogene* 1997; 15: 1815–1822.
CD40 signaling was shown to be able to suppress apoptosis induced by radiation. Many of the previous papers showed that CD40 suppresses apoptosis induced by physiological stimuli and Ca ionophores; however, this is one of the papers that examined the response to DNA damage. The overexpression of Bcl-2 correlated with the consequent survival of cells following irradiation. The combination of both Bcl-2 overexpression and CD40 signaling provided a better survival advantage than either of the two factors alone.

••30. Walker, A Taylor, ST, Hickman JA, Dive C: Germinal center-derived signals act with Bcl-2 to decrease apoptosis and increase clonogenicity of drug-treated human B lymphoma cells. *Cancer Res* 1997; 57: 1939–1945.
Describes the use of both the feeder-layer system and the VCAM-1/Fc, IL-4, and anti-CD40 antibody system to show that CD40, IL-4, and VCAM-1 signaling could suppress drug-induced apoptosis. The important finding is that in combination with the overexpression of Bcl-2, these survival signals promoted a clonogenic survival advantage. This was the first paper to describe that cells in a

survival niche could not only suppress apoptosis induced by physiological stimuli, but also by chemotherapeutic agents.

•31. Liu YJ, Mason DY, Johnson GD, Abbot S, Gregory CD, Hardie DL, Gordon J, MacLennan ICM. Germinal center cells express Bcl-2 protein after activation by signals which prevent their entry into apoptosis. *Eur J Immunol* 1991; 21: 1905–1910.
Linked CD40 signaling to the upregulation of Bcl-2, and thus a link to the suppression of apoptosis. It showed that Bcl-2 was upregulated in GC B-cells, and that these cells did not undergo spontaneous apoptosis in culture.

•32. Choi MSK, Boise LH, Gottschalk AR, Quintans J, Thompson CB, Klaus GGB. Role of Bcl-X(L) in CD40-mediated rescue from anti-Mu-induced apoptosis in WEHI-231 B-lymphoma-cells. *Eur J Immunol* 1995; 25: 1352–1357.
This was one of two papers that showed that CD40 not only regulated Bcl-2 expression in certain circumstances, but that CD40 could also mediate expression of Bcl-X$_L$ another protein that is involved in the suppression of apoptosis. The upregulation of Bcl-X$_L$ was linked to the suppression of antigen-receptor-induced apoptosis.

•33. Wang ZH, Karras JG, Howard RG, Rothstein TL. Induction of Bcl-X by CD40 engagement rescues sIg-induced apoptosis in murine B-cells. *J Immunol* 1995; 155: 3722–3725.
The second of two papers that showed that Bcl-X$_L$ was upregulated following CD40 ligation. The group also showed that the upregulation of Bcl-X$_L$ following CD40 ligation could suppress antigen-receptor-induced apoptosis.

34. Sato T, Irie S, Reed JC. A novel member of the TRAF family of putative signal transducing proteins binds to the cytosolic domain of CD40. *Febs Lett* 1995; 358: 113–118.

35. Cheng GH, Cleary AM, Ye ZS, Hong DI, Lederman S, Baltimore D. Involvement of CRAF1, a relative of TRAF, in CD40 signaling. *Science* 1995; 267: 1494–1498.

36. Tewari M, Dixit VM. Recent advances in tumor-necrosis-factor and CD40 signaling. *Curr Opin Genet Dev* 1996; 6: 39–44.

37. Faris M, Fu SM. CD40 signal-transduction—association of CD40 with Lyn, PI3k, Gap, and PLC-gamma. *Clin Res* 1994; 42: A206–A206.

38. Ren CL, Morio T, Fu SM, Geha RS. Signal-transduction via CD40 involves activation of Lyn kinase and phosphatidylinositol-3-kinase, and phosphorylation of phospholipase C-gamma-2. *J Exp Med* 1994; 179: 673–680.

39. Knox, KA, Gordon, J. Protein-tyrosine phosphorylation is mandatory for CD40-mediated rescue of germinal center B-cells from apoptosis. *Eur J Immunol* 1993; 23: 2578–2584.

40. Knox KA, Johnson GD, Gordon J. Distribution of cAMP in secondary follicles and its expression in B-cell apoptosis and CD40-mediated survival. *Int Immunol* 1993; 5: 1085–1091.

41. Uckun FM, Schieven GL, Dibirdik I, Chandanlanglie M, Tuelahlgren L, Ledbetter JA. Stimulation of protein tyrosine phosphorylation, phosphoinositide turnover, and multiple previously unidentified serine threonine-specific protein-kinases by the pan-B-cell receptor CD40/Bp50 at discrete developmental stages of human B-cell ontogeny. *J Biol Chem* 1991; 266: 17,478–17,485.

42. Berberich I, Shu GL, Clark EA. Cross-linking CD40 on B-cells rapidly activates nuclear factor-kappa-B. *J Immunol* 1994; 153: 4357–4366.

43. Schieven GL, Kirihara JM, Burg DL, Geahlen RL, Ledbetter JA. P72syk tyrosine kinase is activated by oxidizing conditions that induce lymphocyte tyrosine phosphorylation and Ca^{2+} signals. *J Biol Chem* 1993; 268: 16,688–16,692.
44. Baichwal VR, Baeuerle PA. Apoptosis: activate NF-kappa B or die? *Curr Biol* 1997; 7: R94–R96.
45. Sarma V, Lin ZW, Clark L, Rust BM, Tewari M, Noelle RJ, Dixit VM. Activation of the B-cell surface-receptor CD40 induces A20, a novel zinc-finger protein that inhibits apoptosis. *J Biol Chem* 1995; 270: 12,343–12,346.
46. Tewari M, Wolf FW, Seldin MF, O'Shea KS, Dixit VM, Turka LA. Lymphoid expression and regulation of A20, an inhibitor programmed cell death. *J Immunol* 1995; 154: 1699–1706.
47. Rebollo A, Gomez J, Martinez-A C. Lessons from the immunological, biochemical, and molecular pathways of the activation mediated by IL-2 and IL-4. *Adv Immunol* 1996; 63: 127–196.
48. Ning ZQ, Norton JD, Li J, Murphy JJ. Distinct mechanisms for rescue from apoptosis in Ramos human B-cells by signaling through CD40 and interleukin-4 receptor—role for inhibition of an early response gene, Berg36. *Eur J Immunol* 1996; 26: 2356–2363.
49. Koopman G, Keehnen RMJ, Lindhout E, Newman W, Shimizu Y, van Seventer GA, de Groot C, Pals ST. Adhesion through the LFA-1 (CD11a/CD18)-ICAM-1 (CD54) and the VLA (CD49d)-VCAM-1 (CD106) pathways prevents apoptosis of germinal centre B cells. *J Immunol* 1994; 152: 3760–3767.
50. Giancotti FG. Integrin signaling: specificity and control of cell survival and cell cycle progression. *Curr Opin Cell Biol* 1997; 9: 691–700.
51. Kolanus W, Seed B. Integrins and inside-out signal transduction: converging signals from PKC and PIP_3. *Curr Opin Cell Biol* 1997; 9: 725–731.
52. Meredith Jr JE, Schwartz MA. Integrins, adhesion and apoptosis. *Trends Biosci* 1997; 7: 146–150.
53. Singh S, Natarajan K, Aggarwal BB: Capsaicin (8-Methyl-N-Vanillyl-6-Nonenamide) is a potent inhibitor of nuclear transcription factor-κB activation by diverse agents. *J Immunol* 1996; 157: 4412–4420.

15 Chemotherapy, Tumor Microenvironment, and Apoptosis

Constantinos Koumenis, PhD,
Nicholas Denko, PhD, MD
and Amato J. Giaccia, PhD

CONTENTS

ABSTRACT

Minimally transformed cells are more apoptotically sensitive than their untransformed wild-type counterparts to apoptosis induced by chemotherapeutic compounds, or by the microenvironmental stresses

From: *Apoptosis and Cancer Chemotherapy*
Edited by: J. A. Hickman and C. Dive © Humana Press Inc., Totowa, NJ

of the tumor, such as hypoxia. However, most solid-tumor cells have acquired additional genetic alterations that reduce their sensitivity to chemotherapy- and hypoxia-induced apoptosis. One question that has not been explored is how the hypoxic microenvironment of tumors affects the apoptotic response of tumor cells to chemotherapy during treatment. Although it is widely accepted that the decreased proliferative capacity of hypoxic tumor cells makes them more refractory to killing by chemotherapeutic agents that require cell proliferation to be effective, agents now exist that are able to induce apoptosis under growth-inhibiting hypoxic conditions. In addition, new strategies to make tumors more oxic, by inducing apoptosis, should also improve tumor response to chemotherapy.

INTRODUCTION

Evidence is accumulating that the tumor microenvironment affects both the malignant progression of transformed cells and their therapeutic response to chemotherapy and radiotherapy *(1••)*. Although these two consequences, attributable to the adverse conditions in the tumor microenvironment, seem distinct, they are in fact intimately linked by their effect on apoptotic cell death. Of the many factors that fluctuate in the microenvironment of solid tumors, changes in oxygen availability seem to be the most critical. In a landmark study, Thomlinson and Gray *(2••)* proposed that the rapid growth rate of tumors exceeded their ability to develop vasculature to supply tumor cells that were >150 μm from the vessel lumen. Therefore, tumor cells become chronically hypoxic as oxygen diffusion becomes limiting because of cellular metabolism. This concept is further underscored by the highly angiogenic-dependent nature of solid tumor growth. It is now known that a second mechanism exists in which tumor cells can become hypoxic as the result of transient changes in vessel opening and closing *(1••)*. This latter mechanism has been termed transient hypoxia because cells become reoxygenated upon vessel opening. In contrast, diffusion-limited hypoxic cells have less probability of becoming reoxygenated, and will die unless they are able to access a blood supply.

Previous studies on solid tumors have led investigators to champion several hypotheses based on tumor physiology, to explain the refractory nature of solid tumors to chemotherapy. These hypotheses include a physical barrier to delivery of the chemotherapeutic compound by constricted blood vessels *(3)*, cessation of cell proliferation induced by low-oxygen conditions *(4)*, decreased cytotoxic drug activity under low-oxygen conditions *(5)*, and induction of stress proteins that impart resistance to the cytotoxicity of certain chemotherapeutic compounds

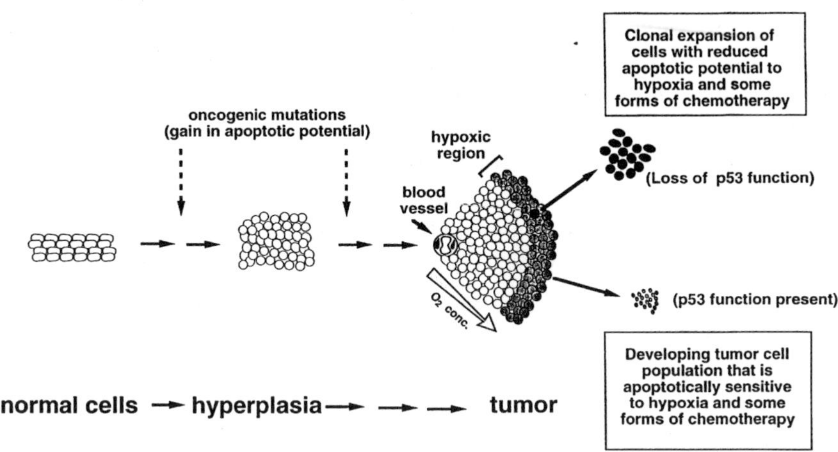

Fig. 1. Model for the role of hypoxia in the clonal selection and expansion of cells with diminished apoptotic sensitivity to chemotherapy.

(6) . Conceptually, all of these factors could be important in increasing the survival of tumor cells to chemotherapy. However, in advanced cervical cancer, changes in the tumor microenvironment brought about by fluctuating oxygen levels have been found to be an independent prognostic factor predicting disease-free survival and local control, regardless of treatment *(7•)*. Additional studies by other groups for squamous cell carcinoma of the head and neck *(8•)* and soft-tissue sarcomas *(9•)* also indicate that patients with hypoxic tumors treated with radiotherapy and/or chemotherapy have a greater chance for metastatic disease, compared to well-oxygenated tumors. These studies, especially those in which tumor oxygenation predicted for a worse outcome in patients treated with surgery alone, suggest that the tumor microenvironment is not just making cells more refractory to killing by therapeutic modalities that require oxygen to be effective, but that it is also making tumor cells more aggressive.

HYPOXIA AND MINIMALLY TRANSFORMED RODENT CELLS

An insight into how low-oxygen conditions may affect the aggressiveness of tumors is that hypoxia initiates apoptosis in oncogenically transformed cells, and can provide a selective pressure for the expansion of populations of oncogenically transformed rodent cells with reduced apoptotic sensitivity. This applies not only to hypoxia (Fig. 1; *10••*), but to chemotherapeutic agents as well *(11••)*. Resistance to chemotherapy will result, because many of the same genes that modulate sensitivity to hypoxia-induced apoptosis also modulate sensitivity to chemotherapy-induced apoptosis *(12)*. Thus, transformed cells that possess mutations in their apoptotic program because of inactivation of the *p53* tumor-suppressor gene or the overexpression of antiapoptotic genes, such as *bcl*-2, will have a survival advantage in a low-oxygen environment. This hypothesis was tested in mixing experiments that demonstrated that small numbers of transformed cells, which lack wild-type p53, possess a survival advantage over isogenic cells that possess wild-type p53, when exposed to multiple rounds of hypoxia followed by aerobic regrowth *(10••)*. In immune-deficient mice, transplanted tumors possessing a wild-type *p53* genotype were found to have apoptotic areas that colocalized with hypoxic areas, further supporting the hypothesis that the tumor microenvironment may play an important role in malignant progression. Some tumors, derived from either p53 wild-type or p53-null cells, can also die by necrosis when exposed to hypoxia for long periods of time, indicating that this form of cell death is possible in both cell types, and does not act as a selective pressure.

HYPOXIA AND HUMAN PAPILLOMAVIRUS

Since human papillomavirus (HPV) infection is strongly associated with cervical neoplasia, and tumor hypoxia has prognostic significance in human cervical carcinomas, the authors examined the relationship between hypoxia and apoptosis in human cervical epithelial cells (ECs) expressing high-risk HPV type 16 oncoproteins. In vitro, hypoxia stimulated both p53 induction and apoptosis in primary cervical ECs infected with the HPV *E6* and *E7* genes *(13•)*. Cervical fibroblasts infected with *E6* and *E7* did not undergo apoptosis when exposed to hypoxia, but, instead, growth arrested. Furthermore, cell lines derived from HPV-associated human cervical squamous cell carcinomas were substantially less sensitive to apoptosis induced by hypoxia, indicating that these cell lines have acquired further genetic alterations that reduced their apoptotic sensitivity. Although the process of long-term cell culturing alone resulted in selection for subpopulations of HPV oncoprotein-expressing cervical ECs with diminished apoptotic potential, exposure of cells to hypoxia greatly accelerated the selection process *(13•)*. These results provide evidence for the role of hypoxia-mediated selection of cells with diminished apoptotic potential in the progression of human tumors, and, in part, offer an explanation why cervical tumors that possess low pO$_2$ values are more aggressive, and potentially more resistant to chemotherapy.

To investigate the relationship between hypoxia and apoptosis in spontaneous human tumors, apoptotic regions were correlated with hypoxic regions in biopsy specimens of patients with squamous cell carcinoma of the cervix, prior to initiation of treatment. Analysis of contiguous histological sections indicated that apoptosis colocalized with hypoxia (C. K. Kim et al., unpublished observations). A direct correlation between clinical stage and incidence of colocalization was suggested by the data, with extensive colocalizations found in patients with advanced stages of the disease. Colocalization of apoptosis and hypoxia occurred in regions distal to blood vessels, similar to that found in experimental rodent tumors. Taken together, both rodent and human studies suggest that the tumor microenvironment can act as a selective pressure for the expansion of transformed cell populations with reduced apoptotic sensitivity to chemotherapeutic agents.

MECHANISM AND IMPLICATIONS OF HYPOXIA-INDUCED APOPTOSIS

What events are required to increase the susceptibility of untransformed cells to stress-induced apoptosis? One event is the deregulated expression of growth-promoting proto-oncogenes, such as *myc,* the adenovirus *E1A* gene, or HPV *E7* gene. How and what targets these oncogenes affect are still under investigation, but all three oncogenes stimulate an increase in p53 protein levels. Thus, the increase in p53 protein levels makes cells with deregulated oncogene expression highly sensitive to changes in their microenvironment, leading to apoptotic cell death rather than growth arrest. Although a number of potential transcriptional targets of p53, such as Bax *(14),* IGFBP-3 *(15),* and p53-inducible genes (PIGs) *(16),* have been identified that might play a role in apoptosis, the requirement of transcription in p53-mediated apoptosis under low-oxygen conditions is unclear. Data indicates that a p53 gene product that lacks ability to induce transcriptional activity is still quite able to induce apoptosis under certain conditions *(17••,18••).* In contrast, induction of a G1/S-phase checkpoint induced by DNA damage requires p53's transcriptional activity.

It has been proposed that p53's transrepressor activity is used in signaling apoptosis. Although hypoxia increases p53 protein accumulation and sequence specific DNA binding activity, known p53 target genes, such as *p21, GADD45, bax,* and *Fas,* are not transcriptionally

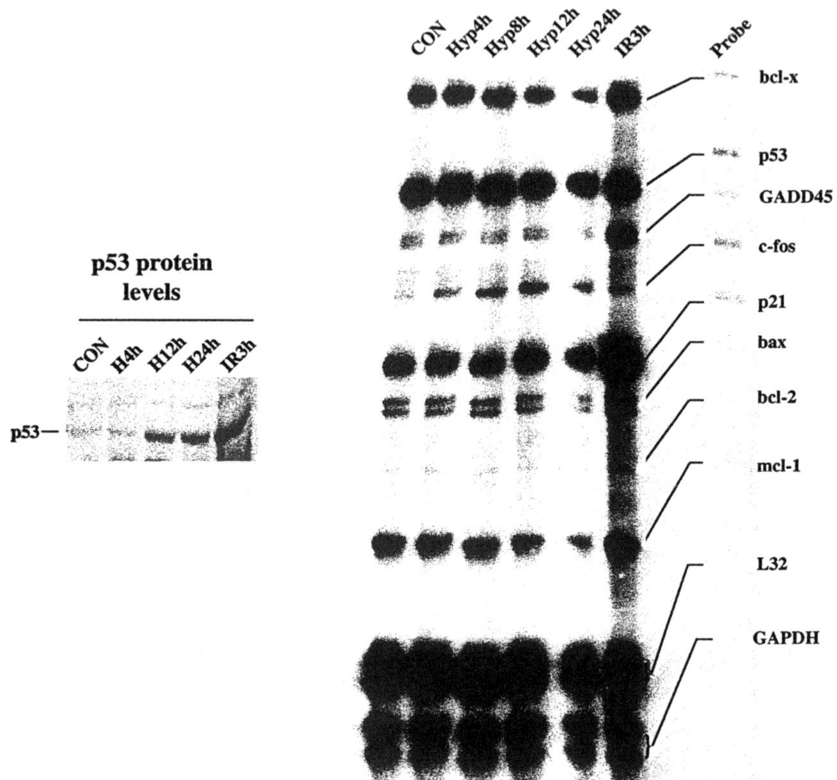

Fig. 2. Ribonuclease protection assay of total RNA from MCF-7 cells exposed to hypoxia (Hyp) or ionizing radiation (IR). The probe fragments, migrate higher than the protected fragments because of extra unprotected bases present in the RNA. Note that hypoxia only induces c-*fos* mRNA expression, but none of the p53-responsive genes (*GADD45, p21,* or *bax*) are induced by hypoxia, in contrast to IR. The L32 ribosomal protein and the glycolytic enzyme GAPDH are used as hybridization and loading controls.

upregulated under hypoxic conditions (Fig. 2; *19*). Furthermore, *bax* knockout cell lines are just as sensitive to hypoxia-induced apoptosis as their wild-type counterparts, indicating that the reason transcriptional induction of p53 is dispensable under hypoxic conditions is not the result of pre-existing high levels of proapoptotic gene expression. Thus, hypoxia represents a model system in which p53's transcriptional activity is uncoupled from its apoptotic-promoting activity. Examination of two genes, *MAP4* and α-tubulin, which have been shown to be transcriptionally repressed in a p53-dependent manner by ultraviolet light, are also repressed under hypoxic conditions in a p53-dependent manner *(20;* C. Koumenis, M. Murphy, and A. Giaccia, unpublished observations). This latter study indicates that, although the activation of p53 is not increased under hypoxic conditions, its ability to transrepress gene expression is increased.

What are the critical upstream and downstream targets of the p53 signal-transduction pathway under hypoxic conditions? Elegant studies have demonstrated that p53 and Ref-1 can interact to increase DNA-binding and transcriptional activity of p53 *(21)*. Although it has been reported that Ref-1 is active under hypoxic conditions *(22),* its interaction with p53 under hypoxia is unproductive, becauseof its diminished role in increasing transcription of downstream effector genes *(19)*. Similarly, proposed interactions between p53 and other hypoxia-inducible transcription factors do not increase the transcriptional activity of downstream p53 effectors present in the genome (although they can activate p53 reporter genes) *(23)*. One critical determinant in p53 regulation is the transcriptional coactivator CREB-binding protein (CBP), which links p53 to RNA polymerase, and which also possesses histone acetyltransferase activity *(24•,25•,26•)*. Thus, a potential explanation for the failure to detect p53-effector transactivation is that CBP becomes limiting under hypoxic conditions, so that a transcriptionally active p53 cannot activate transcription of critical downstream effectors, because it lacks this essential coactivator to link it directly to the transcriptional machinery. Alternatively, p53 transrepressor activity may be induced under hypoxic conditions, and it is possible that CBP may be a target of p53-induced transrepression *(24•)*. Clearly, understanding the posttranslational modifications induced by hypoxia will give better insight into how p53 functions in signaling apoptotic cell death.

CHEMOTHERAPEUTIC AGENTS THAT INDUCE DNA DAMAGE AND KILL HYPOXIC CELLS BY APOPTOSIS

Studies on minimally transformed rodent and human cells, although mechanistically informative, may not reflect of the apoptotic sensitivity of spontaneous human tumors that have accumulated additional genetic alterations compared to the ones introduced in these experimental paradigms. This is discussed in Chapter 1. However, spontaneous human solid tumors, in many instances, still possess intact, although diminished, stress-inducible apoptotic programs. One question that has been inadequately addressed is what chemotherapeutic agents that are presently in clinical protocols are able to kill hypoxic cells by apoptosis?

The answer to this question is not known. One very promising agent, tirapazamine, is a clinically effective hypoxic cell cytotoxin *(1)*. Although tirapazamine can induce apoptotic cell death, tirapazamine can also kill cells by reducing clonogenic survival. In fact, it is quite able to kill a wide variety of human tumor cells that possess highly diminished apoptotic programs.

More than two decades ago, Lin et al. *(27)* proposed that quinone-based compounds, such as mitomycin C, would exhibit cytotoxic activity in hypoxic regions by redox modulation of its benzoquinone ring. Thus, mitomycin C represents a chemotherapeutic compound that is active under hypoxic conditions in experimentally and spontaneously derived tumors. In a different set of experiments, Grau and Overgaard *(28)* examined the cytotoxic effects of a variety of commonly used chemotherapeutic agents, based on an *in situ* local-tumor-control assay. They found that carmustine (BCNU), vincristine (VCR), and etoposide (VP16) all reduced the surviving fraction of hypoxic cells, but 5-fluorouacil (5-FU) and methotrexate (MTX) had no effect on hypoxic tumor cells. Although, in both studies described above, the mechanism of hypoxic cell killing was not vigorously addressed, data is accumulating that some chemotherapeutic agents are active under hypoxic conditions. Recent ongoing experiments by Denko et al. (unpublished) indicate that VP-16 induces apoptotic cell death under hypoxic conditions, in a series of colorectal tumor cell lines. Clearly, much work is needed to investigate the genetic determinants that make hypoxic solid-tumor cells sensitive to chemotherapeutic killing, and how or if these determinants could be exploited therapeutically.

SIGNAL-TRANSDUCTION MODULATORS AND SURVIVAL FACTOR INHIBITORS

As mentioned, oncogenic transformation by *ras/E1A, myc,* or *E6/E7* appears to sensitize cells to hypoxia-induced apoptosis. A common phenotype of cells exposed to hypoxia is reduced macromolecular (protein, RNA, and DNA) synthesis *(29)*. This raises the possibility that hypoxia could be inducing apoptosis by reducing the levels of constitutively active antiapoptotic or survival genes. As a first step to test this hypothesis, the authors investigated whether the specific inhibition of RNA polymerase II (RNAPII) can act as a signal for apoptosis *(30•)*. Untransformed and transformed cells were treated with two highly specific RNAPII inhibitors, 5,6-dichloro-1-β-D-ribofurano-sylbenzimidazole (DRB) and α-amanitin. Both DRB and α-amanitin, which inhibit RNAPII function by two distinct mechanisms, substantially increased apoptosis in a time- and dose-dependent manner in p53$^{+/+}$- and p53$^{-/-}$-transformed mouse embryonic fibroblasts (MEFs) and in HeLa cells under oxic conditions, demonstrating that this type of apoptosis does not require wild-type p53. The specificity of this effect was demonstrated by the fact that engineered expression of an α-amanitin-resistant *RNAPII* gene rendered cells resistant to apoptosis by α-amanitin without affecting their sensitivity to DRB. This indicated that α-amanitin induces apoptosis solely by inhibiting RNAPII function, and not by a nonspecific mechanism. Inhibition of RNAPII for the same periods of time in untransformed cells, such as Rat1 or human AG1522 fibroblasts, did not result in apoptosis, but in growth arrest. In contrast, upregulated expression of c-Myc in Rat1 cells dramatically increased their sensitivity to DRB, suggesting that apoptosis following inhibition of RNAPII function is substantially enhanced by oncogene expression. Apoptosis induced by RNAPII inhibition was independent of the cell cycle or ongoing DNA replication, because DRB induced similar levels of apoptosis in asynchronous cells and cells synchronized by collection at mitosis.

These results provide biochemical and genetic evidence that the specific, nongenotoxic inhibition of RNAPII function can activate a rapid, p53-independent apoptotic program in a variety of oncogenically transformed cells. The differential sensitivity of untransformed and transformed cells to apoptosis elicited by transcriptional inhibition, coupled with the finding that this type of apoptosis is independent of p53 status, suggest that inhibition of RNAPII may be exploited therapeutically for the design of successful antitumor agents. These should be effective under certain stress-inducing conditions, such as hypoxia, in which macromolecular synthesis may already be somewhat compromised. Furthermore, these results are consistent with a model in which p53 inhibits RNAPII function, either globally or at specific promoters, to induce transrepression of antiapoptotic genes (Fig. 3). RNAPII inhibitors induce apoptosis in cells that also undergo apoptosis when p53 function is increased, either by genotoxic damage or by forced p53 overexpression (e.g., transformed MEFs, HeLa, *myc*-expressing Rat1 cells). Thus, we propose that p53 acts, under hypoxic conditions, to amplify the apoptotic signal (perhaps by suppressing the expression of survival genes), which, in the absence of p53, would cause little or no apoptosis. Thus, agents that inhibit transcription can act as a new class of chemotherapeutic agents, exhibiting selectivity toward oncogenically transformed cells, that does not require oxygen to be functional.

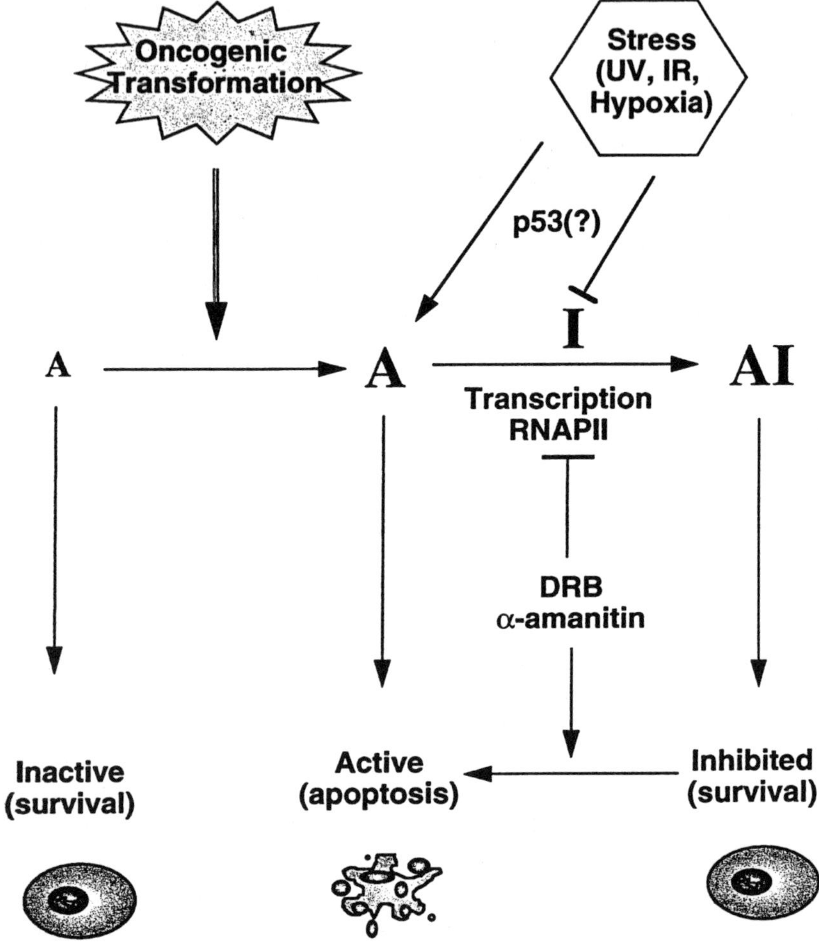

Fig. 3. Model for the role of RNAPII activity in protecting oncogenically transformed cells from undergoing apoptosis. An untransformed cell normally has
very low sensitivity to apoptotic stimuli, and exposure to RNAPII inhibitors or
exogenous stress results in either a cell-cycle arrest or mitotic death. Oncogenic
transformation of the cell results in an increase in its sensitivity to apoptotic
stimuli (A). However, this apoptotic sensitivity is counteracted by the presence
of one or more survival genes, or apoptotic inhibitors (I), whose presence and/or
activity is tightly dependent on ongoing RNAPII activity. Exposure of this cell
to RNAPII inhibitors (and possibly exogenous stress like hypoxia, low serum, or
UV/ionizing irradiation) results in an inhibition of the activity or cellular levels
of (I), which subsequently leads to a shift toward the apoptotic state. p53 may be
transducing the inhibitory effect of these stresses to the RNAPII holoenzyme.

HOW DOES RNAPII INHIBITION SIGNAL FOR APOPTOSIS?

A well-documented effect of inhibition of RNAPII is the induction of changes in chromatin conformation, which include the dissociation and dispersion of the nucleolus into the nucleoplasm *(31,32)*. It is possible that α-amanitin and DRB induce a relaxation of the structure of chromatin, thereby making it more accessible to nucleases that are either constitutively active or become activated by oncogenic transformation. A more plausible target for RNAPII inhibitors, though, is the expression of antiapoptotic genes. For example, the synthesis of proteins of the Bcl-2 family that exhibit antiapoptotic activity (Bcl-2, Bcl-X$_L$) may be the target of these inhibitors. However, the very long half-life of the Bcl-2 protein (about 14 h) *(33)*, and the very rapid onset of apoptosis following DRB or α-amanitin administration, suggests that a different and perhaps yet unidentified member of this family or an entirely different family of apoptotic suppressors may be primarily responsible for maintaining a suppression of apoptosis in oncogenically transformed cells. In addition, Bcl-2, Bcl-X$_L$, and Bax levels do not change under hypoxic conditions, indicating that alterations in their expression are not needed for hypoxia-induced apoptosis. These types of observation are addressed in Chapter 7.

Recently, Mayo et al. *(34•)* demonstrated that NF-κB activation, which accompanies *ras* transformation, is required to suppress p53-independent apoptosis induced by *ras*. This implies that Ras elicits the induction of antiapoptotic proteins in conjunction with an increase in the apoptotic sensitivity of transformed cells. This study also supports the model described in Fig. 3, and identifies specific antiapoptotic gene products that are involved in suppression of oncogene-induced apoptosis. Chu et al. *(35)* have reported that NF-κB activation induces the expression of c-IAP2, a member of the inhibitors of apoptosis family (IAP). This gene family of apoptotic suppressors is homologous to the baculovirus inhibitors of apoptosis that suppress apoptosis of the host cell. In addition to the IAP family, a new family of secreted apoptotic inhibitors was recently identified in rodent cells. These proteins (called secreted apoptosis-related proteins [SARPS]) were isolated from quiescent cells that exhibit increased antiapoptotic activity, compared to cells stimulated with serum to enter the cell cycle *(36)*. However, although SARP-1 expression inhibits apoptosis, expression of SARP-2 appears to increase the apoptotic potential of transfected human MCF-7 cells. Therefore, their role as pro- or antiapoptotic gene products remains to be clarified.

NF-κB inhibition may also prove useful in inducing apoptosis in hypoxic areas of tumors. Koong et al. *(37•,38)* previously demonstrated that hypoxia induces NF-κB activity. Because of the recently demonstrated antiapoptotic role of NF-κB, we can speculate that the role of NF-κB upregulation under hypoxia in some cell types is to prevent apoptosis by the hypoxic stress. Furthermore, because many transformed cell types possess increased NF-κB levels, inhibition of NF-κB may increase the sensitivity of hypoxic tumor cells to die when exposed to chemotherapeutic agents. For example, synthetic peptides that carry the nuclear localization signal for NF-κB have been shown to compete with endogenous NF-κB for nuclear transport, and possibly could act in tumors to inhibit NF-κB activity *(39)*. Other chemical agents that have been shown to inhibit NF-κB activity could also serve as proapoptotic agents. One such agent is caffeic acid phenethyl ester (CAPE), the active component of the folk medicine, propolis. Propolis has long been used for its anti-inflammatory and antiallergenic properties, presumably because of its anti-NF-κB properties *(40)*. CAPE has been found to inhibit NF-κB activity by directly inhibiting DNA binding of NF-κB without affecting IκB activity *(41)*, and has also been shown to prevent the formation of papillomas in mice treated with carcinogens *(42)*. Is the antitumor effect of CAPE caused by inhibition of NF-κB activity? Such a connection has not yet been established, but should be investigated.

STRATEGIES TO MAKE TUMORS MORE OXIC

Another possible strategy to circumvent the therapeutic problems associated with hypoxic tumor cells is to make tumors more oxic. At present, there are several possible strategies that may be used to increase tumor oxygenation. The first strategy can be broadly defined as physiological manipulation of tumor oxygenation. The combination of carbogen breathing and polyethylene glycol-hemoglobin (PEG) hemoglobin increase tissue pO_2, and also increase the effectiveness of radiotherapy and chemotherapy in experimental solid tumor systems *(43)*. Clinical trials are presently ongoing to assess whether these strategies will be beneficial either in improving local control of tumor growth or disease-free survival, compared to matched control populations that do not receive these adjuvants to their therapy. A second strategy is to shrink tumor size by first administering microtubule-inhibiting agents, such as paclitaxel (or its family members) before fractionated radiotherapy *(44••,45)*. Mechanistically, it has been hypothesized that paclitaxel treatment synchronizes cells in the radiosensitive G2/M phases of the cell cycle, and thereby makes the oxic cells more sensitive to radiation-induced killing. In some tumors, the rapid shrinkage of tumor size by paclitaxel-induced apoptosis enhanced tumor responsiveness by inducing reoxygenation of remaining hypoxic tumor cells. Thus, apoptosis can play an indirect role in eradicating hypoxic tumor cells by causing tumor shrinkage that results in reoxygenation of poorly perfused tumor regions. Last, antiangiogenic therapy is the most ardently pursued antitumor strategy that also induces tumor shrinkage and reoxygenation of hypoxic tumor cells *(46)*. In a strategy elegantly pioneered by Folkman et al. *(46)* tumor shrinkage is achieved by eliminating vascular ECs that are critical in composing the tumor microvasculature. Experimental tumors treated with novel antiangiogenic molecules, such as endostatin, exhibit tumor shrinkage that correlates with apoptotic cell death. Therefore, a novel use of chemotherapeutic and antiangiogenic agents that induce apoptotic cell death is to shrink tumor size to increase tumor oxygenation. In this manner, apoptosis-inducing chemotherapeutic agents will be used as tumor debulking agents to increase tumor oxygenation, and need not be, in themselves, the primary treatment modality.

CONCLUSIONS

The tumor microenvironment presents a formidable opponent in successful anticancer therapy, both directly, by inhibiting tumor cell proliferation, and indirectly, by selecting for populations of transformed cells that possess diminished apoptotic programs. The latter result can, in part, explain the strong clinical relationship between tumor hypoxia and tumor control. Although a small percentage of classical chemotherapeutic compounds exhibit some cytotoxicity under hypoxic conditions (with the exception of the highly hypoxic cell cytotoxic bioreductive agent, tirapazamine), the ability to use presently existing or novel chemotherapeutic agents to kill hypoxic cells rapidly by apoptosis will provide two important measures of therapeutic gain. First, the killing of hypoxic tumor cells with apoptosis-inducing chemotherapeutic agents will complement the killing of oxic cells by modalities such as radiotherapy, which exhibit a significant enhancement in effectiveness under oxic conditions. In other words, the two approaches would be highly complementary. Second, the use of chemotherapeutic agents that induce rapid apoptosis will shrink the tumor, and make it more oxic and more responsive to treatment by radiotherapy. Thus, future goals should be to identify apoptosis-inducing compounds or molecules that not only specifically kill transformed tumor cells, but kill transformed tumor cells that are hypoxic. The selective inhibition of endogenous antiapoptotic genes, which work in a p53-independent manner, and provide important survival signals under hypoxic conditions, should receive more attention. Last, the identification of naturally occurring compounds that inhibit EC-survival signals may be highly useful in shrinking solid-tumor size, to increase the effectiveness of radiotherapy. The many ways the adverse conditions of the tumor microenvironment can affect tumor progression and therapeutic outcome have only begun to be appreciated. Because apoptosis plays key roles in tumor progression and resistance to chemotherapy, a more thorough investigation of how tumor hypoxia affects the apoptotic program of transformed cells is needed to better develop newer and more effective strategies to subvert this old foe.

ACKNOWLEDGMENTS

This work was supported by grants from the National Cancer Institute, the American Cancer Society, and the Naomi Vandenberg Research Fund (A.J.G.), by National Institutes of Health postdoctoral fellowship 1F32CA75754-01 (C.K.), and Public Health Service National Research Service Award training grant CA 09302 (N.C.D.).

REFERENCES

• of special interest
•• of outstanding interest

••1. Brown JM, Giaccia AJ. Unique physiology of solid tumors: opportunities (and problems) for cancer therapy. *Cancer Res* 1998; 58: 1408–1416.
   Current review on how the tumor microenvironment influences therapy and potential ways to exploit it.

••2. Thomlinson RH, Gray LH. Histological structure of some human lung cancers and the possible implications for radiotherapy. *Br J Cancer* 1955; 9: 539–549.
   Landmark paper that used morphological analysis to demonstrate that solid tumors possess hypoxic regions, and that these hypoxic regions could be problematic for radiotherapy.

3. Jain RK. Barriers to drug delivery in solid tumors. *Sci Am* 1994; 271: 58–65.

4. Bedford JS, Mitchell JB. Effect of hypoxia on the growth and radiation response of mammalian cells in culture. *Br J Radiol* 1976; 47: 687–696.

5. Durand RE. Influence of microenvironmental factors during cancer therapy. *In vivo* 1994; 8: 691–702.

6. Sakata K, Kwok TT, Murphy BJ, Laderoute KR, Gordon GR, Sutherland RM. Hypoxia-induced drug resistance: comparison to P-glycoprotein-associated drug resistance. *Br J Cancer* 1991; 64: 809–814.

•7. Hockel M, Schlenger K, Aral B, Mitze M, Schaffer U, Vaupel P. Association between tumor hypoxia and malignant progression in advanced cancer of the uterine cervix. *Cancer Res* 1996; 56: 4509–4515.
   This clinical study was one of the first to indicate that hypoxia not only affected

the response of cervical tumors that received radiotherapy, but also tumors that received surgery. Therefore, hypoxia was also somehow affecting the malignant progression of tumors, as well as their response to therapy.

•8. Nordsmark M, Overgaard M, Overgaard J. Pretreatment oxygenation predicts radiation response in advanced squamous cell carcinoma of the head and neck. Radiother Oncol 1996; 41: 31–40.
Clinical study that shows that pretreatment tumor oxygenation acts as an independent prognostic indicator for therapeutic response in head and neck cancer.

•9. Brizel DM, Scully SP, Harrelson JM, Layfield LJ, Bean JM, Prosnitz LR, Dewhirst MW: Tumor oxygenation predicts for the likelihood of distant metastases in human soft tissue sarcoma. *Cancer Res* 1996; 56: 941–943.
Presents data to suggest that tumor hypoxia increases the invasiveness of soft-tissue sarcomas.

••10. Graeber TG, Osmanian C, Jacks T, Housman DE, Koch CJ, Lowe SW, Giaccia AJ. Hypoxia mediated selection of cells with diminished apoptotic potential in solid tumors. *Nature* 1996; 379: 88–91.
p53 mutations are common in many different histological types of solid tumors. This paper proposes that hypoxia may act as a selective pressure to induce apoptosis in transformed cells possessing wild-type p-53, thus resulting in p53 mutation. Thus, transformed cells that have acquired p53 mutations are less susceptible to killing by apoptosis, and possess an increased survival advantage under hypoxic conditions.

••11. Lowe SW, Ruley HE, Jacks T, Housman DE. p53-dependent apoptosis modulates the cytotoxicity of anticancer agents. *Cell* 1993; 74: 957–967.
Landmark study that established the importance of the p53 tumor-suppressor gene in modulating the apoptotic sensitivity of oncogene-transformed cells in response to a wide variety of chemotherapy agents. This paper completely changed how oncologists and basic science researchers thought about sensitivity to anticancer agents.

12. Lotem J, Sachs L. Regulation by bcl-2, c-myc, and p53 of susceptibility to induction of apoptosis by heat shock and cancer chemotherapy compounds in differentiation-competent and -defective myeloid leukemic cells. *Cell Growth Differ* 1993; 4: 41–47.

•13. Kim CY, Tsai MH, Osmanian C, Graeber TG, Lee JE, Giffard RG, et al. Selection of human cervical epithelial cells that possess reduced apoptotic potential to low-oxygen conditions. *Cancer Res* 1997; 57: 4200–4204.
Previous studies indicated that oncogenically transformed rodent cells were sensitive to hypoxia-induced apoptosis. This paper demonstrates that human cervical ECs infected with the HPV E6 and E7 genes are also sensitive to hypoxia-induced apoptosis. This paper also showed that hypoxia could select for HPV-infected human ECs that lost their apoptotic sensitivity to hypoxia.

14. Miyashita T, Reed JC. Tumor suppressor p53 is a direct transcriptional activator of the human bax gene. *Cell* 1995; 80: 293–299.

15. Buckbinder L, Talbott R, Velasco-Miguel S, Takenaka I, Faha B, Seizinger BR, Kley N. Induction of the growth inhibitor IGF-binding protein 3 by p53. *Nature* 1995; 377: 646–649.

16. Polyak K, Xia Y, Zweier JL, Kinzler KW, Vogelstein B. Model for p53-induced apoptosis. *Nature* 1997; 389: 300–305.

•• 17. Chen X, Ko L, Jayaraman L, Prives C. p53 levels, functional domains, and DNA damage determine the extent of the apoptotic response of tumor cells. *Genes Dev* 1996; 10: 2438–2451.
The role of transcription in p53-modulated apoptosis is controversial. Chen et al. investigated how mutations in different regulatory domains of p53 affected apoptosis. The results of this study indicated that inactivation of one of the p53-transactivation domains still permitted apoptosis, suggesting that p53 could signal apoptosis by a mechanism that did not solely rely on transcriptional upregulation of its downstream effectors.

•• 18. Haupt Y, Rowan S, Shaulian K, Vousden K H, Oren M. Induction of apoptosis in HeLa cells by transactivation deficient p53. *Genes Dev* 1995; 9: 2170–2183.
A classic paper demonstrating that apoptosis occurs in cells in which p53 transactivation is inhibited by overexpression of mdm-2.

19. Koumenis C, Derr J, Marquez S, Giaccia AJ. Hypoxia dissociates induction of p53 protein and apoptosis from p53-dependent transactivation. p53 Meeting, Crete, Greece. 1998.

20. Murphy M, Hinman A, Levine AJ. Wild-type p53 negatively regulates the expression of a microtubule-associated protein. *Genes Dev* 1996; 16: 2971–2980.

21. Jayaraman L, Murthy KG, Zhu C, Curran T, Xanthoudakis S, Prives C. Identification of redox repair protein Ref-1 as a potent activator of p53. *Genes Dev* 1997; 11: 558–570.

22. Walker LJ, Craig RB, Harris AL, Hickson ID. Role for the human DNA repair enzyme HAP1 in cellular protection against DNA damaging agents and hypoxic stress. *Nucleic Acids Res* 1994; 22: 4884–4889.

23. An W, Kanekal M, Simon MC, Maltepe E, Blagosklonny MV, Neckers LM. Stabilization of wild-type p53 by hypoxia-inducible factor 1 alpha. *Nature* 1998; 392: 405–408.

• 24. Avantaggiati ML, Ogryzko V, Gardner K, Giodano A, Levine AS, Kelly K. Recruitment of p300/CBP in p53-dependent signal pathways. *Cell* 1997; 89: 1175–1184.
Demonstrated that p53 was signaling to the transcriptional machinery through interaction with the p300/CBP coactivator.

• 25. Gu W, Shi XL, Roeder RG. Synergistic activation of transcription by p53 and CBP. *Nature* 1997; 387: 819–823.
Second important paper demonstrating that p53 and CBP interactions were necessary for transcriptional regulation by p53.

• 26. Scolnick DM, Chehab NH, Stavridi ES, Lien MC, Caruso L, Moran E, Berger SL, Halazonetis TD. CREB-binding protein and p300/CBP-associated factor are transcriptional coactivators of the p53 tumor suppressor gene. *Cancer Res* 1997; 57: 3693–3696.
Third report that demonstrates the in vivo interaction between p53 and CBP.

27. Lin AJ, Cosby L A, Shansky CW, Sartorelli AC. Potential bioreductive alkylating agents. 1. Benzoquinone derivatives. *J Med Chem* 1972; 15: 1247–1252.

28. Grau C, Overgaard J. Effect of etopside, carmustine, vincristine, 5-fluorouracil, or methotrexate on radiobiologically oxic and hypoxic cells in C3H mouse mammary carcinoma in situ. *Cancer Chemother Pharmacol* 1992; 30: 277–280.

29. Kraggerud SM, Sandvik JA, Pettersen EO. Regulation of protein synthesis in human cells exposed to extreme hypoxia. *Anticancer Res* 1995; 15: 683–686.

•30. Koumenis C, Giaccia AJ. Transformed cells require constitutive activity of RNA polymerase II to resist oncogene-induced apoptosis. *Mol Cell Biol* 1997; 17: 7306–7316.
Many oncogenes are thought to promote growth deregulation and increased proliferation in transformed cells. However, the same oncogenes that increase proliferation also induce cell death when RNAPII is inhibited, suggesting that a constitutively active antiapoptotic program is necessary to permit growth deregulation.

31. Haaf T, Ward DC. Inhibition of RNA polymerase II transcription causes chromatin decondensation, loss of nucleolar structure, and dispersion of chromosomal domains. *Exp Cell Res* 1996; 224: 163–173.

32. Damgaard J, Balslev Y, Mollgaard K, Wassermann K. Ongoing activity of RNA polymerase II precludes chromatin collapse and DNA fragmentation in Chinese hamster ovary cells. *Biochem Biophys Res Commun* 1996; 227: 677–683.

33. Kitada S, Miyashita T, Tanaka S, Reed JC. Investigations of antisense oligonucleotides targeted against bcl-2 RNAs. *Antisense Res Dev* 1993; 3: 157–169.

•34. Mayo MW, Wang CY, Cogswell PC, Rogers-Graham KS, Lowe SW, Der C J, Baldwin AS Jr. Requirement of NF-kappa B activation to suppress p53-independent apoptosis induced by oncogenic Ras. *Science* 1997; 278: 1812–1815.
Transcription is needed to suppress ras-*oncogene-signaled apoptosis. This paper demonstrates that one of the potential downstream effectors of* ras *important in preventing apoptosis is the transcription factor NF-κB.*

35. Chu ZL, McKinsey TA, Liu L, Gentry JJ, Malim MH, Ballard DW. Suppression of tumor necrosis factor-induced cell death by inhibitor of apoptosis c-IAP2 is under NF-kappaB control. *Proc Natl Acad Sci USA* 1997; 94: 10,057–10,062.

36. Melkonyan HS, Chang WC, Shapiro JP, Mahadevappa M, Fitzpatrick PA, Kiefer MC, Tomei LD, Umansky SR. SARPs: a family of secreted apoptosis-related proteins. *Proc Natl Acad Sci USA* 1997; 94: 13,636–13,641.

•37. Koong AC, Chen EY, Giaccia AJ. Hypoxia causes the activation of nuclear factor κB through the phosphorylation of IκBα on tyrosine residue. *Cancer Res* 1994; 54: 1425–1430.
This study shows that hypoxia stimulates the activation of NFκB by promoting the phosphorylation and degradation of its inhibitory molecule, IκBα. Therefore, both oxidative stress and a lack of oxygen are able to activate this transcription factor.

38. Koong AC, Chen EY, Mivechi NF, Denko NC, Stambrook P, Giaccia AJ. Hypoxic activation of nuclear factor-kappa B is mediated by a Ras and Raf signaling pathway and does not involve MAP kinase (ERK1 or ERK2). *Cancer Res* 1994; 54: 5273–5279.

39. Lin YZ, Yao SY, Veach RA, Torgerson TR, Hawiger J. Inhibition of nuclear translocation of transcription factor NF-kappa B by a synthetic peptide containing a cell membrane-permeable motif and nuclear localization sequence. *J Biol Chem* 1995; 270: 14,255–14,258.

40. Chen JH, Shao Y, Huang MT, Chin CK, Ho CT. Inhibitory effect of caffeic acid phenethyl ester on human leukemia HL-60 cells. *Cancer Lett* 1996; 108: 211–214.

41. Natarajan K, Singh S, Burke TRJ, Grunberger D, Aggarwal BB. Caffeic acid phenethyl ester is a potent and specific inhibitor of activation of nuclear transcription factor NF-kappa B. *Proc Natl Acad Sci USA* 1996; 93: 9090–9095.

42. Huang MT, Ma W, Yen P, Xie JG, Han J, Frenkel K, Grunberger D, Conney AH. Inhibitory effects of caffeic acid phenethyl ester (CAPE) on 12-O-tetradeca-noylphorbol-13-acetate-induced tumor promotion in mouse skin and the synthesis of DNA, RNA and protein in HeLa cells. *Carcinogenesis* 1996; 17: 761–765.

43. Teicher BA, Ara G, Herbst R, Takeuchi H, Keyes S, Northey D. PEG-hemoglobin: effects on tumor oxygenation and response to chemotherapy. *In Vivo* 1997; 11: 301–311.

••44. Milas L, Hunter NR, Mason KA, Kurdoglu B, Peters LJ. Enhancement of tumor radioresponse of a murine mammary carcinoma by paclitaxel. *Cancer Res* 1994; 54: 3506–3510.
Demonstrated that apoptosis induced by the chemotherapeutic agent, paclitaxel, increased tumor responsiveness to radiotherapy by making the tumor more oxic. The implications of this study are that chemotherapeutic agents, such as paclitaxel, which induce apoptosis, may indirectly increase tumor response to radiotherapy by reducing tumor size, and thereby increasing tumor oxygenation.

45. Milross CG, Mason KA, Hunter NR, Terry NH, Patel N, Harada S, et al. Enhanced radioresponse of paclitaxel-sensitive and-resistant tumors in vivo. *Eur J Cancer* 1997; 33: 1299–1308.

••46. O'Reilly MS, Boehm T, Shing Y, Fukai N, Vasios G, Lane WS, et al. Endostatin: an endogenous inhibitor of angiogenesis and tumor growth. *Cell* 1997; 88: 277–285.
The Achilles heel of tumor growth is vascularization. Inhibition of tumor vascu-larization by naturally occurring small molecules, such as endostatin, induce rapid tumor cell apoptosis and tumor shrinkage in transplanted tumors.

16 Molecular Regulation and Therapeutic Implications of Cell Death in Prostate Cancer

Timothy J. McDonnell, MD, PhD,
Elizabeth M. Bruckheimer, BS,
Bjorn T. Gjertsen, MD, PhD,
Tsuyoshi Honda, MD, PhD,
and Nora M. Navone, MD, PhD

CONTENTS

ABSTRACT

Prostate cancer is typically a low-growth-fraction, indolent malignancy. However, these malignancies not uncommonly progress to become refractory to therapeutic interventions, and ultimately result

From: *Apoptosis and Cancer Chemotherapy*
Edited by: J. A. Hickman and C. Dive © Humana Press Inc., Totowa, NJ

in the death of the patient. Molecular alterations involved in the pathogenesis and progression of prostate cancer frequently include lesions that disrupt normal cell death mechanisms. Available nonsurgical therapies for prostate cancer involve the induction of apoptosis. This chapter describes the current understanding of how cell death regulatory molecules may influence treatment outcomes. Based on the expanding awareness of the molecular regulation of cell death in prostate cancer, novel, more effective treatment strategies may be designed and implemented.

INTRODUCTION

American men have a one in six chance of developing prostate cancer, and current estimates suggest that as many as 40% over the age of 50 yr have prostate cancer. This means that approx 9.7 million Americans, equivalent to the population of Paris, are now living with prostate cancer. Although projections indicate that over 41,000 deaths will be attributable to the disease in 1998 in the United States alone, it is clear that the majority of men with histologic evidence of prostate cancer will not develop clinically significant disease. Selecting which of these patients with localized prostate cancer are likely to benefit from surgical intervention or other treatment modalities, such as radiation and chemotherapy, is frequently difficult.

The pattern of prostate cancer progression is uniquely predictable, characterized by local tumor growth followed by regional lymph node metastases, and subsequently by bone marrow metastases in advanced-stage disease. Many patients with advanced-stage prostate cancer benefit from androgen ablation as a means of achieving tumor control. An initial response rate of about 80–90% can be achieved in patients with metastatic prostate carcinoma by reducing serum testosterone to castrate levels. Although most patients with advanced prostate cancer are initially responsive to hormonal ablation therapy, approx 10–20% of patients are refractory to treatment. Furthermore, patients who exhibit an initial therapeutic response typically relapse within 3 yr, with rapidly fatal, androgen-independent, metastatic carcinoma. Tumor recurrences are thought to represent clonal selection of pre-existing, androgen-independent cells.

ANDROGEN-DEPRIVATION THERAPY
AND APOPTOSIS INDUCTION

The efficacy of androgen-ablation therapy is thought to be associated with the induction of apoptotic cell death in the androgen-dependent prostatic epithelial cells (EC). Although apoptosis induction may contribute to the clinical response in androgen-sensitive prostate cancers, it is not necessarily an invariable outcome of androgen-ablation therapy *(1•)*. Tumor biopsies obtained from a series of 18 prostate cancer patients, sampled prior to and 7 d following castration, demonstrated that the rate of apoptosis remained unchanged following castration in 9 patients, increased in 6 patients, and decreased in 3 patients, whereas the proliferative indices decreased in 15 patients and remained unchanged in 3 patients *(1•)*. These findings indicate that clinical responsiveness of prostate carcinomas to androgen-ablation cannot, in all cases, be attributed to apoptosis induction alone.

An understanding of the molecular pathogenesis of prostate cancer and the biochemical mechanisms of androgen-independent tumor growth are needed to enhance the prediction of disease outcome, and to develop more effective therapeutic interventions. Although substantial effort has been devoted to understanding the positive and negative regulatory events controlling cellular proliferation, the involvement of the deregulation of programmed cell death (PCD) in multistep neoplasia has, until recently, received relatively little attention. It is likely that pathways regulating cell-death susceptibility are disrupted during the development and progression of prostate neoplasia.

Numerous investigators have established in experimental models that androgen ablation results in the death of over 80% of the prostatic ECs within 10 d, by the derepression of an endogenous PCD cascade *(2,3)*. Activation of the PCD cascade in the prostate by androgen ablation has been shown to result in the enhanced expression of specific genes. Specifically, elevations of intracellular calcium in response to castration have been shown to result in early activation of a Ca^{2+}-Mg^{2+}-dependent endonuclease responsible for the endonucleolytic cleavage of DNA, which is the hallmark of apoptosis *(2,4)*. In fact, pharmacologic modulation of intracellular calcium has been suggested as a strategy to induce apoptosis in hormonally unresponsive prostate cancer. Additionally, the upregulation of other genes, such as TGF-β and TRPM-2, has been shown to immediately precede the onset of PCD in the prostate, and has been implicated in the mediation of this process *(5,6)*. More recently, it has been observed that many of the genes that are induced in the prostate by androgen deprivation, and are associated with apoptosis *(fos, myc, hsp-70)*, are also associated with cellular proliferation, leading to the proposal that apoptosis may, in fact, result from defective cell-cycle progression *(7)*.

Examination of the deregulation of cellular proto-oncogenes and tumor-suppressor genes has provided valuable insights into the molecu-

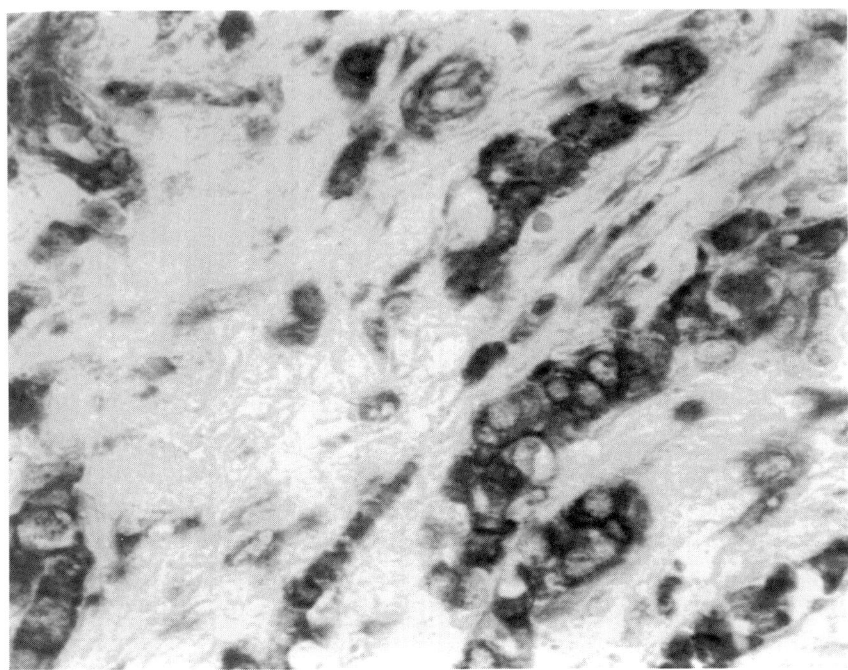

Fig. 1. Immunohistochemical detection of human Bcl-2 in stage-D, hormone-independent prostate cancer. Paraffin-embedded tissue sections were stained with a monoclonal antibody against human Bcl-2, using standard procedures.

lar basis of multistep neoplasia. Several molecular alterations have been implicated in the pathogenesis and progression of prostate cancer, including the deregulated expression of the *bcl-2* proto-oncogene (Fig. 1) and inactivating somatic mutations of the *p53* tumor-suppressor gene (Fig. 2) *(8••)*. Bcl-2 has been shown to inhibit apoptotic cell-death induction in many experimental systems by as-yet poorly understood mechanisms. Additionally, wild-type, but not mutant, forms of p53 have been shown to participate in cell-cycle arrest and apoptosis induction following DNA damage. Additional alterations occur in prostate cancer, such as changes involving the androgen receptor and cell-cycle regulatory genes *(9)*.

ALTERATIONS IN Bcl-2 FOLLOWING TREATMENT

In normal prostate tissue, Bcl-2 expression is restricted to the basal cells of the glandular epithelium, which are resistant to the effects of androgen deprivation *(8••)*. In contrast, the Bcl-2-negative secretory

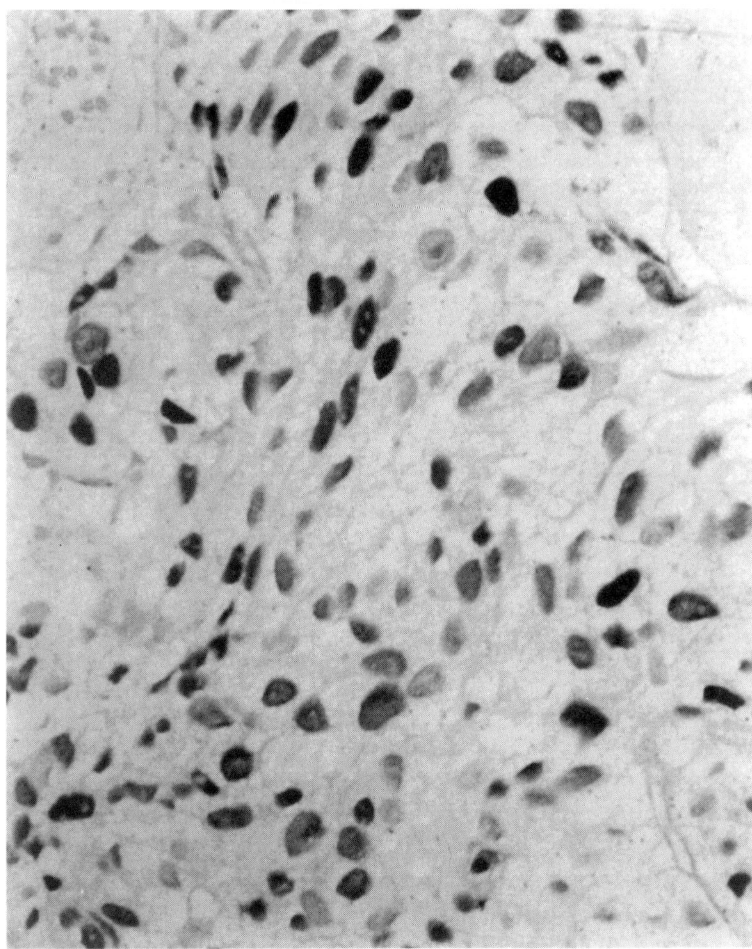

Fig. 2. Immunohistochemical detection of p53 in stage-D, hormone-independent prostate cancer. Paraffin-embedded tissue sections were stained with a monoclonal antibody against p53 standard procedures.

glandular ECs undergo apoptotic cell death in response to androgen deprivation. Therefore, the secretory cells of the prostatic glandular epithelium require testosterone, to remain viable and proliferate. Immunohistochemical studies of prostate cancer suggested that Bcl-2 expression could contribute to tumor progression following androgen-ablation therapy *(8••)*. In this study, Bcl-2 protein levels were undetectable in 13 of 19 cases of androgen-dependent prostate cancers; however, in androgen-independent prostate cancers, 10 of 13 cases displayed intense staining for Bcl-2 protein. Studies by other groups have confirmed the finding that androgen-independent prostate cancers are typically immunoreactive for Bcl-2 protein. Thus, it appears that Bcl-2 may enable prostate cancer cells to remain viable despite castrate levels of androgen, and that hormone-ablation therapy may be selecting for Bcl-2-positive cells that fail to undergo PCD following hormone ablation.

Expression of Bcl-2 has been shown to adversely correlate with clinical outcome in patients undergoing radical prostatectomy for treatment of clinically localized prostate cancer. Bcl-2 expression has also been shown to be an adverse prognostic indicator in patients with locally advanced or metastatic prostate cancer, receiving hormonal therapy. Additionally, expression of Bcl-2 within prostate carcinoma cells is associated with resistance to cell-death induction by various chemotherapeutic agents *(10)*. Members of the Bcl-2 family are potentially important molecular targets for phosphorylation-dependent regulation of apoptosis. Cancer cells treated with the chemotherapeutic drug, taxol, or its analog, paclitaxel, have been proposed to undergo apoptosis through inactivation of Bcl-2 by phosphorylation of a serine residue *(11–13)*. This phosphorylation of Bcl-2 is proposed to involve the serine/threonine protein kinase c-Raf-1, because suppression of c-Raf-1 inhibits both taxol-induced apoptosis and phosphorylation of Bcl-2 *(11)*. Conversely, the antiapoptotic function of Bcl-2 may require phosphorylation of serine 70, one of several residues targeted by the c-Jun N-terminal kinase *(14,15)*. More recently, it has been demonstrated that Bcl-2 phosphorylation, following paclitaxel or vincristine treatment, involves protein kinase A (PKA) *(16)*. The PKA inhibitor Rp diastereomers of cyclic adenosine monophosphate (cAMP) blocked both Bcl-2 hyperphosphorylation and apoptosis. However, it was not determined if the sites of phosphorylation by PKA occur at the same sites as with other previously tested agents.

Direct evidence has recently been provided that the upregulation of Bcl-2 expression, associated with the progression of prostate cancer from androgen-dependence to androgen-independence, is a biologically meaningful event *(17•–19)*. The androgen-sensitive Dunning G and LNCaP prostate cancer cell lines were engineered to express Bcl-2 protein using standard gene-transfer techniques. Tumors from Bcl-2-expressing cells grown as xenografts in male nude mice exhibited a significant growth advantage following castration, but a significant reduction in the rate of growth was observed in the parental control tumors. The rate of apoptosis in control Dunning G tumors exhibited a transient, but significant, decrease following castration, which was accompanied by a corresponding decrease in the proliferative index. In contrast, the spontaneous rate of apoptosis was lower in Bcl-2 transfectant tumors, compared to control tumors, and remained unaltered following castration. Additionally, the proliferative index in Bcl-2 transfectant tumors remained unaltered following castration *(17•,19)*. These findings suggest that acquisition of androgen-independent growth may not be entirely accounted for on the basis of a decreased susceptibility to undergo apoptotic cell death. Several recent observations have demonstrated that the induction of apoptosis in prostate cancers does not invariably occur following castration *(1••)*. The mitotic index and rate of growth of androgen-sensitive R3327PAP prostate cancer explants were significantly reduced following castration, but this was not accompanied by a corresponding increase in the rate of apoptosis *(20)*.

It is intriguing that the level of p53 protein was augmented in both the vector control and LNCaP–Bcl-2 tumors following castration *(19)*. The maximum induction of p53 was observed on postcastration d 3, coincident with the peak in apoptosis in the vector control, but not LNCaP–Bcl-2, tumors. The augmentation in p53 expression was preceded by a transient elevation in the level of c-Myc protein in both vector control and LNCaP–Bcl-2 tumors. Enforced expression of c-Myc has been shown to result in apoptosis induction in some cell types, and also to transcriptionally activate the p53 promoter. In fact, it has been suggested that c-myc-mediated apoptosis may in part depend on the availability of wild-type p53 *(21)*. It has also been demonstrated that Bcl-2 can inhibit apoptosis induction mediated by either c-myc *(22)* or wild-type p53 *(23)*.

RESISTANCE TO THERAPY: POTENTIAL ROLE OF Bax

The augmentation in the level of Bax protein observed in both vector control and LNCaP–Bcl-2 tumors may also be a relevant event in the induction of apoptosis in prostate cancer cells following androgen ablation *(19)*. Bax, a member of the *bcl-2* gene family, has been shown to function as a dominant-acting cell death effector protein. The *bax* promoter also possesses a functional p53-binding element, and *bax* has been shown to be transactivated by wild-type p53 protein. These observations suggest that *bax* may be a transcriptionally regulated target of p53 and an important mediator of p53-dependent apoptosis. Recent experimental evidence in support of this contention has been provided *(24)*. It is thought that an important determinant of the susceptibility of a cell to undergo cell death induction is the relative amount of

Bcl-2 and Bax within a cell *(25)*. If this interpretation is correct, then the protection of LNCaP–Bcl-2 tumors from apoptosis induction following castration could be attributed to the abrogation of Bax function by Bcl-2.

Although a role for Bax in the mediation of cell death in prostate cancer cells in response to androgen ablation has not yet been directly demonstrated, several observations suggest that this is plausible. First, the time-course of Bax induction in vector control LNCaP tumors corresponds to the time-course of apoptosis following castration. Additionally, we have previously demonstrated that Bax protein is also induced in the rodent ventral prostate following castration and again corresponds, temporally, with apoptosis induction *(17•)*. Conversely, the androgen-sensitive Dunning G prostate carcinoma cell line does not exhibit an increase in either the rate of apoptosis or Bax protein following castration *(1•)*. It may also be relevant in this context that, among the most common molecular alterations observed during prostate cancer progression, p53 inactivation and upregulation of Bcl-2 are potential modulators of Bax function. In LNCaP prostate carcinoma cells the acquisition of androgen-independent growth may not be exclusively a consequence of selection for an apoptosis-resistant phenotype, but may also involve the ability to supplant the requirement for androgen-stimulated proliferation. Together, these findings suggest a model for a cascade of events involving the sequential upregulation of c-myc, p53, and Bax that may be important mediators of the apoptotic response in androgen-dependent prostatic ECs following androgen ablation (Fig. 3).

ALTERATIONS IN CALCIUM HOMEOSTASIS AND SIGNALING PATHWAYS MEDIATING APOPTOSIS INDUCTION

A modest and sustained increase in intracellular Ca^{2+} is considered to be important in triggering apoptosis in benign and malignant prostatic ECs. The ability of Bcl-2 to modulate intracellular Ca^{2+} was examined in Dunning G prostate carcinoma cells following apoptosis induction, and has recently been described *(26•)*. Adriamycin and thapsigargin, an endoplasmic reticulum Ca^{2+}-pump inhibitor, were effective inducers of apoptosis in control, but not Bcl-2-transfected, cells. These treatments were accompanied by a sustained rise in cytoplasmic and nuclear Ca^{2+} in control cells, but only modest increases were observed in Bcl-2-expressing cells. Furthermore, Ca^{2+} was excluded from nuclei isolated from Bcl-2-expressing cells, but was sequestered in control nuclei, following the addition of ATP. These findings suggest that Bcl-2 may

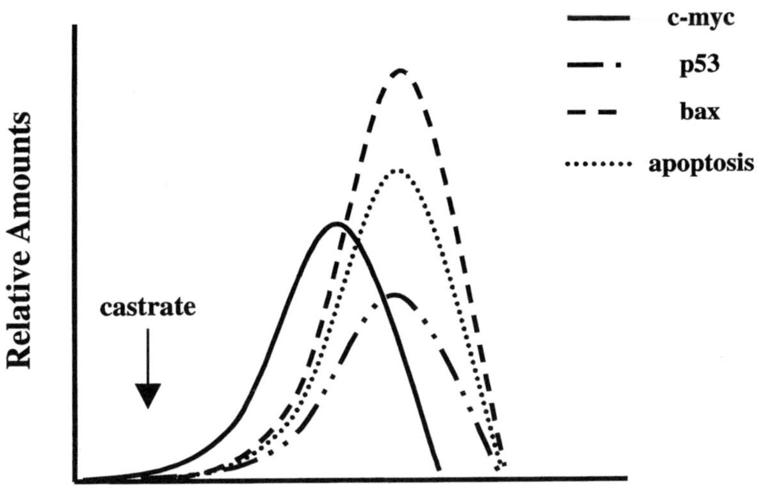

Fig. 3. Molecular alterations following hormone ablation in LNCaP tumor xenografts. Maximum induction of c-Myc protein occurs on d 1 following castration, preceding the maximum induction of p53 and Bax protein, and the peak rate of apoptosis, which occurs on d 3.

regulate levels of intranuclear Ca^{2+} independently of cytosolic Ca^{2+} levels, and that the ability of Bcl-2 to inhibit sustained increases in intracellular Ca^{2+} may provide the basis for resistance to cell death induction in these cells. The importance of alterations in intracellular Ca^{2+} in the control of apoptotic cell death in some experimental systems seems well established. What remains unclear is the specific events that these alterations are regulating. It is noteworthy that several of the enzymes implicated in the mediation of apoptosis are Ca^{2+}-dependent, such as the Ca^{2+}-Mg^{2+}-dependent endonuclease, transglutaminase, and ced-4. Alterations in intracellular Ca^{2+} may also be anticipated to influence nucleocytoplasmic trafficking and the regulation of transcription.

THE p53 CONTROVERSY

The *p53* tumor-suppressor gene is a transcription factor that functions in cell-cycle regulation, DNA repair, and PCD. In response to DNA damage, p53 functions to inhibit the cells from entering the cell cycle, allowing the cell to either repair the DNA or trigger an apoptotic signal that mediates the selective deletion of the affected cell. The identification of p53-responsive genes that mediate these outcomes is a matter of widespread interest. It appears that the p53-dependent transcriptional regulation of p21$^{WAF1/CIP1}$ is an important event for mediating G1 cell-cycle arrest in response to genotoxic damage. It has been shown that wild-type p53 is able to transcriptionally upregulate the cell-death effector of the Bcl-2 family, Bax, and to potentially downregulate Bcl-2 itself *(27)*. Furthermore, Bcl-2 protein expression was increased specifically in prostatic glandular ECs in p53-deficient mice [27]. Because the ratio of Bcl-2 to Bax is believed to be an important determinant of cell death susceptibility, it may be considered that p53 regulates apoptosis by modulating the expression of Bcl-2 family members. It has also been suggested that a deficiency of Bax promotes drug resistance and oncogenic transformation by attenuating p53-dependent apoptosis *(28)*.

Somatic mutations in p53 have been identified in diverse types of cancer. The induction of p53 protein following irradiation may exhibit cell-type specificity in cells isolated from human prostate tissue. ECs derived from normal prostate, benign prostatic hyperplasia (BPH), and adenocarcinoma showed no evidence of p53 accumulation following radiation; however, in stromal cells derived from the same patients, increased levels of p53 protein were observed. Additionally, increased expression of p21$^{WAF1/CIP1}$ was observed only in the stromal cells. The significance of these observations with respect to the radiation responsiveness of prostate cancer has yet to be determined.

The *p53* gene is located on the short arm of chromosome 17, and it has been reported that loss of heterozygosity (LOH) on chromosome 17 occurs in approx 20% of prostate cancers. Somatic mutations within p53 most commonly occur in the DNA-binding domain, but have also been identified in the carboxyl terminal nuclear localization domain, transactivation domain, and the oligomerization domain. Mutations in the *p53* gene typically result in an increased half-life of the p53 protein, which then enables its detection by immunohistochemical methods; wild-type p53 protein is difficult to detect under these same conditions *(29,30)*. These mutations usually result in the loss of function of p53 and consequent reduction in cell-cycle control and increased instability in genomic integrity. The reported incidence of p53 mutations in localized prostate cancer varies considerably, ranging from 0 to 80%. It is not surprising, therefore, that the significance of p53 mutations in localized prostate cancer is controversial. Recent immunohistochemical evidence suggests that the frequency of p53 mutations in prostate cancer from radical prostatectomy specimens varies between 47 and 80%. Additionally, Yang et al. *(31)* have demonstrated that clustered p53 immunoreactivity in localized prostate cancer is correlated with a higher incidence of tumor recurrence, compared to similar lesions that lack evidence of p53 mutations.

Arguably, there is more consistent evidence that p53 mutations contribute to the progression of prostate cancer. It has been shown that p53 protein expression is correlated with higher Gleason score, nuclear grade, pathological stage, and proliferation in localized primary prostate carcinomas, and that there is a increase in p53 mutations in advanced prostate cancer, with the highest incidence occurring in androgen-independent tumors *(30)*. Several other studies have demonstrated that p53 protein expression serves as an adverse prognostic indicator in prostate cancer *(32)*. The contribution of p53 inactivation to the resistance of prostatic ECs to the effects of androgen-deprivation is incompletely understood. The significance of the augmentation in the level of p53 following castration may be related to DNA repair processes, and may not necessarily be associated with the molecular events of cell-cycle reentry. Subsequent observations from p53 knockout mice indicate that postcastration elevations of p53 enhance rates of cell death in the prostate, but are not essential for cell-death induction. The potential contribution of other members of the *p53* gene family, most notably the p73 protein, to cell death regulation in the prostate is unknown.

On the basis of these observations, it might be anticipated that the inactivation of p53 and the expression of Bcl-2 may each confer a growth advantage to prostate cancer cells, as well as resistance to therapeutic cell death induction. The relationship between the occurrence of p53 mutations and upregulation of Bcl-2 in stage-D, androgen-independent bone marrow metastases has recently been examined *(30)*. In case-matched tissue samples, the expression of Bcl-2 inversely correlated with p53 protein accumulation ($p < 0.01$). The presence or absence of p53 protein accumulation and/or Bcl-2 expression did not correlate with tumor burden or patient survival in stage-D, androgen-independent prostate cancer bone marrow metastases. It should be noted that the ability to detect a modest impact on survival is limited by the small numbers of patients, and by currently available therapeutic interventions. The data imply that p53 mutations and expression of Bcl-2 are independent genetic processes in advanced prostate cancer. The data also indicated that these molecular alterations, although common, are not necessary for fatal progression of prostate cancer, in that a subset of these cases had no demonstrable evidence of Bcl-2 expression or p53 mutations.

Experimental evidence indicates that Bcl-2 can block p53-mediated apoptosis, and that inactivation of p53 in cells that overexpress Bcl-2 does not genetically complement during in vivo tumorigenesis *(23)*. It may be that Bcl-2 and p53 serve repressor and effector functions, respectively, of a common cell-death pathway; therefore, the inverse correlation between Bcl-2 expression and p53 mutation would be expected. Although Bcl-2 is able to inhibit p53-dependent apoptosis induction following genotoxic damage, the mechanistic basis of this inhibition has not been elucidated. Recently, however, it has been demonstrated that the nuclear import of wild-type p53 protein, following DNA damage, was significantly inhibited in Bcl-2-transfected, but not control prostate cancer cells *(33)*. Furthermore, enforced expression of Bcl-2 significantly inhibited the ability of p53 to transactivate a p53-responsive promoter element in cotransfection assays.

CONCLUSIONS

It is now widely appreciated that the deregulation of cell death, as well as cell proliferation, can contribute to the malignant phenotype. Further, it is recognized that the molecular regulators of cell death share features with the regulators of the cell cycle. More recently, it has become apparent that the efficacy of many therapeutic interventions may be a consequence of their common feature of effective cell-death induction, selectively, in cancer cells. An extension of this realization is that the refractoriness of many types of cancer to available anticancer therapy is the selection, or the acquisition, of molecular mechanisms to suppress cell-death induction. These mechanisms of resistance frequently involve the inappropriate expression of a cell-death repressor, such as the *bcl*-2 proto-oncogene, or the inactivation of a cell-death effector such as bax, or the *p53* tumor-suppressor gene. Strategies involving restoration of these cell death regulatory pathways are now being designed and implemented. Currently, at the University of Texas M.D. Anderson Cancer Center, neoadjuvant adenoviral-mediated, wild-type *p53* gene transfer is being evaluated in patients with locally advanced prostate cancer, prior to prostatectomy. It is anticipated that the efficacy of this therapy may depend on context-dependent variable such as the p53 status in individual tumors and the expression levels of Bcl-2. Treatment strategies designed to overcome Bcl-2-associated therapeutic resistance are also being developed, including adenoviral-mediated *bax* gene transfer and Bcl-2 antisense. Thus, recent insights into the molecular basis of cancer have provided the rationale for the consideration of novel approaches to treat cancer involving the restoration of cell death signaling pathways. These strategies provide the potential for improving long-term disease-free survival.

REFERENCES

• of special interest
•• of outstanding interest

•1. Westin P, Stattin P, Damber J-E, Bergh A. Castration therapy rapidly induces apoptosis in a minority and decreases cell proliferation in a majority of human prostatic tumors. *Am J Path* 1995; 146: 1368–1375.
   Demonstrated that there was a heterogeonous response in the induction of apoptosis in prostate cancer following hormone ablation therapy.

2. English HF, Kyprianou N, Isaacs JT. Relationship between DNA fragmentation and apoptosis in the programmed cell death in the rat prostate following castration. *Prostate* 1989; 15: 233–250.

3. Kyprianou N, Isaacs JT. Activation of programmed cell death in the rat ventral prostate after castration. *Endocrinology* 1988; 122: 552–562.

4. Kyprianou N, English HF, Isaacs JT. Activation of a Ca^{2+}, Mg^{2+} dependent endonuclease as an early event in castration-induced prostatic cell death. *Prostate* 1988; 13: 103–117.

5. Montpetit M, Lawless KR, Tenniswood M. Androgen-repressed messages in the rat ventral prostate. *Prostate* 1986; 8: 25–36.

6. Kyprianou N, Isaacs JT. Expression of transforming growth factor-β in the rat ventral prostate during castration-induced programmed cell death. *Mol Endocrinol* 1989; 3: 1515–1522.

7. Colombel M, Olsson CA, Ng P-Y, Buttyan R. Hormone-regulated apoptosis results from reentry of differentiated prostate cells onto a defective cell cycle. *Cancer Res* 1992; 52: 4313–4319.

••8. McDonnell TJ, Troncoso P, Brisbay SM, Logothetis C, Chung LW, Hsieh JT, Tu SM, Campbell ML. Expression of the proto-oncogene bcl-2 in the prostate and its association with emergence of androgen-independent prostate cancer. *Cancer Res* 1992; 52: 6940–6944.
   First direct evidence that upregulation of the cell-death suppressing bcl-2 *proto-oncogene was associated with progression from androgen-dependent to androgen-independent prostate cancer.*

9. Bruckheimer EM, Gjertsen BT, McDonnell TJ. Implications of cell death regulation in the pathogenesis and treatment of prostate cancer. *Semin Oncol,* in press.

10. Tu SM, McConnell K, Marin MC, Campbell ML, Fernandez A, von Eschenbach AC, McDonnell TJ. Combination adriamycin and suramin induces apoptosis in bcl-2 expressing prostate carcinoma cells. *Cancer Lett* 1995; 93: 147–155.

11. Blagosklonny MV, Schulte T, Nguyen P, Trepel J, Neckers LM. Taxol-induced apoptosis and phosphorylation of Bcl-2 protein involves c-Raf-1 and represents a novel c-Raf-1 signal transduction pathway. *Cancer Res* 1996; 56: 1851–1854.

12. Haldar S, Chintapalli J, Croce CM. Taxol induces bcl-2 phosphorylation and death of prostate cancer cells. *Cancer Res* 1996; 56: 1253–1255.

13. Beham A, Vogel M, McDonnell TJ, Farkas S, Furst A, Jauch KW. Apoptosis after genotoxic damage is enhanced by taxol-induced phosphorylation of bcl-2 and correlates with nuclear import of p53. *Proc Am Assoc Cancer Res* 1998; 39: 463.

14. Ito T, Deng X, Carr B, May WS. Bcl-2 phosphorylation required for antiapoptosis function. *J Biol Chem* 1997; 272: 11,671–11,673.

15. Maundrell K, Antonsson B, Magnenat E, Camps M, Muda M, Chabert C, et al. Bcl-2 undergoes phosphorylation by c-Jun N-terminal Kinase/stress-activated protein kinases in the presence of the constitutively active GTP-binding protein rac1. *J Biol Chem* 1997; 272: 25,238–25,242.

16. Srivastava RK, Srivastava AR, Korsmeyer SJ, Nesterova M, Cho-Chung YS, Longo DL. Involvement of microtublues in the regulation of bcl-2 phosphorylation and apoptosis through cyclic AMP-dependent protein kinase. *Mol Cell Biol* 1998; 18: 3509–3517.

17. Westin P, Lo P, Marin CM, Fernandez A, Sarkiss M, Tu SM, et al. Bcl-2 expression confers androgen independence in androgen sensitive prostatic carcinoma. *Int J Oncol* 1997; 10: 113–118.

•18. Beham AW, Sarkiss M, Brisbay S, Tu SM, von Eschenbach AC, McDonnell TJ. Molecular correlates of bcl-2 enhanced growth following androgen-ablation in prostate carcinoma cells in vivo. *Int J Mol Med* 1998; 1: 953–959.
Experimental evidence directly implicating upregulation of the proapoptotic Bax protein as a mediator of androgen-sensitivity in prostate cancer cells.

•19. Raffo AJ, Perlman H, Chen MW, Day ML, Streitman JS, Buttyan R. Overexpression of bcl-2 protects prostate cancer cells from apoptosis in vitro and confers resistance to androgen depletion in vivo. *Cancer Res* 1995; 55: 4438–4445.
Provides the first experimental evidence that enforced Bcl-2 expression provides an in vivo growth advantage to prostate cancer cells following androgen ablation.

20. Brandstrom A, Westin P, Bergh A, Cajander S, Damber JE. Castration induces apoptosis in the ventral prostate but not in an androgen-sensitive prostatic adenocarcinoma in the rat. *Cancer Res* 1994; 54: 3594–3601.

21. Wagner AJ, Kokontis JM, Hay N. Myc-mediated apoptosis requires wild-type p53 in a manner independent of cell cycle arrest and the ability of p53 to induce p21waf1/cip1. *Genes Dev* 1994; 23: 2817–2830.

22. Marin MC, Hsu B, Stephens LC, Brisbay S, McDonnell TJ. Functional basis of c-myc and bcl-2 complementation during multistep lymphomagenesis in vivo. *Exp Cell Res* 1995; 217: 240–247.

23. Marin MC, Hsu B, Meyn RE, Donehower LA, El-Naggar AK, McDonnell TJ. Evidence that p53 and bcl-2 are regulators of a common cell death pathway important for *in vivo* lymphomagenesis. *Oncogene* 1994; 9: 3107–3112.

24. Yin C, Knudson CM, Korsmeyer SJ, Van Dyke T. Bax suppresses tumorigenesis and stimulates apoptosis in vivo. *Nature* 1997; 385: 637–640.

25. Korsmeyer SJ, Shutter JR, Veis DJ, Merry DE, Oltvai ZN. Bcl-2/Bax: a rheostat that regulates an anti-oxidant pathway and cell death. *Semin Cancer Biol* 1993; 4: 327–332.

•26. Marin MC, Fernandez A, Bick RJ, Brisbay S, Buja LM, Snuggs M, et al. Apoptosis suppression by bcl-2 is correlated with the regulation of nuclear and cytosolic Ca^{2+}. *Oncogene* 1996; 12: 2259–2266.
Provides a basis for the possible mechanism of Bcl-2-mediated apoptosis. The ability of Bcl-2-expressing cells to regulate intranuclear Ca^{2+} provides evidence for the pore-forming properties of the Bcl-2 family members.

27. Miyashita T, Krajewski S, Krajewksa M, Wang HG, Lin HK, Liebermann DA, Hoffman B, Reed JC. Tumor suppressor p53 is a regulator of bcl-2 and bax gene expression in vitro and in vivo. *Oncogene* 1994; 9: 1799–1805.

28. McCurrach ME, Connor TM, Knudson CM, Korsmeyer SJ, Lowe SW. Bax-deficiency promotes drug resistance and oncogenic transformation by attenuating p53-dependent apoptosis. *Proc Natl Acad Sci USA* 1997; 94: 2345–2349.

29. Navone NM, Troncoso P, Psiters LL, Goodrow TL, Palmer JL, Nichols WW, von Eschenbach AC, Conti CJ. p53 protein accumulation and gene mutation in the progression of human prostate carcinoma. *J Natl Cancer Inst* 1993; 85: 1657–1659.

30. McDonnell TJ, Navone NM, Troncoso P, Pisters LL, Conti C, von Eschenbach AC, Brisbay S, Logothetis CJ. Expression of bcl-2 oncoprotein and p53 protein accumulation in bone marrow metastases of androgen independent prostate cancer. *J Urol* 1995; 157: 569–574.

31. Yang G, Stapleton AMF, Wheeler TM, Truong LD, Timme TL, Scardino PT, Thompson TC. Clustered p53 immunostaining: a novel pattern associated with prostate cancer progression. *Clin Cancer Res* 1996; 2: 399–401.

32. Thomas DJ, Robinson M, King P, Hasan T, Charlton R, Martin J, Carr TW, Neal DE. p53 expression and clinical outcome in prostate cancer. *Br J Urol* 1993; 72: 778–781.

33. Beham A, Marin MC, Fernandez A, Herrmann J, Brisbay S, Tari AM, et al. Bcl-2 inhibits p53 nuclear import following DNA damage. *Oncogene* 1997; 15: 2767–2772.

17 Chemotherapy and Apoptosis in the Ovary
Cancer Treatment Comes with a Price

Jonathan L. Tilly, PhD
and Alan L. Johnson, PhD

CONTENTS

ABSTRACT

The female gonad represents a unique organ when the efficacy and actions of chemotherapeutic drugs are evaluated, because a finite and irreplaceable stockpile of germ cells (oocytes) housed within the ovaries unfortunately suffer the same consequence of anticancer-drug exposure that is deemed clinically desirable in a tumor mass, namely, the induction of apoptosis. Only recently has emphasis been placed on dissecting the intracellular machinery responsible for committing germ cells to mass suicide when the ovary is exposed to

From: *Apoptosis and Cancer Chemotherapy*
Edited by: J. A. Hickman and C. Dive © Humana Press Inc., Totowa, NJ

chemotherapy. Thus, the possibility of developing novel drugs or approaches that spare these precious germ cells while still providing efficient cancer cell killing can begin to be explored. This new area of investigation, coupled with the fact that ovarian cancer is the leading cause of mortality in women who develop gynecologic tumors, combine to make the ovary a clinically important, but very complex, organ for further in-depth study of the relationships between chemotherapy and apoptosis.

INTRODUCTION

Ovarian cancer accounts for approx 4% of the total cancers diagnosed per year, and is the seventh most common cause of tumors in women *(1,2•)*. The majority of patients diagnosed with ovarian cancer are between 50 and 60 yr of age, and 60–70% of these women present with advanced stages of the disease (stage III or IV) at the time of diagnosis. It is widely believed that tumors of the surface epithelium represent the vast majority (85–90%) of ovarian cancers, which are then subcategorized as serous cystadenocarcinoma (42%), mucinous cystadenocarcinoma (12%), endometrioid carcinoma (15%), undifferentiated carcinoma (17%), and clear cell carcinoma (6%). The remaining forms of ovarian tumors are thought to arise from the transformation of granulosa, stromal, or germ cells *(1,2•)*. It should be pointed out, however, that recent work has challenged currently accepted concepts regarding the etiology ovarian cancer. It has been argued that cells within Müllerian duct remnants, originating from the paraovarian and ovarian hilar regions, as opposed to cells of the mesothelial lining of the ovarian surface (Fig. 1), may actually be the progenitors of at least some ovarian cancers *(3)*.

Whatever the case, of great concern is the severity of ovarian cancer and the poor prognosis for survival of these patients, primarily because no symptoms are manifested during early stages of the disease. Although several serum parameters, such as α-fetoprotein, CA-125, or carcinoembryonic antigen, have been used as ovarian tumor markers, no reliable end point yet exists to enable early screening. As such, ovarian cancers represent the most common cause of death in women who develop gynecologic tumors *(1,2•)*. Of further concern, the efficacy of ovarian cancer therapies is limited because of the inherent resistance of these tumors to currently available chemotherapeutic drugs jand/or the propensity of ovarian cancer cells to quickly develop resistance to treatment. Although ovarian cancers, in general, exhibit relatively high response rates to first-line chemotherapy regimens (e.g., up to 85% of

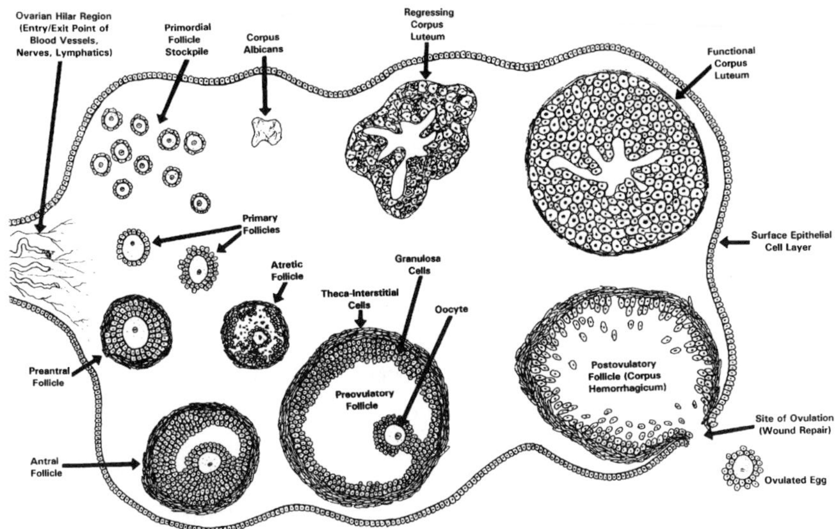

Fig. 1. Schematic diagram of the structure of the adult human ovary.

primary surface epithelial cell lesions initially react well to chemotherapy), they are extremely prone to recurrence and 80% of such recurrent cancers exhibit resistance to a subsequent round of chemotherapy *(4,5)*.

Another important consideration when evaluating chemotherapy and the ovary concerns the anticancer-drug-mediated destruction of innocent or bystander cells present within the female gonads, i.e., germ cells *(6–8)*. Unlike males, who continually produce mature germ cells (sperm) capable of fertilization throughout adult life, females do not share the luxury of a renewable germ cell pool. At birth, females of most, if not all, vertebrate species are endowed with a finite number of irreplaceable oocytes, each enclosed by a single layer of somatic (granulosa) cells in a structure referred to as a primordial follicle (Fig. 1; *9•,10*). This pool of follicles serves as the stockpile from which all eggs released at ovulation in adulthood are derived. Throughout the life of the female, this germ cell reserve is gradually depleted, so that near-exhaustion of the follicle pool occurs about the fifth decade of life, leading to the menopause *(10,11)*. Of central relevance to discussions of chemotherapy and the ovary, it is currently believed that many anticancer drugs accelerate this process of normal germ cell loss, and thus hasten the time to menopause *(6–8)*. Until recently, however, the mechanisms underlying this undesirable side effect of many chemotherapy regimens were ill-defined *(12••)*.

Therefore, this chapter will cover the present understanding of chemotherapy as it relates to apoptosis in ovarian tumors, followed by an evaluation of new data implicating apoptosis and its signaling molecules in the context of chemotherapy-induced infertility in women treated for both gynecologic and nongynecologic cancers.

ETIOLOGY OF OVARIAN CANCER

A significant hereditary component of ovarian cancer that may account for as many as 5–10% of ovarian tumors has been described *(2•,13,14)*. Currently, the most studied hereditary factor is the tumor-suppressor gene, *BRCA1,* which is located on the long arm of chromosome 17 (17q12–21). Although the function of the *BRCA1* gene product has yet to be fully defined, mutations within this gene may account for 70–80% of familial ovarian cancers. Unfortunately, screening for *BRCA1* gene mutations as a marker for ovarian neoplasia will likely prove problematic because the gene is extremely large (>100 kb), and more than 40 distinct mutations have been identified among affected families. A second tumor-suppressor gene, *BRCA2,* has recently been characterized and localized to chromosome 13q12–q13. Thus far, however, mutations of this gene have been found to confer only a low risk for an affected individual to develop ovarian cancer. Similarly, despite the fact that spontaneous or mutagen-induced alterations of other tumor-suppressor genes (i.e., *p53, Rb*), as well as several proto-oncogenes (i.e., K-*ras,* c-*myc,* c-*erb* B2, c-*fms*), collectively occur in at least 90% of ovarian cancers, relatively little is known of the significance of these mutations in the initial neoplastic transformation of ovarian cells (Table 1).

The oncogenic potential of the ovarian surface epithelium has also been causally linked to the high rate of normal cell proliferation that occurs with repair of the surface of the ovary following each ovulation (Fig. 1). It has been proposed that disruption of regular ovulatory cycles, such as that occurring with multiple pregnancies or with the use of oral contraceptives, may reduce the risk of development of ovarian cancer *(15•).* In keeping with this notion, women with Turner's syndrome, who are anovulatory, rarely develop epithelial ovarian cancer. Conversely, ovulation-inducing treatments (e.g., the use of exogenous gonadotropins for fertility control) may increase the risk of developing ovarian cancer *(15•).* At present, however, it is not clear if suppressing ovulation, specifically for reducing the risk of ovarian tumorigenesis, is justified by the limited clinical data available, or if mutations in the above-described genes combine with postovulatory surface epithelial cell proliferation to produce an optimum growth environment for ovarian neoplasia.

Table 1
Relationship Between Mutation of Tumor-Suppressor Genes
or Overexpression of Proto-Oncogenes and Ovarian Cancer

Gene (mutation/ overexpression)	Frequency of observation %	Prognosis/clinical outcome
BRCA1 (mutation)	5–10	Conveys genetic predisposition
BRCA2 (mutation)	<2	Not determined; lower risk of ovarian, as compared with breast cancer development
c-*myc* (gene amplification)	30–50 of malignant tumors	Most frequent in advanced serous carcinomas
c-*erb*B1/H*er*1 (overexpression)	36–54	Correlates with more aggressive tumors
c-*erb*B2/H*er*2/*neu* (overexpression)	20–35	Lower survival rate; prognostic value controversial
c-*erb*B3/H*er*3 (lack kinase activity)	Benign: 61 Borderline: 100 Malignant: 89	Role in transformation?; prognostic value unknown
c-*fms* (overexpression; unusual transcripts)	Stage III invasive: 78	Overexpression in stages III/IV (high tumor grade)
K-*ras* (point mutation, amplification)	20–50	More common in mucinous vs serous carcinomas; commonly found in borderline cancers; prognostic significance uncertain
AKT2 (amplification)	12	Associated with aggressive tumors; poor prognosis
p53 (loss-of-function mutation)	Stages I/II: 10–15 Stages III/IV: 40–50 Borderline tumors: 4	Decreased patient survival time
Retinoblastoma/Rb (mutation)	Unknown	Prognostic significance unknown

RELATIONSHIP OF APOPTOSIS TO DIAGNOSIS
AND PROGNOSIS OF OVARIAN CANCER

In ovarian carcinomas, a high apoptotic index is directly correlated
with a high mitotic index, severe histologic grade of tumor, and an
overall poor prognosis for long-term survival *(16)*. One obvious inter-

pretation of these data is that mitotically active cancerous cells in aggressive malignant tumors are also more prone to cell-cycle anomalies and, hence, apoptotic cell death. From a genetic standpoint, a number of disruptions within cellular pathways that result in programmed death of normal cells have been identified in ovarian carcinomas, and altered expression of one or more of these genes within cancerous ovarian tissues has been linked to chemotherapy resistance. Thus, a recent focus of interest has been to determine if abnormal profiles of these specific proteins and enzymes, known to either mediate or suppress apoptotic cell death, can serve as predictive indicators for remission and long-term survival in ovarian cancer patients.

For example, the expression pattern of the antiapoptotic protein, Bcl-2, varies with individual forms of ovarian cancer, but elevated Bcl-2 is weakly associated with prolonged patient survival *(17–19)*. This apparent paradox may be explained by the observation that cells expressing high levels of Bcl-2 exhibit a reduced rate of proliferation, and thus a slower growth of the solid tumor would be predicted *(20)*. More recent studies have documented elevated levels of the related death-suppressor protein, Bcl-X$_L$, in cancerous, when compared to normal, ovarian surface epithelium *(21)*. However, the usefulness of Bcl-X$_L$ expression patterns in predicting ovarian tumor severity or patient survival has not been established. In a cohort of 215 patients with ovarian cancer, an increased level of the proapoptotic Bcl-2-related protein, Bax, when found in the absence of Bcl-2, paradoxically, was again found to be related to poor clinical prognosis *(19)*. Although more work is clearly needed to clarify the relationship between expression of Bcl-2 family members and ovarian cancer treatment outcome, these initial observations suggest that a disrupted balance or ratio of Bax to Bcl-2/Bcl-X$_L$ may impact on patient survival by deregulating apoptosis in the tumor. Consequently, this alters sensitivity of ovarian cancer cells to antineoplastic drugs.

Mutations in the gene encoding the p53 tumor-suppressor protein represent the most common form of aberration leading to human cancers, in general; one-half of patients with ovarian carcinomas overexpress mutant p53 *(15•,22)*. Mutations in p53 have been detected more frequently in stage III/IV ovarian cancers (50%), compared with stage I/II tumors (15%), and mutant p53 is highly expressed in the most aggressive forms of ovarian carcinomas. Consequently, of all the data generated to date regarding female gonadal tumors, there is strong consensus that disruption of normal p53 function, either by allelic losses or mutations, is clearly one of the critical events responsible for development of advanced-stage ovarian carcinoma, and possibly the initial neoplastic event (Table 1). This proposal would be consistent with the well-known bifunctional role of p53 as a prominent inhibitor of cell-cycle progression, as well as an inducer of cell death *(23••,24);* thus, inactivating mutations in p53 would allow for rapid tumor growth resulting from a loss of proliferative control concomitant with a lack of cell elimination via apoptosis.

DEVELOPMENT OF DRUG RESISTANCE IN OVARIAN CANCERS

Ovarian tumors initially demonstrate relatively high response rates to chemotherapy: Up to 85% of primary ovarian epithelial carcinomas respond to first-line therapy, and 60% may show complete clinical remission. Chemotherapeutic agents, such as platinum compounds, doxorubicin and taxanes, used alone or in combination with each other or with alkylating agents (e.g., cyclophosphamide), have the greatest antitumor actions in ovarian carcinoma *(25,26).* As discussed earlier, however, unfortunately, ovarian tumors are prone to frequent recurrence and to the development of resistance to subsequent rounds of chemotherapy. As in other forms of cancer, chemotherapeutic drug resistance in ovarian tumors is undoubtedly multifactorial, although a unifying feature of resistance may be an alteration in the expression or function of one or more apoptosis-related genes or their products. This hypothesis, in the context of ovarian cancer, will be further evaluated with the following two examples.

First-line therapy based on platinum compounds (i.e., cisplatin, carboplatin) is relatively successful in advanced cases of ovarian cancer: Approximately 70% of patients treated with cisplatin-based therapy achieve initial tumor remission *(26);* however, the 10-yr survival rate is still only about 13%. At least some of the antitumor effects of cisplatin are via the effect of platinum-DNA crosslinks on G2-stage cell-cycle arrest leading to apoptotic cell death *(27),* with the latter event involving caspase-3 activation *(28).* Development of ovarian cancer cell resistance to platinum therapy has been linked to a reduced susceptibility to drug-induced apoptosis, which is believed to be a consequence of loss of wild-type p53 function. The resultant inability of mutant p53 to promote cell death has been further proposed to involve a loss of p53-mediated *bax* gene transactivation *(29).* Furthermore, the propensity for ovarian cancer cells to overexpress cell-death-suppressing proteins, such as Bcl-2 and Bcl-X_L, may provide further protection from cisplatin-induced apoptosis.

However, not all results are consistent with this interpretation of the role of p53 mutations in development of resistance to chemotherapy; other studies have reported that cisplatin retains cytotoxic effects in ovarian cancer cells expressing mutant p53 *(30)*. The reasons for these discrepancies remain to be clarified, although one likely answer is that DNA damage induced by platinum-based compounds is sensed by the cancer cell through multiple pathways *(31)*. Amelioration of just one sensor (e.g., p53) may not fully abrogate the cell's ability to die, and thus development of a drug-resistant phenotype probably depends on multiple hits in the cellular genotoxic-response pathways, leading to apoptosis.

Cytotoxicity following chemotherapy with taxanes (i.e., paclitaxel, docetaxel) is primarily the result of the ability of these agents to promote tubulin assembly into stable microtubules, thus preventing tubulin de-polymerization and cell division *(32)*. Paclitaxel treatment results in a marked increase in the percentage of cells in the late G2 to M phase of the cell cycle, followed by apoptotic cell death. Unlike platinum compounds and doxorubicin, expression of wild-type p53 is not neces-sarily a prerequisite for cytotoxicity induced by paclitaxel in ovarian cancer *(33,34)*. Although loss of normal p53 function has been reported to further sensitize some cell types to paclitaxel-induced cell death *(35)*, the emergence of taxane-resistant tumor cells, which can ultimately lead to treatment failure, does not appear dependent on p53 mutations. Consequently, first-line therapy using paclitaxel in combination with cisplatin may prove as effective, if not more so, than other cisplatin-combinatorial regimens because of the different mechanisms of action of each drug *(22)*.

In addition, paclitaxel has received considerable attention as a salvage treatment for ovarian cancer after patients become refractory to, or relapse from, platinum-based therapy, but objective responses occur in only 20–30% of patients, and are usually of short duration *(36)*. Mechanisms leading to taxane treatment failure have yet to be estab-lished, but such resistance may be associated with overexpression of either the *mdr*-1 or *mrp* gene *(37)*, or from altered expression of tubulin isotypes that affect the binding of taxanes to microtubules, and thus decrease the overall levels of tubulin polymerization *(38)*. Although *bax* overexpression *per se* does not appear to alter the incidence of apoptosis in several ovarian cancer cell lines, paclitaxel-induced cyto-toxicity is markedly enhanced by elevated levels of Bax *(39)*. This observation may be explained by the ability of paclitaxel to induce phosphorylation of Bcl-2, an event that can result in increased intracellular levels of unbound Bax, and thus facilitate Bax:Bax homodimer formation, leading to enhanced cell-death susceptibility. In further support of this, overexpression of Bcl-2 or Bcl-X$_L$ has been shown to inhibit paclitaxel-induced apoptosis in several tumor cell lines *(40,41)*. Such results indicate that monitoring the relative levels of Bcl-2 and related proteins in individual patients, prior to first-line and subsequent rounds of chemotherapy, may be beneficial for prescribing the most efficacious combinatorial anticancer-drug-treatment regimen.

CHEMOTHERAPY AND OVARIAN GERM CELL APOPTOSIS

As advances are made in understanding the mechanisms of action of chemotherapeutic drugs, and in using this information for the development of new drugs capable of more efficient tumor cell killing, investigators and clinicians must also be cognizant of the severity of side effects of such therapies in cancer patients. This point becomes even more critical as long-term survival rates for a number of cancers are improved, and thus steps to ensure a high quality of posttherapy life should be taken, when possible, in providing optimum patient care. Of the many drawbacks faced by cancer patients given chemotherapy, one of the most worrisome is the loss of fertility, and, in the case of women, the acceleration of the time to menopause *(6–8)*. For both outcomes, premature depletion of the oocyte stockpile is thought to be the underlying mechanism, although, until recently, little effort was placed on defining the specific events activated in oocytes by anticancer drugs that lead to their demise and, hence, ovarian failure. Aside from infertility following chemotherapy, young girls and women who lose ovarian function early in life because of anticancer drug treatment, also prematurely face the general health risks observed in normal postmenopausal women who do not receive hormone replacement therapy, including osteoporosis, cardiovascular disease, and neurodegenerative disorders.

Recent efforts to clarify the molecular biology of ovarian failure in female cancer patients arose from a foundation of earlier work directed at elucidation of the genes and pathways responsible for the normal depletion of germ cells from the ovary throughout life. As one would expect, apoptosis is the program used in all vertebrate species, and perhaps invertebrate species *(42)*, to normally delete germ cells and follicles from the ovary *(10,11)*. As one would further expect, Bcl-2 family members, p53, and caspases comprise the basic ovarian cell-death machinery *(43,44•)*. For example, apoptosis in the rodent and human ovary is well correlated with increased expression of the pro-apoptotic molecule, Bax *(45,46),* and knockout of the *bax* gene in mice leads to dramatic defects in both ovarian germ cell and somatic (granulosa) cell death *(47••,48).* Conversely, ablation of functional Bcl-2 by gene targeting in mice yields females with a compromised endowment of germ cells in postnatal life *(49).* Nuclear translocation of p53 has also been demonstrated to occur in ovarian granulosa cells destined for apoptosis *(50),* possibly as a consequence of DNA damage induced by reactive oxygen species *(51),* and serving to increase expression of the *bax* gene *(45,52••).* Consistent with the role of caspases in further disrupting homeostasis and dismantling the cell, leading to its fragmentation during apoptosis *(53••),* members of this family of proteases probably function as the downstream executioners of ovarian cell death. This is evidenced by both gene-expression studies *(46,54)* and substrate-cleavage experiments *(55),* as well as targeted disruption of the *caspase-2 (Ich-1)* locus in mice, which endows females with a surfeit of oocyte-containing primordial follicles in their starting stockpile shortly after birth *(56••).*

With these data in mind, and working under the presumption that oocytes, like tumor cells, would respond to chemotherapeutic drugs with inappropriate engagement of the specific components of the intracellular cell-death machinery discussed above, the actions of one specific anti-cancer drug, doxorubicin (DXR; also referred to as 14-hydroxydaunomycin), in murine oocytes, were recently explored in detail (Fig. 2; *12••*). Results from this series of in vivo and in vitro studies revealed that eggs, upon exposure to therapeutically relevant doses of DXR, rapidly undergo apoptosis, as defined by cytoplasmic condensation (retraction of the oocyte plasma membrane away from the zona pellucida protein coat that surrounds the egg), membrane budding, DNA cleavage, and, finally, fragmentation of the oocyte into apoptotic bodies of unequal sizes (Fig. 2; *12••,56••*). This set of morphological and biochemical events is observed both in vitro and in vivo, supporting the use of cultured oocytes as a model for dissecting the events responsible for apoptosis induction in the female germline.

Consistent with the ability of DXR and related compounds to trigger apoptosis in tumor cells via generation of ceramide, through activation of either ceramide synthase *(57•)* or sphingomyelinase *(58),* female germ cells appear to use ceramide as the initial death signal. This is evidenced by the fact that pretreatment of oocytes with sphinogosine-1-phosphate, a component of the sphingolipid cycle that effectively suppresses ceramide-promoted stress kinase activation and apoptosis

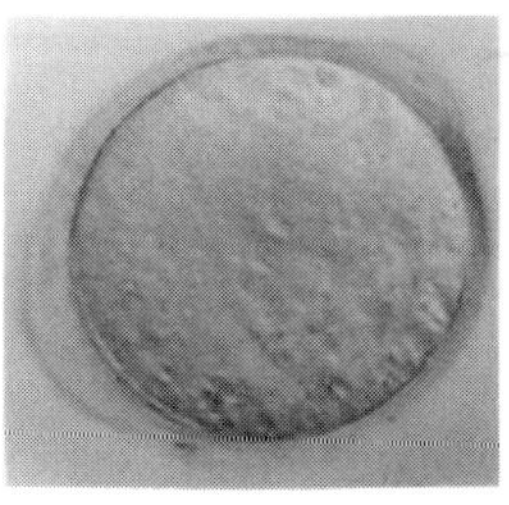

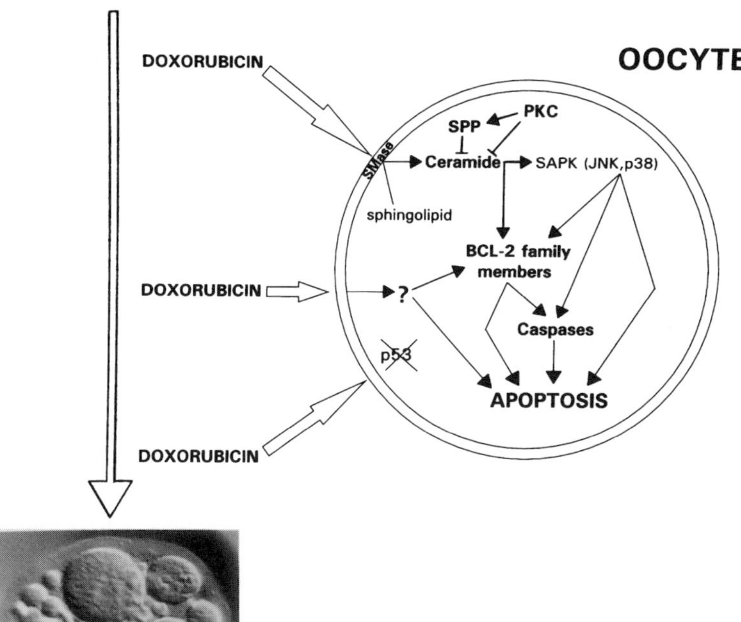

Fig. 2. Proposed model of chemotherapy-induced apoptosis in female germ cells. The *upper panel (left)* shows the morphology of a healthy murine oocyte; the *lower panel (left)* reveals the cellular budding and fragmentation observed to rapidly occur in oocytes treated with anticancer drugs. The *center panel* depicts the intracellular-death pathways believed to be activated oocytes following exposure to DXR, with key control checkpoints highlighted in bold letters (compiled from refs. *12, 43,* and *49*). Abbreviations: SMase, sphingomyelinase; PKC, protein kinase C; SPP, sphingosine-1-phosphate; SAPK, stress-activated protein kinases; JNK, c-jun N-terminal kinase. Note that although SMase is indicated within the membrane, this does not imply the specific involvement of neutral, as opposed to acidic, sphingomyelinase in DXR-induced signaling in oocytes.

(59•), abolishes DXR-induced apoptosis *(12•)*. Moreover, preliminary data have shown that prior activation of protein kinase C in oocytes conveys resistance to DXR-induced death *(60)*, in agreement with the reported ability of protein kinase C activators to inhibit signaling events and apoptosis initiated by ceramide in various somatic cell lineages *(61••)*.

The downstream targets of ceramide, or other yet-undescribed early signals activated by anticancer drugs, in germ cells remain to be fully elucidated, although it is known that functional Bax protein is of absolute necessity for apoptosis to occur in oocytes exposed to DXR in vivo or in vitro *(12••)*. This observation is consistent with data above indicating that Bax also plays a central role in normal cell loss from the ovary in both rodent models *(45,47••,48••)* and humans *(46)*. Finally, Bax probably activates the caspase-effector arm of the cell-death pathway in oocytes as a means to commit DXR-exposed germ cells to apoptosis. Not only have peptide inhibitors of caspases been reported to block apoptotic events in oocytes treated with DXR *(12••)*, but recent data derived from analysis of germ cells in caspase-2 (Ich-1)-deficient female mice support a central role for this specific caspase family member in chemotherapy-induced oocyte loss *(56••)*. Collectively, current data argue that ovarian germ cells exposed to anticancer drugs respond with the same discrete and orderly series of intracellular events normally used to eliminate oocytes and follicles from the ovary *(43,44••)*. Thus, manipulation of key checkpoints in the oocyte apoptosis pathway prior to and during chemotherapy may protect this precious and irreplaceable cohort of eggs from certain death.

CONCLUSIONS

From the data discussed herein, it is clear that apoptosis, and its regulatory genes and signals, are probably fundamental to understanding and improving the efficacy of the cytotoxic actions of chemotherapeutic drugs in the female gonads. Regarding the status of chemotherapy and ovarian cancer, more work is certainly needed to expand on the limited clinical data presented, as well as the hypotheses raised in the preceding paragraphs. There should be particular emphasis on the development of molecular markers to screen for ovarian neoplastic transformation and early stages of the disease. Further exploration is also needed into the possibility that the profile of expression of apoptosis-regulatory genes in individual patients should be accounted for, when determining the best course of adjuvant chemotherapy. The survival rates in patients with other forms of cancer following high-dose anticancer drug treatment is dramatically influenced by the pre-existing genetic environment within the tumor *(62)*. Regarding chemotherapy and premature ovarian failure, additional studies to define the molecular events underlying germ cell loss in women treated for various forms of cancer may one day allow development of oocyte-sparing drugs that retain tumoricidal activity. Although exactly how this will be accomplished is at present unknown, research in this field has not only defined commonalities between oocytes and cancer cells in the context of chemotherapy-initiated proapoptotic signaling, but differences as well, such as the lack of p53 involvement in DXR-induced germ cell demise (Fig. 2; *12••*). This final area of investigation is critical for ensuring that the clinical outcome of any chemotherapeutic regimen is not only long-term remission, but also to provide the highest quality of life for cancer survivors posttherapy.

ACKNOWLEDGMENTS

Studies conducted by the authors and described herein were supported by the United States National Institutes of Health (R01-AG12279, R01-HD34226, R01-HD36095, R01-ES08430, Office of Research on Women's Health), the United States National Science Foundation (IBN94-19613), the Walther Cancer Institute (Indianapolis, IN), and the Vincent Memorial Research Fund (Boston, MA).

REFERENCES

• of special interest
•• of outstanding interest

1. *Cancer Facts and Figures.* American Cancer Society: Atlanta, GA. 1997.
•2. Langston AA, Ostrander EA. Hereditary ovarian cancer. *Curr Opin Obstet Gynecol* 1997; 9: 3–7.
   Concise analysis of past and current literature on the etiology and treatment of human ovarian cancer.
3. McCluskey LL, Dubeau L. Biology of ovarian cancer. *Curr Opin Oncol* 1997; 9: 465–470.
4. Mitch DG, Williams S. Biology of epithelial ovarian cancer. *Clin Obstet Gynecol* 1994; 37: 406–422.
5. Rabinerson D, Kaplan B, Levavi H, Neri A. The biology of ovarian cancer of epithelial origin. *Isr J Med Sci* 1996; 32: 1128–1133.
6. Waxman J. Chemotherapy and the adult gonad: a review. *J Roy Soc Med* 1983; 76: 144–148.
7. Reichman BS, Green KB. Breast cancer in young women: effect of chemotherapy on ovarian function, fertility and birth defects. *Monogr Natl Cancer Inst* 1994; 16: 125–129.

8. Forman R, Gilmour-White S, Forman N, eds. Cancer chemotherapy. In: *Drug-Induced Infertility and Sexual Dysfunction.* Cambridge University Press: Cambridge (UK). 1996; pp. 52–71.

• 9. Gougeon A. Regulation of ovarian follicular development in primates: facts and hypotheses. *Endocr Rev* 1996; 17: 121–155.
Excellent and comprehensive review of primate follicular development, with particular emphasis on human ovarian biology.

10. Tilly JL, Ratts VS. Biological and clinical importance of ovarian cell death. *Contemp Obstet Gynecol* 1996; 41: 59–86.

11. Tilly JL. Apoptosis and ovarian function. *Rev Reprod* 1996; 1: 162–172.

••12. Perez GI, Knudson CM, Leykin L, Korsmeyer SJ, Tilly JL. Apoptosis-associated signaling pathways are required for chemotherapy-mediated female germ cell destruction. *Nature Med* 1997; 3: 1228–1232.
First study to delineate the role of apoptosis in oocyte destruction caused by chemotherapy, the genetic and biochemical pathways responsible for activating germ cell death, and the possibility for preservation of ovarian function post-therapy.

13. Boyd J, Rubin SC. Hereditary ovarian cancer: molecular genetics and clinical implications. *Gynecol Oncol* 1997; 64: 196–206.

14. Katso RMT, Manek S, O'Byrne K, Playford MP, Le Meuth V, Ganesan TS. Molecular approaches to diagnosis and management of ovarian cancer. *Cancer Metas Rev* 1997; 16: 81–107.

•15. Berchuck A, Carney M. Human ovarian cancer of the surface epithelium. *Biochem Pharmacol* 1997; 54: 541–544.
Excellent review of the etiology, genetics, and clinical management of human ovarian cancer.

16. Yamasaki F, Tokunaga O, Sugimori H. Apoptotic index in ovarian carcinoma: correlation with clinicopathologic factors and prognosis. *Gynecol Oncol* 1997; 66: 439–448.

17. Henriksen R, Wilander E, Oberg K. Expression and prognostic significance of Bcl-2 in ovarian tumours. *Br J Cancer* 1995; 72: 1324–1329.

18. Herod JJ, Eliopoulos AG, Warwick J, Niedobitek G, Young LS, Kerr DJ. The prognostic significance of Bcl-2 and p53 expression in ovarian carcinoma. *Cancer Res* 1996; 56: 2178–2184.

19. Marx D, Binder C, Meden H, Lenthe T, Ziemek T, Hiddemann T, Kuhn W, Schauer A. Differential expression of apoptosis-associated genes *bax* and *bcl-2* in ovarian cancer. *Anticancer Res* 1997; 17: 2233–2240.

20. Pietenpol JA, Papadopoulos N, Markowitz S, Willson JK, Kinzler KW, Vogelstein B. Paradoxical inhibition of solid tumor growth by Bcl-2. *Cancer Res* 1994; 54: 3714–3717.

21. Witty JP, Jensen RA, Johnson AL. Expression and localization of Bcl-2 related proteins in human ovarian cancers. *Anticancer Res* 1998; 18: 1223–1230.

22. Strasser A, Huang, DCS, Vaux DL. The role of the *bcl-2/ced*-9 gene family in cancer and general implications of defects in cell death control for tumourigenesis and resistance to chemotherapy. *Biochim Biophys Acta* 1997; 1333: F151–F178.

••23. Ko LJ, Prives C. p53: puzzle and paradigm. *Genes Dev* 1996; 10: 1054–1072.
Outstanding review of the actions of p53 in cell proliferation and death.

24. Oren M. Lonely no more: p53 finds its kin in a tumor suppressor haven. *Cell* 1997; 90: 829–832.

25. Goldspiel BR. Clinical overview of the taxanes. *Pharmacotherapy* 1997; 17: 110S–125S.
26. Martoni A, Cacciari N, Angelelli B, Zamagni C, Pannuti F. Chemotherapy of advanced ovarian cancer. *Frontiers Biosci* 1997; 2: G20–G26.
27. Sorenson CM, Eastman A. Mechanism of cis-diamminedichloroplatinum(II)-induced cytotoxicity: role of G_2 arrest and DNA double-strand breaks. *Cancer Res* 1998; 48: 4484–4488.
28. Chen Z, Naito M, Mashima T, Tsuruo T. Activation of actin-cleavable interleukin 1β-converting enzyme (ICE) family protease CPP-32 during chemotherapeutic agent-induced apoptosis in ovarian carcinoma cells. *Cancer Res* 1996; 56: 5224–5229.
29. Perego P, Giarola M, Righetti SC, Supino R, Caserini C, Delia D, et al. Association between cisplatin resistance and mutation of p53 gene and reduced *bax* expression in ovarian carcinoma cell systems. *Cancer Res* 1996; 56: 556–562.
30. De Feudis P, Debernardis D, Beccaglia P, Valenti M, Graniela SE, Arzani D, et al. DDP-induced cytotoxicity is not influenced by p53 in nine human ovarian cancer cell lines with different p53 status. *Br J Cancer* 1997; 76: 474–479.
31. Brown L, McCarthy N. A sense-abl response? *Nature* 1997; 387: 450,451.
32. Long BH, Fairchild CR. Paclitaxel inhibits progression of mitotic cells to G_1 phase by interference with spindle formation without affecting other microtubule functions during anaphase and telephase. *Cancer Res* 1994; 54: 4355–4361.
33. Debernardis D, Sire EG, De Feudis P, Vikhanskaya F, Valenti M, Russo P, et al. p53 status does not affect sensitivity of human ovarian cancer cell lines to paclitaxel. *Cancer Res* 1997; 57: 870–874.
34. Lanni JS, Lowe SW, Licitra EJ, Liu JO, Jacks T. p53-independent apoptosis induced by paclitaxel through an indirect mechanism. *Proc Natl Acad Sci USA* 1997; 94: 9679–9683.
35. Wahl AF, Donaldson KL, Fairchild C, Lee FY, Foster SA, Demers GW, Galloway DA. Loss of normal p53 function confers sensitization to Taxol by increasing G_2/M arrest and apoptosis. *Nature Med* 1996; 2: 72–79.
36. Bishop JF, Macarounas-Kirchman K. The pharmacoeconomics of cancer therapies. *Semin Oncol* 1997; 24(Suppl 19): 106–111.
37. Mutch DG, Williams S. Biology of epithelial ovarian cancer. *Clin Obstet Gynecol* 1994; 37: 406–422.
38. Kavallaris M, Kuo DYS, Burkhart CA, Regl DL, Norris MD, Haber M, Horwitz SB. Taxol-resistant epithelial ovarian tumors are associated with altered expression of specific beta-tubulin isotypes. *Clin Invest* 1997; 100: 1282–1293.
39. Strobel T, Swanson L, Korsmeyer SJ, Cannistra SA. BAX enhances paclitaxel-induced apoptosis through a p53-independent pathway. *Proc Natl Acad Sci USA* 1996; 93: 14,094–14,099.
40. Ibrado AM, Liu L, Bhalla K. Bcl-X_L overexpression inhibits progression of molecular events leading to paclitaxel-induced apoptosis of human acute myeloid leukemia HL-60 cells. *Cancer Res* 1997; 57: 1109–1115.
41. Tang C, Willingham MC, Reed JC, Miyashita T, Ray S, Ponnathpur V, et al. High levels of p26 BCL-2 oncoprotein retard taxol-induced apoptosis in human pre-B leukemia cells. *Leukemia* 1994; 8: 1960–1969.
42. Gumienny TL, Hengartner MO. *C. elegans* as a model system for germ cell death. In: *Cell Death in Reproductive Physiology*, Tilly JL, Strauss JF, Tenniswood M, eds. Springer-Verlag: New York. 1997; pp. 8–18.

43. Tilly JL, Tilly KI, Perez GI. Genes of cell death and cellular susceptibility to apoptosis in the ovary: a hypothesis. *Cell Death Differ* 1997; 4: 180–187.

••44. Tilly JL. Cell death and species propagation: molecular and genetic aspects of apoptosis in the vertebrate female gonad. In: *When Cells Die. A Comprehensive Evaluation of Apoptosis and Programmed Cell Death,* Lockshin RA, Zakeri Z, Tilly JL, eds. Wiley-Liss: New York. 1998; pp. 431–452.
In-depth evaluation and integration of the literature regarding the molecular biology and genetics of apoptosis in the ovary, with several hypotheses offered for future testing.

45. Tilly JL, Tilly KI, Kenton ML, Johnson AL. Expression of members of the *bcl-2* gene family in the immature rat ovary: equine chorionic gonadotropin-mediated inhibition of apoptosis is associated with decreased *bax* and constitutive *bcl-2* and *bcl-x*$_{long}$ messenger ribonucleic acid levels. *Endocrinology* 1995; 136: 232–241.

46. Kugu K, Ratts VS, Piquette GN, Tilly KI, Tao X-J, Martimbeau S, et al. Analysis of apoptosis and expression of *bcl-2* gene family members in the human and baboon ovary. *Cell Death Differ* 1998; 5: 67–76.

••47. Knudson CM, Tung KSK, Tourtellote WG, Brown GAJ, Korsmeyer SJ. Bax-deficient mice with lymphoid hyperplasia and male germ cell death. *Science* 1995; 270: 96–99.
First report of the generation and phenotypes of mice with a targeted disruption in the bax *gene.*

••48. Perez GI, Robles R, Knudson CM, Flaws JA, Korsmeyer SJ, Tilly JL. Prolongation of ovarian lifespan into advanced chronological age by Bax deficiency. *Nature Genet* 1999; 21: 200–203.
First study to show that the mouse equivalent of menopause can be prevented.

49. Ratts VS, Flaws JA, Kolp R, Sorenson CM, Tilly JL. Ablation of *bcl-2* gene expression decreases the numbers of oocytes and primordial follicles established in the post-natal female mouse gonad. *Endocrinology* 1995; 136: 3665–3668.

50. Tilly KI, Banerjee S, Banerjee PP, Tilly JL. Expression of the p53 and Wilms' tumor suppressor genes in the rat ovary: gonadotropin repression *in vivo* and immunohistochemical localization of nuclear p53 protein to apoptotic granulosa cells of atretic follicles. *Endocrinology* 1995; 136: 1394–1402.

51. Tilly JL, Tilly KI. Inhibitors of oxidative stress mimic the ability of follicle-stimulating hormone to suppress apoptosis in cultured rat ovarian follicles. *Endocrinology* 1995; 136: 242–252.

••52. Miyashita T, Reed JC. Tumor suppressor p53 is a direct transcriptional activator of the human *bax* gene. *Cell* 1995; 80: 293–299.
Elegant series of experiments providing an important link between p53 and apoptosis induction: transcriptional regulation of the bax *gene.*

••53. Cryns VL, Yuan J. The cutting edge: caspases in apoptosis and disease. In: *When Cells Die. A Comprehensive Evaluation of Apopotosis and Programmed Cell Death,* Lockshin RA, Zakeri Z, Tilly JL, eds. Wiley-Liss: New York. 1998; pp. 177–210.
Thorough review of the biochemistry of caspases, their intracellular targets following activation, and the implications of manipulating caspase activity in treating various disease states involving abnormal apoptosis.

54. Flaws JA, Kugu K, Trbovich AM, DeSanti A, Tilly KI, Hirshfield AN, Tilly JL. Interleukin-1β-converting enzyme-related proteases (IRPs) and mammalian cell death: dissociation of IRP-induced oligonucleosomal endonuclease activity

from morphological apoptosis in granulosa cells of the ovarian follicle. *Endocrinology* 1995; 136: 5042–5053.

55. Maravei DV, Trbovich AM, Perez GI, Tilly KI, Talanian RV, Banach D, Wong WW, Tilly JL. Cleavage of cytoskeletal proteins by caspases during ovarian cell death: evidence that cell-free systems do not always mimic apoptotic events in intact cells. *Cell Death Differ* 1997; 4: 707–712.

••56. Bergeron L, Perez GI, Mcdonald G, Shi L, Sun Y, Jurisicova A, et al. Defects in regulation of apoptosis in caspase-2-deficient mice. *Genes Dev* 1998; 12: 1304–1314.
First report of the generation and phenotypes of caspase-2 (Ich-1)-deficient mice.

•57. Bose R, Verheij M, Haimovitz-Friedman A, Scotto K, Fuks Z, Kolesnick R. Ceramide synthase mediates daunorubicin-induced apoptosis: an alternative mechanism for generating death signals. *Cell* 1995; 82: 405–414.
Interesting report providing evidence for ceramide generation as a novel pro-apoptotic signal in cancer therapy.

58. Jaffrezou JP, Levade T, Bettaieb A, Andrieu N, Bezombes C, Maestre N, et al. Daunorubicin-induced apoptosis: triggering of ceramide generation through sphingomyelin hydrolysis. *EMBO J* 1996; 15: 2417–2424.

•59. Cuvillier O, Pirianov G, Kleuser B, Vanek P, Coso O, Gutkind S, Spiegel S. Suppression of ceramide-mediated programmed cell death by sphingosine-1-phosphate. *Nature* 1996; 381: 800–803.
Intriguing series of experiments characterizing the novel function of sphingosine-1-phosphate as a counteracting intracellular signal to ceramide-promoted stress kinase activation and apoptosis.

60. Perez GI, Tilly JL. Oocyte apoptosis following anti-cancer drug exposure is prevented by activation of protein kinase-C. *J Soc Gynecol Invest* 1998; 5(Suppl): 46A.

••61. Spiegel S, Foster D, Kolesnick RN. Signal transduction through lipid second messengers. *Curr Opin Cell Biol* 1996; 8: 159–167.
Outstanding review of the biochemistry of lipid signaling molecules, and how these second messengers act and interact to determine cell fate in response to external stimuli and stressors.

62. Muss HB, Thor AD, Berry DA, Kute T, Liu ET, Koerner F, et al. c-*erb*B-2 expression and response to adjuvant therapy in women with node-positive early breast cancer. *New Engl J Med* 1994; 330: 1260–1266.

18 Role of Bcl-2 Family Members in Homeostasis, in Toxicology in Normal Intestine, and for Prognosis of Colonic Carcinomas

D. Mark Pritchard, BSc, MBChB, MRCP

CONTENTS

ABSTRACT

INTRODUCTION

EXPRESSION AND EFFECTS OF BCL-2 FAMILY
  PROTEINS IN NORMAL INTESTINAL EPITHELIUM

BCL-2 FAMILY OF MOLECULES IN COLORECTAL
  CANCER

BAX MUTATIONS OCCUR IN COLON CARCINOMAS
  THAT DISPLAY MICROSATELLITE INSTABILITY

ARE EFFECTS OF THERAPY IN INTESTINE AFFECTED
  BY EXPRESSION OF BCL-2 FAMILY PROTEINS?
    IN VITRO STUDIES
      ANIMAL STUDIES IN NORMAL INTESTINAL
        EPITHELIUM
      CLINICAL STUDIES IN COLORECTAL CARCINOMA
      CHEMOPROTECTIVE STUDIES

CONCLUSIONS

ACKNOWLEDGMENTS

REFERENCES

ABSTRACT

The influence and implications of Bcl-2 and Bax expression upon the responses of both normal and malignant intestinal epithelium to chemotherapeutic drugs are discussed with reference to in vitro,

From: *Apoptosis and Cancer Chemotherapy*
Edited by: J. A. Hickman and C. Dive © Humana Press Inc., Totowa, NJ

animal, and clinical studies. The functions of members of the Bcl-2 family during homeostasis of the normal intestinal epithelium are described by referring to immunohistochemical studies and to experiments involving knockout mice. Of particular importance is Bcl-2 itself, which appears to be a critical regulator of cell survival in the crucial stem-cell zone at the base of colonic crypts. Changes in Bcl-2 expression appear to be important during colorectal tumorigenesis; the protein is most abundant at the early adenoma stage, and Bcl-2 positivity is an independent predictor of good prognosis. Frameshift mutations in bax *appear to be common in those colorectal carcinomas that display microsatellite instability.*

INTRODUCTION

Colonic carcinoma is a common malignancy in the Western world, and the multistep hypothesis of carcinogenesis *(1)* has recently led to increased understanding of the genetic changes that occur during the progression from normal mucosa to adenoma to carcinoma. Moreover, the crypt–villus axis in normal intestinal epithelium provides an excellent model system for studying the apoptosis of cells at different points along a differentiation hierarchy *(2••)*. This chapter will discuss the roles of various members of the Bcl-2 family of proteins during homeostasis of the normal intestinal epithelium, and in colon carcinogenesis.

Treatment of advanced colorectal carcinoma by chemotherapy is currently not particularly successful *(3)*, and many chemotherapeutic agents also cause dose-limiting toxicity to the normal intestine. Because chemotherapeutic drugs induce apoptotic cell death, it has been suggested that the expression of various apoptosis-regulating Bcl-2 family proteins might influence the responses of both normal and malignant intestinal epithelium to chemotherapy. Animal models have proved helpful in elucidating the effects of Bcl-2 and Bax expression on chemotherapy-induced apoptosis in the normal intestine, but clinical studies of the responses of human tumors have not yet shown any clear-cut effects. However, these clinical studies are difficult to interpret, and the reasons for this are discussed.

EXPRESSION AND EFFECTS OF Bcl-2 FAMILY PROTEINS IN NORMAL INTESTINAL EPITHELIUM

The crypt–villus axis in normal intestinal epithelium permits study of the effects of expression of various proteins on cells at different stages in the pathway of differentiation *(2••)*. Briefly, cells originate from a putative stem-cell zone located at or near the crypt base, prolifer-

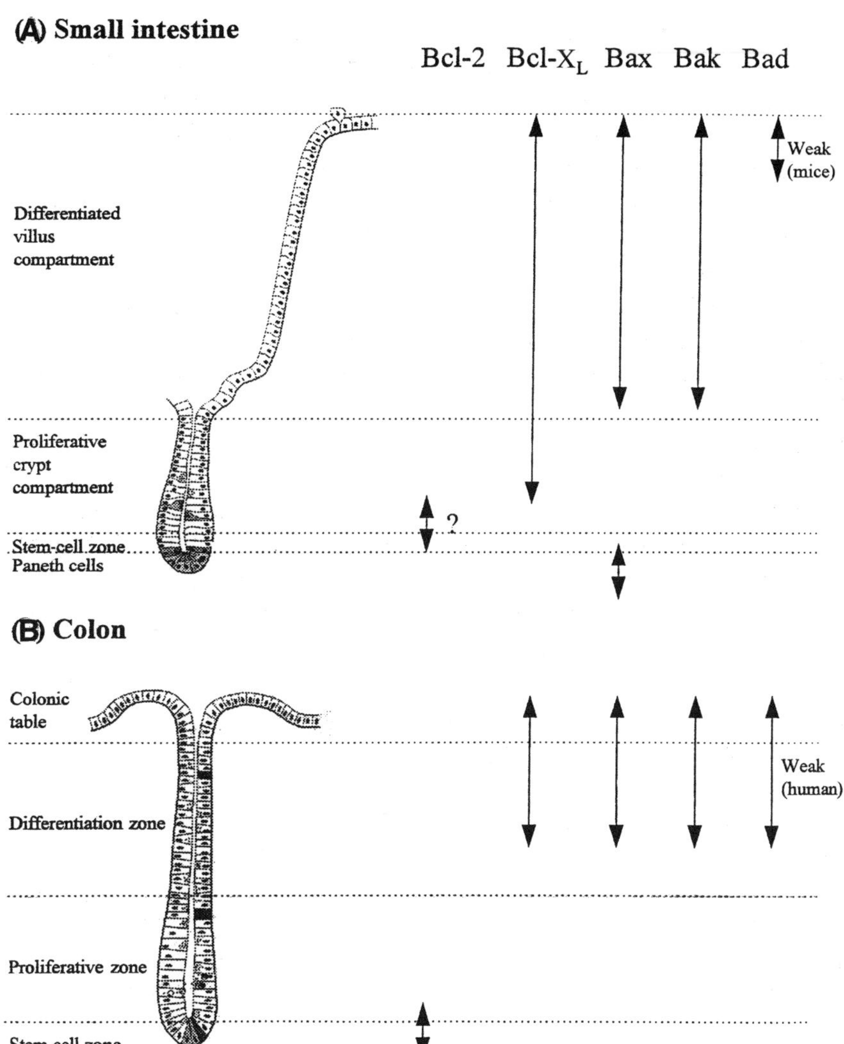

Fig. 1. Immunohistochemical expression of various members of the Bcl-2 family within small intestinal and colonic epithelia. Bcl-2 expression in the small intestine is controversial and is illustrated by "?". Data are taken from refs. *5–7,13,16,17,20,21.*

ate rapidly within the crypt compartment, and then differentiate as they continue to pass upward toward the villus (small intestine) or table (colon). Cells are shed from the surface of the epithelium by a process distinct from classical apoptosis (Fig. 1).

The immunohistochemical expression of members of the Bcl-2 family therefore provides evidence of their function within this epithelium. Further analysis of function comes from studying the phenotype and intestinal apoptosis of mice rendered homozygously null for each particular gene. Apoptotic cells can be readily detected in the crypt compartment of intestinal epithelia by light microscopy of conventional hematoxylin and eosin sections, and low levels of spontaneous apoptosis can be observed *(2••)*.

Initial publications of the immunohistochemical distribution of Bcl-2 in the human intestine demonstrated expression in the lower parts of the crypts of both small and large intestine, with little or no expression in more differentiated cells *(4•)*. The colonic expression pattern has been confirmed by a number of other investigators *(5,6)*. However, the small intestinal expression has been questioned by some observers, especially in murine specimens *(7••)*. A 30-kDa phosphorylated form of Bcl-2 protein has also been detected in normal human colonic mucosa *(8)*. Homozygously *bcl*-2-null mice have been generated independently by three groups *(9–12)*. The phenotype of all these mice was identical in most tissues, but, although the first two groups found no intestinal phenotype *(9–11)*, Kamada et al. *(12)* did report abnormalities in the shape and length of small intestinal villi in their *bcl*-2-null mice. Using the mice generated by Nakayama et al. *(9,10)*, Merritt et al. *(7••)* found no difference in the level of spontaneous apoptosis in the small intestine of *bcl*-2-null mice, compared with their wild-type counterparts. However, the level of spontaneous apoptosis was greatly elevated in the colon of *bcl*-2-null mice, and this occurred specifically at the base of colonic crypts, where immunohistochemical Bcl-2 expression had previously been detected in wild-type animals. This suggests that, irrespective of whether immunoreactivity is detected in the small intestine, the major function of Bcl-2 is in controlling the level of apoptosis in the putative stem-cell zone at the base of colonic crypts.

The other antiapoptotic family members studied have been Bcl-X$_\text{L}$ and Bcl-w. Bcl-X$_\text{L}$ expression has been detected in all the cells within the crypt–villus axis of the small intestine except Paneth cells. In colonic crypts, the intensity of staining for Bcl-X$_\text{L}$ was more intense apically than at the base of the crypts, in contrast to Bcl-2 *(13•)*. *Bcl-X$_\text{L}$*-null mice died at embryonic d 13, so study of any intestinal phenotype has not been possible *(14)*. Bcl-w has only been detected by Northern blot analysis in colonic tissue, and no published data exist on its expression pattern within the epithelium *(15)*.

Of the proapoptotic Bcl-2 family members, there is most data on Bax. Bax immunoreactivity in the mouse small intestine was strongest in the Paneth cells, with also some weak staining in villus enterocytes. Immunoreactivity in the colon was strongest in epithelial cells in the upper half of crypts *(16•,17•)*. No change in Bax immunoreactivity was observed in *bcl*-2-null mice *(17•)*. Homozygously *bax*-null mice showed no overt intestinal phenotype *(18)*, and no differences in the levels of spontaneous small intestinal or colonic apoptosis were seen when compared with wild-type counterparts *(19)*.

Bak expression in human intestine has been detected in the villus enterocytes of the small intestine, and in the more differentiated cells at the luminal surface of the colon *(20•)*. There are no publications to date documenting *bak*-null mice. Bad expression could not be detected in human small intestine, and only weak expression was seen in the colonic mucosa. Strong immunoreactivity was seen in smooth muscle cells of the colon *(21•)*. Another immunohistochemical study on mice showed weak cytoplasmic Bad immunostaining in the small intestine, particularly near the villus tip, and negligible staining was seen in the colon *(17•)*. Increased levels of Bad staining were observed in the intestinal mucosa of *bcl*-2-null mice *(17•)*. There have been no publications to date documenting *bad*-null mice. Immunohistochemical patterns of expression of the known Bcl-2 family members are summarized in Fig. 1.

Bcl-2 FAMILY OF MOLECULES
IN COLORECTAL CANCER

Since apoptosis is often deregulated during neoplasia, is there any evidence to suggest that apoptosis-modulating members of the Bcl-2 family are involved in colorectal carcinogenesis? Most work in this area has involved immunohistochemical studies of sporadic human colonic tumors, rather than model systems.

The immunohistochemical expression of Bcl-2 in human colonic neoplasms has been studied by a number of groups. Within the conventional adenoma to carcinoma pathway, adenomas have generally been found to express more Bcl-2 than carcinomas. However, methodological differences, particularly the criteria used to assign a specimen as either Bcl-2-positive or -negative, and the number of tissue samples studied, mean that the percentage of tumors in each category varies widely among papers *(5,6,22–29)*. Table 1 gives details of those studies performed up to 1997. In one of these studies, there appeared to be

Table 1
Summary of Individual Studies of Immunohistochemical Bcl-2 Expression During Progression
from Colonic Adenoma to Carcinoma

Author and ref. no.	No. of adenomas	% positive	No. of carcinomas	Characteristics of carcinomas	% positive	Criteria for positivity	Effect of Bcl-2 positivity on prognosis
Hague (5)	13	92	22	All stages	86	Any staining classified positive	Not tested
Ofner (22)	0	–	104	All stages	47	Any staining classified positive, but subclassified further by %	Improved prognosis
Bosari (23)	49	100	205	All stages	47	Classified positive if >1% cells stained. Subclassified further as focal or diffuse	No correlation with prognosis
Bronner (6)	26	100	24	All stages	92	Any staining classified positive	Not tested
Sinicrope (24)	24	71	21	Mostly Dukes' C	67	Classified positive if 0.5% cells stained. Intensity of staining also scored	Not tested

Sinicrope (25)	0	–	110	Stage 2 only	65	Classified positive if >5% cells stained. Intensity of staining also scored	Improved prognosis
Watson (26)	19	63.2	53	All stages	36.5	Classified as diffuse if >50% cells stained, focal if areas contained staining >50% of epithelial cell cytoplasm	Not tested
Baretton (27)	44	86	95	All stages	67	Any staining classified positive	Improved prognosis
Langlois (28)	0	–	74	All stages, patients <45 yr	5	Classified positive if >10% cells stained	No correlation with prognosis
Manne (29)	0	–	134	All stages	50	Classified positive if score, based on intensity and distribution of staining >0.5	Improved prognosis

Some of these have also studied prognostic significance. The table lists those studies published up to 1997. Different studies used varying levels of stringency when assigning Bcl-2 status to tumors, and the sample numbers also varied considerably; these variables may account for the different percentages of Bcl-2-positive tumors claimed by individual authors.

reciprocity of expression of Bcl-2 and p53 in the small proportion of neoplasms that dual-stained for both proteins *(26)*. This study *(26)* also demonstrated higher levels of Bcl-2 expression in normal colonic crypts adjacent to carcinomas, compared to normal crypts more than 5 cm from the neoplasm. These patterns of Bcl-2 staining suggest either that changes in Bcl-2 expression occur at the earliest stages of colorectal tumorigenesis, or that tumors arise from the Bcl-2-positive putative stem cells at the colonic crypt base (i.e., the apparent increase in Bcl-2 expression simply reflects cell of origin).

Ilyas et al. *(30)* found no difference in Bcl-2 immunoreactivity between sporadic colorectal carcinomas with different levels of microsatellite instability, but did find that carcinomas arising in patients with ulcerative colitis had a much lower level of Bcl-2 expression, suggesting that these tumors arose via a different pathway.

Some of the studies quoted above have attempted to correlate Bcl-2 status with other tumor parameters, and with prognosis. Studies that have assessed the effect of Bcl-2 status on the response to chemotherapy will be discussed later. As shown in the table, four studies have shown Bcl-2 positivity to be associated with a favorable clinical prognosis in colonic carcinoma *(22,25•,27,29••)*, and only one large study has shown no correlation *(23)*. The study of Sinicrope et al. *(25•)* was particularly robust in assessing prognosis within a group of patients with the same stage of tumor. Those colon carcinomas that one would suppose should undergo less apoptosis, because of Bcl-2 expression, actually had an improved prognosis; this contrasts with findings in other tumors, such as lymphoma and carcinomas of the prostate and breast, in which Bcl-2 confers a poorer prognosis. The reasons for this correlation have not been fully established, but the finding that colonic adenomas, in general, express more Bcl-2 than carcinomas suggests that those carcinomas that still express Bcl-2, although within the same Dukes' stage, may actually represent tumors at an earlier stage in their overall development, and therefore have a better prognosis.

Changes in the immunohistochemical expression of various other Bcl-2 family members have also been studied in a series of 30 colonic adenocarcinomas and 24 adenomas in which the expression levels were compared with those in adjacent normal mucosa *(31)*. The intensity of Bcl-X_L staining was seen to increase in about 60% of carcinoma cases, in a pattern reciprocal to Bcl-2 staining. Bak expression decreased in about 90% of cases, and no change in Bax immunoreactivity was observed. The relevance of these findings to the role played by these proteins during colorectal tumorigenesis remains to be clarified, but the authors speculate that Bcl-X$_L$ may be the predominant antiapoptotic Bcl-2 family member in advanced colon cancer, contributing to both tumor progression and chemoresistance.

Bax MUTATIONS OCCUR IN COLON CARCINOMAS THAT DISPLAY MICROSATELLITE INSTABILITY

Frameshift mutations have been demonstrated in a tract of eight deoxyguanosines within *bax,* in about 50% of primary microsatellite-mutator phenotype colonic adenocarcinomas *(32••).* Similar mutations were found in a number of microsatellite-mutator phenotype colorectal carcinoma cell lines *(32••).* The same *bax* mutations have also been found in 41% of microsatellite-unstable colorectal cancers in another cohort *(33),* as well as in 33% of gastric cancers and 12% of endometrial cancers. Colonic tumors that display microsatellite instability are less likely to have *p53* mutations, and the authors speculate that *bax* mutations, by affecting the coupling of p53 to apoptosis, eliminate the selective pressure on *p53* mutations during colorectal tumorigenesis. However, experiments detailing the effects of γ-radiation on intestinal apoptosis in *bax*-null mice (discussed in Animal Studies in Normal Intestinal Epithelium) demonstrate that the extent to which this apoptosis is dependent on Bax is much less than on p53 *(19).* This suggests that p53-driven Bax expression is not crucial in the normal intestine, hence, the full significance of *bax* mutations in colonic carcinomas remains to be clarified. However, this finding means that immunohistochemical studies for members of the Bcl-2 family need to be interpreted carefully, because such experiments cannot assess whether the protein being detected is functionally active.

ARE EFFECTS OF THERAPY IN INTESTINE AFFECTED BY EXPRESSION OF Bcl-2 FAMILY PROTEINS?

It has been hypothesized that the failure of a mutated stem cell to be deleted by apoptosis is one of the earliest events during colorectal tumorigenesis. A number of dietary (e.g., fiber) and chemopreventive agents (e.g., nonsteroidal anti-inflammatory drugs) are thought to prevent colonic tumor formation by promoting such apoptosis. Current chemotherapeutic agents used for the treatment of colorectal carcinoma also induce apoptotic cell death in both normal and malignant cells. Because the Bcl-2 family of proteins are known regulators of apoptosis, is there any evidence to suggest that these proteins regulate the effects of pharmacological agents in the intestine, with effects on both tumor formation and tumor therapy?

In Vitro Studies

Butyrate is generated in the colon as a result of bacterial fermentation of dietary fiber, and the ability of butyrate to induce apoptosis in colonic cells has been proposed as one of the possible mechanisms by which dietary fiber is protective against colorectal carcinoma. Hague et al. *(34)* have studied butyrate-induced apoptosis in one colorectal carcinoma and three adenoma cell lines. In one adenoma line that lacked Bcl-2, butyrate-induced apoptosis appeared to be mediated by Bak, and, in other lines, Bcl-2 appeared to protect against this type of apoptosis. These in vitro studies, therefore, suggest that Bcl-2 family members may influence the carcinogenic potential of dietary components in the intestine.

bcl-2 has been transfected into a number of colorectal carcinoma cell lines and resulted in growth inhibition *(35)*. The effects of such *bcl*-2 transfection on the sensitivity of colorectal carcinoma cell lines to various chemotherapeutic agents have not been widely studied. However, Bcl-2 expression has been shown to affect the sensitivity of other transformed cells to the class of drugs most commonly used to treat advanced colon cancer, namely, thymidylate synthase inhibitors *(36)*.

Animal Studies in Normal Intestinal Epithelium

As mentioned above, neither homozygously *bcl*-2 nor *bax*-null mice have any major intestinal phenotype. Assessment of the spontaneous levels of apoptosis in the intestine showed an increased level concentrated at the colonic crypt base in *bcl*-2 null mice *(7••)*, but no differences in the crypts of *bax*-null mice *(19)*. More informative data on the role played by these proteins in controlling apoptosis in the intestinal epithelium was obtained when these animals were stressed either by administration of γ-radiation or the chemotherapeutic agent, 5-fluorouracil (5-FU).

bcl-2-null mice showed elevated levels of apoptosis 3 h after 8 Gy of γ-radiation: This was again concentrated specifically at the base of colonic crypts. No differences were seen in the level of γ-radiation-induced apoptosis in the small intestine when *bcl*-2-wild-type mice were compared with null mice. Similar results were obtained with 5-FU. 4.5 h after a 40 mg/kg intraperitoneal injection of 5-FU, *bcl*-2-null mice showed greatly increased levels of apoptosis specifically at the colonic crypt base. However, by 24 h, the time-point of peak apoptosis after 5-FU, there was no difference in the level of colonic apoptosis between *bcl*-2 wild-type and null mice, suggesting that *bcl*-2 at this site normally functions to delay apoptosis, rather than to prevent it completely *(19)*. Again, Bcl-2 expression had little effect on the small intestinal apoptosis induced by 5-FU.

The consequences of administering agents that induce intestinal apoptosis have also been studied in homozygously *bax*-null mice. No differences in intestinal apoptosis were observed between *bax* wild-type and null mice 4.5 h following 1 or 8-Gy γ-radiation or 4.5 h following 40-mg/kg 5-FU *(19)*. However, *bax*-null mice showed significantly less small intestinal and midcolonic apoptosis 24 h following 40-mg/kg 5-FU *(19)*. In contrast to *bcl*-2-null mice, the reduction in apoptotic yield in *bax*-null mice following 5-FU occurred throughout the crypt, rather than being concentrated at any particular cell positions. Although immunohistochemical studies have only demonstrated Bax expression in Paneth cells and differentiated enterocytes *(16•,17•)*, the data from *bax*-null mice suggests that there is sufficient functional Bax protein at other crypt cell positions to affect their capacity to undergo apoptosis in response to some types of cytotoxic insult.

These studies in mice suggest that expression of genes such as *bcl*-2 and *bax* can indeed act as arbiters of the cell death induced by radiation and chemotherapeutic drugs in the normal intestinal epithelium. The important questions are whether their expression levels also influence either the toxicity induced by cumulative doses of these agents in the normal intestine or the chemosensitivity of tumors. The influence upon toxicology has yet to be formally tested, and the issue of tumor chemosensitivity is discussed below.

Clinical Studies in Colorectal Carcinoma

One group has studied whether Bcl-2 expression affected the response of 5-FU-based chemotherapy (modulated with interferon) in 135 patients with advanced colorectal adenocarcinoma *(37)*. No difference in overall survival was seen, based on tumor Bcl-2 status. However, such studies are very difficult to interpret, because of such factors as intra- and interpatient heterogeneity, and the inherent general chemoresistance of colonic carcinomas. Either very large and detailed clinical studies or an alternative experimental approach are needed to obtain a definitive answer to this question. To date, no clinical studies of the effect of Bcl-2 on adjuvant chemotherapy in colon cancer have been published.

Chemopreventive Studies

Numerous lines of evidence suggest that nonsteroidal anti-inflammatory drugs (NSAIDs) are chemopreventive agents in colorectal cancer, and recent studies have suggested that this effect is caused by inhibition of one particular isoform of the cyclo-oxygenase enzyme (COX-2). It now appears that COX-2 expression modulates intestinal apoptosis via changes in Bcl-2 expression. Tsuji and Dubois *(38•)* first demonstrated that forced expression of the COX-2 gene in rat intestinal epithelial cells resulted in resistance of these cells to undergoing butyrate-induced apoptosis, and the cells showed increased levels of Bcl-2 protein. Thus, overexpression of COX-2 led to phenotypic changes in the cells that could enhance tumorigenic potential. The same group has advanced these studies by showing that treatment of the COX-2-overexpressing colon carcinoma cell line, HCA-7, with the prostaglandin E_2 (PGE_2), resulted in increased levels of Bcl-2 protein, again suggesting that COX-2 expression leads to inhibition of apoptosis by modulating Bcl-2 expression *(39)*. Hence, by inhibition of COX-2, and thus decreasing the expression of Bcl-2, NSAIDs promote intestinal apoptosis. This may be sufficient to prevent tumor formation.

CONCLUSIONS

Members of the Bcl-2 family of proteins appear to have important effects in controlling apoptosis in normal intestinal epithelium. Bcl-2 expression is strongest in the presumed stem cells at the base of colonic crypts, where it plays a crucial role in protecting these cells from toxic insult. This protection from toxins may allow the stem cells at this site to tolerate mutations. This may explain why colorectal tumors are relatively common, and why the earliest colonic tumors show highest levels of Bcl-2 expression. Bcl-2 expression in colorectal carcinoma seems to be associated with tumors that have a better prognosis. This may be because these cancers actually represent tumors at an earlier point in the carcinogenic pathway, which will thus have an inherently better prognosis. Recent studies have also suggested that *bax* mutations may play an important role in the development of those colonic tumors that show microsatellite instability.

The expression levels of members of the Bcl-2 family of proteins also influence the apoptosis induced by a number of pharmacological agents in the intestine. It is therefore possible that the expression of various members of this family of proteins will influence the response of normal intestinal epithelium to both chemopreventive and chemotherapeutic agents, and also the response of colonic tumors to current treatment regimens. If these hypotheses prove to be correct, then pharmacological agents targeted at members of the Bcl-2 family, although not necessarily having any therapeutic effect themselves, may improve the efficacy, or reduce the toxicity of current chemotherapeutic drugs, by affecting the relative propensities of tumor and normal tissue to undergo apoptosis.

ACKNOWLEDGMENTS

I am grateful for fellowships from the British Digestive Foundation and the Medical Research Council.

REFERENCES

- of special interest
- of outstanding interest

1. Kinzler KW, Vogelstein B. Lessons from hereditary colorectal cancer. *Cell* 1996; 87: 159–170.
••2. Pritchard, DM, Watson AJM. Apoptosis and gastrointestinal pharmacology. *Pharmacol Ther* 1996; 72: 149–169.
   Review on apoptosis in the normal gastrointestinal tract, especially in response to various pharmacological agents.
3. Van Triest B, van Groeningen SJ, Pinedo HM. Current chemotherapeutic possibilities in the treatment of colorectal cancer. *Eur J Cancer* 1995; 31A: 1193–1197.
•4. Hockenberry DM, Zutter M, Hickey, W, Nahm M, Korsmeyer SJ. Bcl-2 protein is topographically restricted in tissues characterized by apoptotic cell death. *Proc Natl Acad Sci USA* 1991; 88: 6961–6965.
   Hockenberry et al. were the first to describe the immunohistochemical localization of Bcl-2 in normal tissues.
5. Hague A, Moorghen M, Hicks D, Chapman M, Paraskeva C. Bcl-2 expression in human colorectal adenomas and carcinomas. *Oncogene* 1994; 9: 3367–3370.
6. Bronner MP, Culin C, Reed JC, Furth E. bcl-2 proto-oncogene and the gastrointestinal epithelial tumor progression model. *Am J Pathol* 1995; 146: 20–26.
••7. Merritt AJ, Potten CS, Watson AJM, Loh DY, Nakayama K, Nakayama K, Hickman JA. Differential expression of bcl-2 in intestinal epithelia—correlation with attenuation of apoptosis in colonic crypts and the incidence of colonic neoplasia. *J Cell Sci* 1995; 108: 2261–2271.
   This paper elegantly describes spontaneous and radiation-induced apoptosis in bcl-2-null mice.
8. Guan RJ, Moss SF, Arber N, Krajewski S, Reed JC, Holt PR. 30kda phosphorylated form of bcl-2 protein in human colon. *Oncogene* 1996; 12: 2605–2609.
9. Nakayama K, Nakayama K, Negishi I, Kuida K, Shinkai Y, Louie MC, et al. Disappearance of the lymphoid system in Bcl-2 homozygous mutant chimeric mice. *Science* 1993; 261: 1584–1588.
10. Nakayama K, Nakayama K, Negishi I, Kuida K, Sawa H, Loh DY. Targeted disruption of bcl-2ab in mice: occurrence of gray hair, polycystic kidney disease, and lymphocytopenia. *Proc Natl Acad Sci USA* 1994; 91: 3700–3704.

11. Veis DI, Sorenson CM, Dhutter JR, Korsmeyer SJ. Bcl-2 deficient mice demonstrate fulminant lymphoid apoptosis, polycystic kidneys and hypopigmented hair. *Cell* 1993; 75: 229–240.

12. Kamada S, Shimono A, Shito Y, Tsujimura T, Takahashi T, Noda T, et al. Bcl-2 deficiency in mice leads to pleiotropic abnormalities: accelerated lymphoid cell death in thymus and spleen, polycystic kidney, hair hypopigmentation, and distorted small intestine. *Cancer Res* 1995; 55: 354–359.

•13. Krajewski S, Krajewska M, Shabaik A, Wang HG, Irie S, Fong L, Reed JC. Immunohistochemical analysis of in vivo patterns of Bcl-X expression. *Cancer Res* 1994; 54: 5501–5507.
The first to describe the immunohistochemical localization of Bcl-X$_L$ in normal tissues.

14. Motoyama N, Wang F, Roth KA, Sawa H, Nakayama K, Nakayama K, et al. Massive cell death of immature haematopoietic cells and neurons in Bcl-x-deficient mice. *Science* 1995; 267: 1506–1510.

15. Gibson L, Holmgreen SP, Huang DCS, Bernard O, Copeland NG, Jenkins NA, et al. Bcl-w, a novel member of the bcl-2 family, promotes cell survival. *Oncogene* 1996; 13: 665–675.

•16. Krajewski S, Krajewska M, Shabaik A, Miyashita T, Wang HG, Reed JC. Immunohistochemical determination of in vivo distribution of Bax, a dominant inhibitor of bcl-2. *Am J Pathol* 1994; 115: 1323–1336.
Describes the immunohistochemical localization of Bax in normal tissues.

•17. Wilson JW, Potten CS. Immunohistochemical localization of BAX and BAD in the normal and BCL-2 null gastrointestinal tract. *Apoptosis* 1996; 1: 183–190.
Describes the immunohistochemical localization of Bax and Bad in murine intestine.

18. Knudson CM, Tung KSK, Tourtellotte WG, Brown GAJ, Korsmeyer SJ. Bax-deficient mice with lymphoid hyperplasia and male germ cell death. *Science* 1995; 270: 96–99.

19. Pritchard DM, Potten CS, Korsmeyer SJ, Hickman JA. 5-fluorouracil-induced apoptosis in intestinal epithelia from bcl-2-null and bax-null mice: implications for the determinants of organ-specific toxicity. 1998; submitted.

•20. Krajewski S, Krajewska M, Reed JC. Immunohistochemical analysis of in vivo patterns of bak expression, a proapoptotic member of the bcl-2 protein family. *Cancer Res* 1996; 56: 2849–2855.
Describes the immunohistochemical localization of Bak in normal tissues.

•21. Kitada S, Krajewska M, Zhang X, Scudiero D, Zapata JM, Wang H-G, et al. Expression and location of pro-apoptotic bcl-2 family protein BAD in normal human tissues and tumor cell lines. *Am J Path* 1998; 152: 51–61.
The first to describe the immunohistochemical localization of Bad in normal human tissues.

22. Ofner D, Riehemann K, Maier H, Riedmann B, Nehoda H, Totsch M, et al. Immunohistochemically detectable bcl-2 expression in colorectal carcinoma: correlation with tumour stage and patient survival. *Br J Cancer* 1995; 72: 981–985.

23. Bosari S, Moneghini L, Graziani D, Lee AKC, Murray JJ, Coggi G, Viale G. bcl-2 oncoprotein in colorectal hyperplastic polyps, adenomas and adenocarcinomas. 1995; 26: 534–540.

24. Sinicrope FA, San Bao Ruan, Cleary KR, Stephens LC, Lee JJ, Levin B. bcl-2 and p53 oncoprotein expression during colorectal tumorigenesis. *Cancer Res* 1995; 55: 237–241.

•25. Sinicrope FA, Hart J, Michelassi F, Lee JJ. Prognostic value of bcl-2 oncoprotein expression in stage II colon carcinoma. *Clin Cancer Res* 1: 1103–1110.
One of the most robust studies of the prognostic significance of Bcl-2 in colon carcinoma.

26. Watson AJM, Merritt AJ, Jones LS, Askew JN, Anderson E, Becciolini A, et al. Evidence for reciprocity of bcl-2 and p53 expression in human colorectal adenomas and carcinomas. *Br J Cancer* 1996; 73: 889–895.

27. Baretton GB, Diebold J, Christoforis G, Vogt M, Muller C, Dopfer K, et al. Apoptosis and immunohistochemical bcl-2 expression in colorectal adenomas and carcinomas. *Cancer* 1996; 77: 255–264.

28. Langlois NEI, Lamb J, Eremin O, Heys SD. Apoptosis in colorectal carcinoma occurring in patients aged 45 years and under: relationship to prognosis, mitosis and immunohistochemical demonstration of p53, c-myc and bcl-2 protein products. *J Pathol* 1997; 182: 392–397.

••29. Manne U, Myers RB, Moron C, Poczatek RB, Dillard S, Weiss H, et al. Prognostic significance of bcl-2 expression and p53 nuclear accumulation in colorectal adenocarcinoma. *Int J Cancer* 1997; 74: 346–358.
This paper, as well as describing its own cohort of patients, also reviews most of the previous prognostic studies of p53 and Bcl-2 in colon carcinoma.

30. Ilyas M, Tomlinson IPM, Hanby AM, Yao T, Bodmer WF, Talbot IC. Bcl-2 expression in colorectal tumors, evidence of different pathways in sporadic and ulcerative-colitis-associated carcinomas. *Am J Pathol* 149: 1719–1726.

31. Krajewska M, Moss S, Krajewski S, Song S, Holt PR, Reed JC. Elevated expression of bcl-x and reduced bak in primary colorectal adenocarcinomas. *Cancer Res* 1996; 56: 2422–2427.

••32. Rampino N, Yamamoto H, Ionov Y, Li Y, Sawai H, Reed JC, Perucho M. Somatic frameshift mutations in the BAX gene in colon cancers of the microsatellite mutator phenotype. *Science* 1997; 275: 967–969.
This important paper was the first to describe Bax mutations in colon cancers with microsatellite instability.

33. Ouyang H, Furukawa T, Abe T, Kato Y, Horii A. BAX gene, the promoter of apoptosis, is mutated in genetically unstable cancers of the colorectum, stomach and endometrium. *Clin Cancer Res* 1998; 4: 1071–1074.

34. Hague A, Diaz GD, Hicks DJ, Krajewski S, Reed JC, Paraskeva C. bcl-2 and bak may play a pivotal role in sodium butyrate-induced apoptosis in colonic epithelial cells; however overexpression of bcl-2 does not protect against bak-mediated apoptosis. *Int J Cancer* 1997; 72: 898–905.

35. Pietenpol JA, Papadopoulos N, Markowitz S, Willson JKV, Kinzler KW, Vogelstein B. Paradoxical inhibition of solid tumor cell growth by bcl2. *Cancer Res* 1994; 54: 3714–3717.

36. Fisher TC, Milner AE, Gregory CD, Jackman AL, Aherne GW, Hartley JA, Dive C, Hickman JA. bcl-2 modulation of apoptosis induced by anticancer drugs: resistance to thymidylate stress is independent of classical resistance pathways. *Cancer Res* 1993; 53: 3321–3326.

37. Schneider HJ, Sampson SA, Cunningham D, Norman AR, Andreyev HJN, Tilsed JVT, Clarke PA. Bcl-2 expression and response to chemotherapy in colorectal adenocarcinomas. *Br J Cancer* 1997; 75: 427–431.

•38. Tsuji M, DuBois R. Alterations in cellular adhesion and apoptosis in epithelial cells overexpressing prostaglandin endoperoxide synthase 2. *Cell* 1995; 83: 493–501.

This was the first paper to discuss the interrelationships between nonsteroidal anti-inflammatory drugs, cyclooxygenase-2, and Bcl-2, implicating Bcl-2 in colon carcinogenesis.

39. Sheng H, Shao J, Morrow, JD, Beauchamp RD, DuBois RN. Modulation of apoptosis and bcl-2 expression by prostaglandin E2 in human colon cancer cells. *Cancer Res* 1998; 58: 362–366.

19 Apoptosis and Breast Cancer

Nancy E. Davidson, MD,
Hillary A. Hahm, MD, PhD,
and Deborah K. Armstrong, MD

CONTENTS

ABSTRACT

Apoptosis is an integral part of normal mammary gland development, differentiation, and function. Numerous descriptive and mechanistic studies of programmed cell death pathways in normal and malignant mammary cells have been presented. This chapter reviews the evidence that established breast cancer treatments act via apoptotic pathways, and how ongoing laboratory work defining these pathways might lead to the development of new therapeutic ap-

From: *Apoptosis and Cancer Chemotherapy*
Edited by: J. A. Hickman and C. Dive © Humana Press Inc., Totowa, NJ

proaches. Also, the possibility that evaluation of apoptotic-related molecules or events might serve as prognostic or predictive factors is discussed.

INTRODUCTION

Activation of programmed cell death (PCD) plays a role in nearly all phases of mammary gland development and function. Studies using human tissues have shown that normal breast tissue goes through a sequence of proliferation and apoptosis through the course of the menstrual cycle. A mitotic peak is noted morphologically at d 25, concurrent with the peak levels of progesterone and estrogen in the luteal phase. This is followed by an apoptotic peak at d 28, in association with the drop in steroid hormones marking the end of the 28-d cycle *(1)*. Classic studies in animal models have shown that the extensive cell death and tissue remodeling that occur during postlactational involution are also a consequence of PCD *(2•)*. It is increasingly evident that hormone-responsive breast cancer retains a similar PCD pathway that is triggered by hormonal manipulation. Also, similar or identical pathways exist in endocrine-unresponsive breast cancer cells, even though endocrine manipulations can no longer initiate the pathways. Indeed, a majority of the standard chemotherapeutic agents used in breast cancer treatment appear to trigger apoptotic responses. Thus, careful dissection of the component parts of these pathways has become a goal that might have several practical implications for breast cancer management. It is the aim of this chapter to examine two such areas. First, the evidence that established breast cancer treatments act via apoptotic pathways, as well as how definition of the component parts of the process might provide opportunities for development of new therapeutic approaches, will be examined. Second, the possibility that characterization of apoptotic-related molecules or events might provide prognostic information about patient outcome, or predictive information about likelihood of response to a particular intervention, will be reviewed. These and other topics about PCD in normal and malignant mammary epithelium are reviewed in greater detail elsewhere *(3•,4,5••)*.

PROGRAMMED CELL DEATH IN HORMONE-DEPENDENT BREAST CANCER CELLS

Human breast cancer cell lines have been the primary model system used in vitro and in vivo to examine effects of cancer therapy on PCD pathways in breast cancer. These models have permitted dissection of the role that PCD plays in breast cancer cell response to endocrine interventions. These studies are critical, because hormonal therapy has been a mainstay of breast cancer therapy for over 100 yr. About two-thirds of primary breast cancers express estrogen receptor (ER) and/or progesterone receptor (PR) at diagnosis, and these tumors often respond to hormonal treatments, such as ovarian ablation, antiestrogens, and aromatase inhibitors. Although these therapies have traditionally been considered to be cytostatic rather than cytotoxic, their use often results in at least partial tumor regression, suggesting that a net decrease in cell number results. This clinical observation suggests that hormonal therapy might both inhibit cell proliferation and promote cell death.

Estrogen Withdrawal

Kyprianou et al. *(6•)* first reported that regression of estrogen-dependent MCF-7 breast tumors in nude mice was associated with apoptosis. Decreased proliferation, increased apoptosis, oligonucleosomal DNA fragmentation, and enhanced expression of several apoptosis-related mRNAs, including transforming growth factor-β1 (TGF-β1), sulfated glycoprotein-2, and c-*myc,* characterized this process. This study showed that the malignant hormone-responsive MCF-7 cells retain the ability to activate a PCD pathway in response to estrogen deprivation. The issue is controversial, though, because other investigators have not been able to demonstrate apoptosis of MCF-7 tumors under similar conditions, perhaps because of variation in MCF-7 lines and/or experimental design *(7).*

The findings have been confirmed in vitro, however. Warri et al. *(8)* showed that MCF-7 cells growing in tissue culture demonstrated apoptotic morphology with estrogen depletion; Wilson et al. *(9)* observed high-mol-wt DNA cleavage in estrogen-deprived MCF-7 cells. The importance of *bcl*-2 in this process is suggested by several studies showing that treatment of MCF-7 cells with 17β-estradiol in tissue culture leads to a time- and dose-dependent increase in *bcl*-2 mRNA levels, whereas *bax* and *bcl*-x_L mRNA expression is not affected *(10,11).*

Antiestrogen Treatment

The apoptotic effects of antiestrogens have also been established via a series of laboratory studies. Bardon et al. *(12)* first showed that tamoxifen, its metabolite 4-hydroxytamoxifen, and the antiprogestin RU 486, induced morphological changes of apoptosis in ER-positive human breast cancer cell lines. Warri et al. *(8)* confirmed this finding using the antiestrogen toremifene in ER-positive ZR-75-1 and MCF-7 cells. Also, treatment of MCF-7 cells with the pure antiestrogen, ICI 182,780 led to decreased cell number in conjunction with high-mol-wt DNA cleavage, but in the absence of apoptotic morphological changes *(9)*. A functional role for TGF-β1 in this process is implicated by a study in which treatment of MCF-7 cells with an anti-TGF-β1 antibody blocked tamoxifen-induced DNA fragmentation *(13)*. Further, involvement of the Bcl-2 protein is supported by the observation that treatment with tamoxifen or the pure antiestrogen, ICI 164,384, significantly inhibited estrogen-induced increases in Bcl-2 in MCF-7 cells *(10,11)*. Thus, in aggregate, studies using human breast cancer cells growing in tissue culture, or in xenograft models, support the hypothesis that hormonal interventions, such as estrogen withdrawal or antiestrogen administration, work through both inhibition of cell proliferation and induction of PCD.

PROGRAMMED CELL DEATH IN MODELS OF HORMONE-INDEPENDENT BREAST CANCER

Although the majority of tumors in women with newly diagnosed breast cancer possess ER, only a fraction actually respond to endocrine maneuvers. In addition, many breast cancers either initially lack ER or evolve to a hormone-unresponsive state over time. This raised the question of whether these hormone-independent cancers still retained the ability to undergo PCD, even though endocrine therapy could not initiate the pathway(s). Studies using a variety of conventional and novel agents have addressed this question.

Peptide Growth Factors and Their Receptors

The importance of peptide growth factor pathways in breast cancer cell growth and death is increasingly evident. In particular, the receptor tyrosine kinases, epidermal growth factor receptor (EGFR) and HER-2/neu/c-erbB2 (hereafter called HER-2), are overexpressed in up to one-third of breast cancers. Their overexpression has been associated with poor clinical outcome in several studies. Autocrine or paracrine loops involving these two growth factors are attractive therapeutic targets. In vitro studies using human breast cancer cell lines have shown that either ligand binding or antibody treatment against these types of receptors may induce apoptosis, depending on experimental conditions. For example, EGF induces PCD in the ER-negative MDA-MB-468 cell line, which overexpresses EGFR *(14)*. This process is characterized by concentration-dependent cytotoxicity, apoptotic morphologic changes, DNA fragmentation, and activation of several downstream caspases. Similar studies have shown that either ligand binding or antibodies directed against the extracellular domain of HER-2 can inhibit proliferation and induce apoptosis in breast cancer cells that overexpress that protein *(5••,15•)*. These findings have direct implications for ongoing clinical efforts to use monoclonal antibodies directed against EGFR or HER-2, either alone or in combination with conventional cytotoxics for management of human breast cancer.

Chemotherapeutic Agents

Numerous studies have demonstrated that treatment of ER-negative (and -positive) human breast cancer cell lines with established chemotherapeutic agents triggers PCD. The authors' initial studies of treatment of MDA-MB-468 cells with fluoropyrimidine inhibitors of thymidylate synthase showed that these agents lead to growth inhibition, loss of clonogenic capacity, apoptotic morphologic changes, oligonucleosomal DNA fragmentation, and induction of TGF-β1 mRNA *(16)*. The ability to induce PCD in these cells is not limited to fluoropyrimidines.

Several studies have provided evidence that treatment of a variety of human breast cancer cell lines with paclitaxel (a drug that stabilizes microtubules and does not damage DNA) also results in growth inhibition, apoptotic morphology, and characteristic DNA fragmentation *(17,18)*. Activation of several caspases is also a feature of the death pathway for both of these drugs. In vitro work also suggests that paclitaxel results in Bcl-2 phosphorylation, which may result in loss of Bcl-2 function, perhaps via interference with its binding to the proapoptotic Bax protein *(19)*. The topoisomerase II inhibitor, etoposide, also induces PCD, confirming that multiple chemotherapeutic agents with diverse mechanisms of action induce PCD in ER-positive and -negative human breast cancer cell lines *(18,20)*. These models can be used to dissect the molecular changes that characterize the initiation and execution phases of PCD in breast cancer cells. The implications of the observed changes can then potentially be explored in mechanistic studies. For example, MCF-7 cells transfected with *bcl*-x$_S$ show increased sensitivity to etoposide and paclitaxel *(18)*. Also, intratumoral injection of an adenoviral vector containing *bcl*-x$_S$ into MCF-7 tumors growing in nude mice led to partial tumor regression in conjunction with apoptotic morphologic changes at the injection site *(21•)*. Thus, this is but one example of how in vitro studies might focus work on chemotherapy sensitivity and resistance toward particular molecular targets, which may play a functional role in determining cell death.

Novel Approaches to Induce Apoptosis

Descriptive studies of PCD in model systems have also permitted the identification of new inducers of apoptosis that might have therapeutic potential. Of the many avenues under exploration, only a few are mentioned here as representative of the diverse types of ongoing studies: polyamine analogs, fatty acid synthase inhibitors, and gene therapy.

A potential relationship between polyamines and PCD is suggested by the findings that spermidine and spermine stabilize chromatin, and that polyamine-depleted cells undergo changes in chromatin and DNA structure. As a consequence, the authors have investigated a class of compounds that target the polyamine metabolic pathways. In particular, the ability of the unsymmetrically alkylated polyamine analog, N^1-ethyl-N^{11}-((cyclopropyl)methyl)-4,8-diazaundecane, to deplete intracellular polyamines, inhibit growth, and induce high-mol-wt and oligonucleosomal DNA fragmentation in human breast cancer cell lines has been established *(22)*. Multiple analogs that disrupt the polyamine metabolic pathways are under development, and hold promise as a way to decrease cell proliferation and increase cell death. In addition, in vitro studies of the ability of these agents to potentiate chemotherapy-induced cell death are warranted.

A fraction of human breast, ovarian, endometrial, colorectal, and prostate cancers express elevated levels of fatty acid synthase, the major enzyme necessary for endogenous fatty acid biosynthesis. Cancer cells that express high levels of fatty-acid-synthesizing enzymes use endogenously synthesized fatty acids for membrane synthesis; normal cells preferentially use dietary lipids. Thus, inhibition of fatty acid synthase with agents like cerulenin, a noncompetitive inhibitor of fatty acid synthase, could be selectively toxic for cancer cells with increased fatty acid biosynthesis. Indeed, cerulenin treatment of ZR-75-1 human breast cancer cells resulted in dose-dependent reduction in clonogenic potential, morphologic evidence of apoptosis, and high-mol-wt DNA fragmentation *(23)*. Thus, the difference in fatty acid synthesis between malignant and normal cells is potentially an exploitable target for induction of PCD in breast cancer cells.

The identification of crucial modulators of PCD sequences might make possible efforts to use those molecules as triggers of PCD via gene therapy strategies. Already noted above is work showing that a replication-deficient adenovirus expressing *bcl*-x$_S$ can induce apoptosis in breast cancer xenografts growing in nude mice *(21•)*. Other investigators *(24)* have used a liposome–*p53* complex for the in vivo delivery of wild-type p53 to MDA-MB-435 tumors, which are known to possess mutant p53. This treatment led to reduction in size in some tumors, with concomitant evidence of apoptotic morphology in treated, but not untreated, tumors. These studies provide proof of the principle that components of the PCD pathway can be targeted, although the problem of delivery is, of course, enormous.

CLINICAL STUDIES OF APOPTOSIS AND BREAST CANCER

A key question is how observations made in experimental model systems currently pertain to treatment of women with breast cancer. Two major types of translational studies have been pursued to test this link. The first type tests the prognostic implications of a particular marker via characterization of tumors obtained at a single point in time, usually time of diagnosis; the second examines the predictive implications of the biomarker via serial tissue sampling during a course of therapy to evaluate sequential changes associated with tumor response or resistance to the treatment.

Apoptotic Markers as Prognostic Factors

Many investigators have evaluated apoptotic and proliferative indices and/or expression of critical apoptotic markers like p53 or Bcl-2 family members as prognostic or predictive markers for women with newly diagnosed breast cancer. Ideally, studies of this type should be carried out in tissue banks derived from a large cohort of women who received defined therapy, e.g., women enrolled in a clinical trial. In practice, of course, this is extremely difficult to do, and the discrepant results that are summarized below probably reflect in part heterogeneity of tumor types and therapy, as well as a lack of standardization of the assays used to evaluate the parameter under study.

In a study of 288 invasive breast cancers, high apoptotic index correlated with increased grade, tumor necrosis, aneuploidy, high S-phase fraction and mitotic index, negative ER and PR, and mutant p53. High mitotic index was significantly associated with decreased survival in lymph-node-negative and -positive patients, but had no independent prognostic value *(25)*.

Aberrant p53 expression is found in 20–50% of breast cancers, depending on the method of detection used. Most studies suggest that the presence of mutant p53 is correlated with other poor prognostic factors, such as high S-phase fraction, high tumor grade, and lack of steroid receptors *(5••)*. Its value as an independent prognostic factor for recurrence or survival is uncertain. One large study of 1400 women with node-negative breast cancer, treated only with local therapy, showed that immunohistochemical expression of p53 protein was associated with negative steroid receptors, older age, tumor size >2 cm, and higher recurrence rate. In multivariate analysis, p53 expression correlated with relapse and survival, as did all of the factors mentioned above *(26)*.

The prognostic value of Bcl-2 and Bax protein expression in breast cancer has also been extensively studied, and is reviewed elsewhere *(5••)*. In sum, Bcl-2 expression is associated with the presence of ER, wild-type p53, lower-grade histology, low proliferative index, and Bax expression. It is inversely correlated with EGFR, HER-2, and p21 expression. Studies linking Bcl-2 expression with lymph node status have given mixed results. Work to date does not adequately address whether Bcl-2 and/or Bax expression is an independent prognostic factor for relapse or death. A single study suggests that absence of Bcl-2 or Bax expression, or especially both, predicts the later development of metastatic disease with lymph node-negative breast cancer.

Apoptosis as a Predictive Factor

Although preclinical work outlined above strongly suggests that both endocrine and chemotherapy approaches for breast cancer work, in part, via induction of apoptotic pathways, support for this hypothesis from human studies has only recently emerged. Ellis et al. *(27••,28••)* have used the setting of preoperative systemic therapy for operable breast cancers to demonstrate that administration of both chemotherapy and antiestrogens induces apoptosis. As part of a trial of two types of anthracycline-containing regimens, core needle biopsies were obtained from 27 patients with large operable breast cancers, 7 d before chemotherapy and 24 h after first chemotherapy administration. Sufficient tissue was available in both samples from 19 patients to assess apoptotic index via an *in situ* end-labeling assay, proliferative index using Ki67 expression, and Bcl-2 protein expression by immunohistochemistry. The median apoptotic index before and after treatment was significantly increased from 0.46% pretreatment to 1.02% after treatment ($p =$

0.009). Median Ki67 scores before and after treatment were 19.2 and 15.8% respectively, which was a nonsignificant decrease ($p = 0.16$). Median Bcl-2 expression also did not change (60% pretreatment and 72% posttreatment, $p = 0.84$). Nine of 13 patients who manifested a clinical response to chemotherapy showed a >50% increase in apoptotic index; 3 of 4 nonresponders showed no change in apoptotic index. This small study provides preliminary support for the concept that chemotherapy treatment does induce apoptosis in human breast cancer, and that induction of apoptosis or lack thereof may predict the likelihood of subsequent clinical response *(27••)*.

A similar study design was used to assess apoptotic effects of tamoxifen and ICI 182,780 in women with operable breast cancer *(28••)*. In two companion trials, women were randomly assigned to tamoxifen or placebo for 1–7 wk before surgery, or to ICI 182,780 or placebo for 1 wk before surgery. Pretreatment samples were obtained before drug administration, and posttreatment specimens were taken at the time of definitive surgery. In the tamoxifen trial, adequate pairs of samples were available from 17 patients taking tamoxifen, and from 22 patients given placebo. Tamoxifen recipients had a median apoptotic index of 0.45% before treatment and 0.54% after treatment; the corresponding indices were 0.53 and 0.67%, respectively, in placebo-treated women: Neither of these differences was significant. However, a similar analysis confined to women with positive ER showed a significant increase in apoptotic index from 0.34 to 0.47% ($p = 0.005$) with tamoxifen treatment. No difference was seen in ER-negative tumors, or in treated or control women with ER-negative tumors. Also, no significant modulation of Ki67 was observed in this study.

Analysis of paired tumors from 13 ICI 182,780 patients and 9 placebo patients also demonstrated that ICI 182,780, but not placebo, treatment led to increased apoptotic index. Median apoptotic index increased from 0.34% before treatment to 0.97% after ICI 182,780 administration ($p = 0.003$); no significant changes were noted in the placebo group. Evaluation by ER status could not be performed because of the small numbers of patients. These studies provide preliminary clinical data that antiestrogens also induce apoptosis in human breast cancers in patients. The short duration of therapy precludes any assessment of change in apoptotic index as a predictor of subsequent clinical response.

Because experimental studies suggest a function for p53 in chemotherapy sensitivity, the role of p53 expression as a predictor for response to antineoplastic therapy has also been studied in several small preoperative chemotherapy studies. Results have been mixed. Two studies suggested that patients whose tumors express mutant p53 are more likely to demonstrate progressive disease during chemotherapy, but two other studies failed to substantiate any relationship between p53 status and chemotherapy response *(5••)*. Also, a large cooperative group trial evaluating p53 status and clinical outcome, after observation or adjuvant cyclophosphamide, methotrexate, 5-fluorouracil, and prednisone therapy, failed to confirm any relationship between p53 expression and clinical outcome *(29•)*. Thus, the importance of p53 status as a predictive factor is also uncertain; the need for studies on large numbers of clinical samples, from women of defined status and standardization of methodology for p53 assessment, is only too obvious.

Three studies have examined the relationship between Bcl-2 expression and response to endocrine therapy *(5••)*. Together, they suggest that Bcl-2 expression is associated with enhanced response to hormonal therapy. Patients whose tumors expressed both ER and Bcl-2 derived the greatest benefit from therapy. This may be explained, in part, by laboratory findings that Bcl-2 is an estrogen-regulated protein; thus, its presence, like that of PR, implies a functional estrogen response pathway within the cell that would be susceptible to the effects of hormonal intervention.

CONCLUSIONS

Studies from preclinical model systems and human tissues have shown that apoptosis plays a role in normal mammary gland development and function. Work carried out using mostly human breast cancer cells growing in tissue culture or nude mice suggests that the critical components of this apoptotic pathway(s) are preserved in malignant cells, and that appropriate stimuli, such as hormonal manipulation or chemotherapy administration, can trigger PCD. That these approaches also induce PCD in tumors in women is suggested by preliminary studies of primary endocrine therapy or chemotherapy. Thus, the current challenge is to delineate the components of these pathways, their activators, and their inhibitors, because their identification could provide opportunities for the development of new therapeutic approaches. Several examples of novel agents and their targets are described here and elsewhere *(3•,4,5••)*. In addition, tailoring of therapy to the individual tumor phenotype is an important goal. Therefore, studies that define the potential of apoptotic markers to serve as prognostic or predictive factors in breast cancer treatment are critical pursuits as well.

ACKNOWLEDGMENTS

Support from the National Cancer Institute, United States Army Research and Material Command, and Susan G. Komen Breast Cancer Foundation for some of the work described here is gratefully acknowledged.

REFERENCES

• of special interest
•• of outstanding interest

1. Ferguson DJP, Anderson TJ. Morphological evaluation of cell turnover in relation to the menstrual cycle in the 'resting' human breast. *Br J Cancer* 1981; 44: 177–181.

•2. Strange R, Li F, Saurer S, Burkhardt A, Friis RR. Apoptotic cell death and tissue remodelling during mouse mammary gland involution. *Development* 1992; 115: 49–58.
   Classic report of apoptosis after cessation of lactation in the mouse mammary gland.

•3. McCloskey DE, Armstrong DK, Jackisch C, Davidson NE. Programmed cell death in human breast cancer cells. *Recent Prog Horm Res* 1996; 51: 493–508.
   Review that places special emphasis on the therapeutic aspects of PCD in breast cancer cells.

4. Denmeade SR, McCloskey DE, Joseph IBJK, Hahm HA, Isaacs JT, Davidson NE. Apoptosis in hormone-responsive malignancies. *Adv Pharmacol* 1997; 41: 553–583.

••5. Hahm HA, Davidson NE. Apoptosis in the mammary gland and breast cancer. *Endocr-Related Cancer* 1998; 10: 475–478.
   Extensive review of the literature on apoptosis in normal and malignant mammary tissues.

•6. Kyprianou N, English HF, Davidson NE, Isaacs JT. Programmed cell death during regression of the MCF-7 human breast cancer following estrogen ablation. *Cancer Res* 1991; 51: 162–166.
   Initial description of PCD after an endocrine manipulation in an animal model of human breast cancer.

7. Osborne CK, Hobbs K, Clark GM. Effect of oestrogens and antioestrogens on the growth of breast cancer cells in athymic nude mice. *Cancer Res* 1985; 45: 584–590.

8. Warri AM, Huovinen RL, Laine AM, Martikainen PM, Harkonen PL. Apoptosis in toremifene-induced growth inhibition of human breast cancer cells *in vivo* and *in vitro*. *J Natl Cancer Inst* 1993; 85: 1412–1418.

9. Wilson JW, Wakeling AE, Morris ID, Hickman JA, Dive C. MCF-7 human mammary adenocarcinoma cell death *in vitro* in response to hormone-withdrawal and DNA damage. *Int J Cancer* 1995; 61: 502–508.

10. Wang TTY, Phang JM. Effects of estrogen on apoptotic pathways in human breast cancer cell line MCF-7. *Cancer Res* 1995; 55: 2487–2489.

11. Teixeira C, Reed JC, Pratt MAC. Estrogen promotes chemotherapeutic drug resistance by a mechanism involving bcl-2 proto-oncogene expression in human breast cancer cells. *Cancer Res* 1995; 55: 3902–3907.

12. Bardon S, Vignon F, Montcourrier P, Rochefort H. Steroid receptor-mediated cytotoxicity of an anti-estrogen and an anti-progestin in breast cancer cells. *Cancer Res* 1987; 47: 1441–1448.

13. Perry RR, Kang Y, Greaves BR. Relationship between tamoxifen-induced transforming growth factor β1 expression, cytostasis, and apoptosis. *Br J Cancer* 1995; 72: 1441–1446.

14. Armstrong DK, Kaufmann SH. Ottaviano YL, Furaya Y, Buckley JA, Isaacs JT, Davidson NE. Epidermal growth factor-mediated apoptosis of MDA-MB-468 human breast cancer cells. *Cancer Res* 1994; 54: 5280–5283.

•15. Kita Y, Tseng J, Horan T, Wen J, Philo J, Chang D, et al. ErbB receptor activation, cell morphology changes, and apoptosis induced by anti-HER2 monoclonal antibodies. *Biochem Biophys Res Commun* 1996; 226: 59–69.
   Of interest because of recent clinical trials using Herceptin (Genentech, South San Francisco, CA), a monoclonal antibody targeted against HER-2.

16. Armstrong DK, Isaacs JT, Ottaviano YL, Davidson NE. Programmed cell death in an estrogen-independent human breast cancer cell line, MDA-MB-468. *Cancer Res* 1992; 52: 3418–3424.

17. McCloskey DE, Kaufmann SH, Prestigiacomo LJ, Davidson NE. Paclitaxel induces programmed cell death in MDA-MB-468 human breast cancer cells. *Clin Cancer Res* 1996; 2: 847–854.

18. Sumantran VN, Ealovega MW, Nunez G, Clarke MF, Wicha M. Overexpression of bcl-x$_S$ sensitizes MCF-7 cells to chemotherapy-induced apoptosis. *Cancer Res* 1995; 55: 2507–2510.

19. Haldar S, Jena N, Croce C. Inactivation of bcl-2 by phosphorylation. *Proc Natl Acad Sci USA* 1995; 92: 4507–4511.

20. Sokolova IA, Cowan KH, Schneider E. Ca^{2+}/Mg^{2+}-dependent endonuclease activation is an early event in VP-16-induced apoptosis of human breast cancer MCF7 cells *in vitro*. *Biochim Biophys Acta* 1995; 1266: 135–142.

•21. Ealovega MW, McGinnis PK, Sumantran VN, Clarke MF, Wicha MS. Bcl-x$_S$ gene therapy induces apoptosis of human mammary tumors in nude mice. *Cancer Res* 1996; 56: 1965–1969.
   Demonstration of how gene therapy using apoptotic molecules can be applied in an in vivo *model.*

22. McCloskey DE, Casero RA Jr, Woster PM, Davidson NE. Induction of programmed cell death in human breast cancer cells by an unsymmetrically alkylated polyamine analogue. *Cancer Res* 1995; 55: 3233–3236.

23. Pizer ES, Jackisch C, Wood FD, Pasternack GR, Davidson NE, Kuhajda FP. Inhibition of fatty acid synthesis induces programmable death in human breast cancer cells. *Cancer Res* 1996; 56: 2745–2747.

24. Lesoon-Wood LA, Kim WH, Kleinman HK, Weintraub BD, Mixson AJ. Systemic gene therapy with p53 reduces growth and metastases of a malignant breast cancer in nude mice. *Hum Gene Ther* 1995; 6: 395–405.

25. Lipponen P, Aaltomaa S, Losma V-M, Syrjanen K. Apoptosis in breast cancer as related to histopathological characteristics and prognosis. *Eur J Cancer* 1994; 30A: 2068–2073.

26. Silvestrini R, Daidone MG, Benini E, Faranda A, Tomasic G, Boracchi P, Salvado B, Veronesi U. Validation of p53 accumulation as a predictor of distant metastasis at 10 years of follow-up in 1400 node-negative breast cancers. *Clin Cancer Res* 1996; 2: 2007–2013.

••27. Ellis PA, Smith IE, McCarthy K, Detra S, Salter J, Dowsett M. Preoperative chemotherapy induces apoptosis in early breast cancer. *Lancet* 1997; 349: 849.
Demonstration that the apoptotic effects of chemotherapy in preclinical models are also observed in the setting of primary chemotherapy adminstration to women with breast cancer.

••28. Ellis PA, Saccani-Jotti G, Clarke R, Johnston SRD, Anderson E, Howell A, et al. Induction of apoptosis by tamoxifen and ICI 182780 in primary breast cancer. *Int J Cancer* 1997; 72: 608–613.
Apoptotic effects of antiestrogen observed in tissue culture models are also seen in tumors taken from women receiving primary hormonal therapy for breast cancer.

•29. Elledge RM, Gray R, Mansour E, Yu Y, Clark GM, Ravdin P, et al. Accumulation of p53 protein as a possible predictor of response to adjuvant combination chemotherapy with cyclophosphamide, methotrexate, fluorouracil, and prednisone for breast cancer. *J Natl Cancer Inst* 1995; 87: 1254–1256.
Example of the type of study required to evaluate the prognostic and predictive value of a molecular marker, in this case, p53.

20 Role of Apoptosis in Human Neuroblastomas

Garrett M. Brodeur, MD
and Valerie P. Castle, MD

CONTENTS

ABSTRACT

Programmed cell death (PCD) probably plays a critical role in the clinical behavior of human neuroblastomas. This tumor has the highest propensity for spontaneous regression, and there is evidence

From: *Apoptosis and Cancer Chemotherapy*
Edited by: J. A. Hickman and C. Dive © Humana Press Inc., Totowa, NJ

that this regression may be the consequence of apoptosis induced by neurotrophin deprivation. Indeed, several ligand-receptor pathways have been identified that appear to be important in regulating neuronal survival or cell death. Furthermore, analysis of gene expression in neuroblastomas suggests that the bcl-2 *family genes and the caspases may also play important roles in regulating neuronal survival. Finally, like many tumors, apoptosis may be a common pathway of cell death in response to chemotherapeutic agents. Therapeutic approaches aimed at selectively activating PCD in neuroblastomas are under development, and may lead to a more effective and less toxic approach to the treatment.*

INTRODUCTION

Neuroblastomas are heterogeneous in terms of genetic features and clinical behavior. A remarkable feature of these tumors is the propensity to undergo spontaneous differentiation. Furthermore, some tumors are very responsive to chemotherapy, but others are quite resistant. This chapter evaluates the data regarding mechanisms of inducing apoptosis in neuroblastoma cell lines and primary tumors. The authors review ligand–receptor interactions that either induce or protect against apoptosis, and also examine the expression and function of genes involved in promoting or preventing apoptosis within the cell, such as the *bcl*-2 family and caspase families. Finally, the role of these genes in chemotherapy-induced cell death is reviewed.

CLINICAL HETEROGENEITY OF NEUROBLASTOMA

Neuroblastoma is the most common solid tumor of childhood *(1)*. This tumor has the remarkable propensity to undergo spontaneous regression, particularly in infants. In some older patients, the tumor may differentiate into a benign ganglioneuroma. Unfortunately, in the majority of patients, neuroblastomas are metastatic at diagnosis, and most of these children ultimately die of their disease, especially if they are over 1 yr of age at diagnosis *(1)*. The reason for such divergent and age-dependent clinical outcomes appears to be strongly related to genetic features of the tumor at diagnosis.

GENETIC HETEROGENEITY OF NEUROBLASTOMA

Several genetic features have been found to be characteristic of subsets of neuroblastomas, including hyperdiploidy with whole chromosome gains, amplification of the *MYCN* oncogene, and deletions of the short arm of chromosome 1 (1p) *(2)*. Hyperdiploidy is characteristic of tumors in infants, particularly those who are prone to undergo spontaneous regression or who respond well to chemotherapy. However, the most aggressive tumors, particularly those that are metastatic in older patients, frequently have *MYCN* amplification, and the majority of these also have 1p deletion *(2)*. Partial trisomy for 17q is also a common genetic change in neuroblastomas, and its presence in the karyotype is generally associated with a worse outcome. Other deletions and rearrangements are also found, such as deletion of 11q or 14q, but thus far these features have not been consistently associated with clinical behavior.

ROLE OF THE *Trk* GENE FAMILY IN NEUROBLASTOMA SURVIVAL AND DIFFERENTIATION

Neuroblastomas are derived from sympathetic neurons. The survival and differentiation of these cells are dependent on the presence of nerve growth factor (NGF), which is a member of a family of neurotrophins that includes brain-derived neurotrophic factor (BDNF), neurotrophin-3 (NT3), and NT4. The receptor for NGF is TrkA, the receptor for BDNF and NT4 is TrkB, and the receptor for NT3 is TrkC. It appears that each of these ligand–receptor pathways may be important for subsets of neurons, as well as subsets of neural tumors.

In neuroblastomas, the expression of TrkA and TrkC has been associated with a favorable outcome; expression of the full-length TrkB receptor is associated with *MYCN* amplification and a poor outcome *(3,4••,5••,6)*. Tumors expressing TrkB consistently express the ligand BDNF (Fig. 1); favorable neuroblastomas rarely express the ligands for TrkA or TrkC *(6,7)*. TrkA-expressing, favorable neuroblastomas appear to be dependent on exogenous NGF for survival, at least in vitro. In the absence of NGF, the cells undergo programmed cell death (PCD), and die within 5–7 d; in the presence of NGF, the cells differentiate into ganglion cells, and can survive for months in vitro (Fig. 1;*4••*). Interruption of either the autocrine loop of TrkB and BDNF in unfavorable neuroblastomas, or the TrkA/TrkC pathways in favorable tumors, may lead to PCD in the tumors.

OTHER RECEPTORS AND APOPTOSIS IN NEUROBLASTOMAS

Low-affinity neurotrophin receptor *(LNTR)* is a gene that encodes a protein, p75, which binds all the NGF family neurotrophins with low affinity *(8–10•,11•)*. This gene is a member of the tumor necrosis factor receptor *(TNFR)* family of transmembrane receptors, which have been

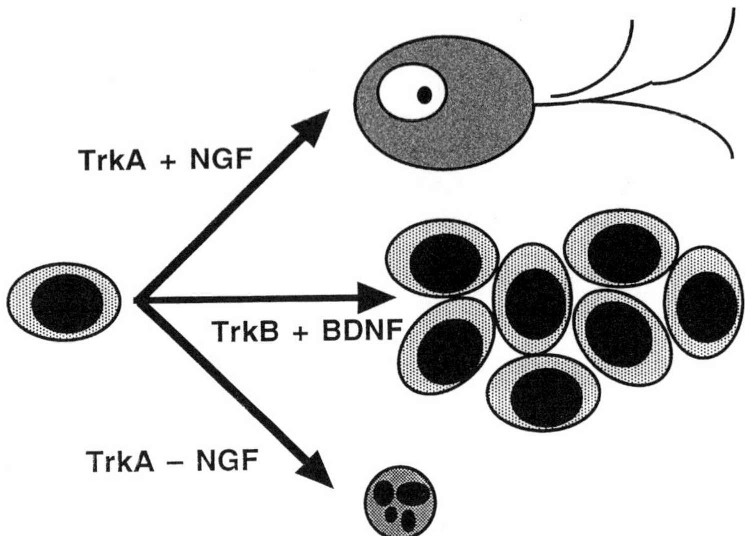

Fig. 1. Effects of neurotrophins and their receptors on neuroblastoma cells. At left is an undifferentiated neuroblastoma cell. Favorable neuroblastoma cells generally express TrkA. Cells expressing TrkA in the presence of NGF will differentiate into a ganglion cell, and the corresponding tumor would become a benign ganglio-neuroma. In the absence NGF, the TrkA-expressing cells eventually undergo apoptosis. Unfavorable neuroblastomas, particularly those with MYCN amplification, express TrkB and its ligand BDNF. This provides an autocrine survival pathway that favors continued proliferation.

associated with triggering PCD. Sometimes apoptosis is triggered in the presence of ligand; in other cases, it is the absence of ligand that leads to PCD. A neuroblastoma cell line was transfected with *LNTR*, and then showed a high level of cell death, especially in the absence of serum *(11•)*. This increased tendency to undergo PCD was inhibited by NGF. There is evidence that NGF binding of p75 may lead to activation of NF-κB *(10•)*. NGF was shown to induce neuronal cell death in another cell line after transfection with *LNTR (8):* Thus, this receptor may mediate either survival or cell death, depending on the cell type or the stage of neuronal differentiation *(9).*

CD95/Fas/APO1 is also a member of the *TNFR* family of receptors. These receptors have been associated with the induction of PCD in the presence of Fas ligand (CD95L), and this cell death is mediated through a death domain. Current evidence suggests that engagement of CD95 with CD95L is a common pathway for induction of apoptosis in a number of cell systems (Fig. 2). Indeed, chemotherapy-induced

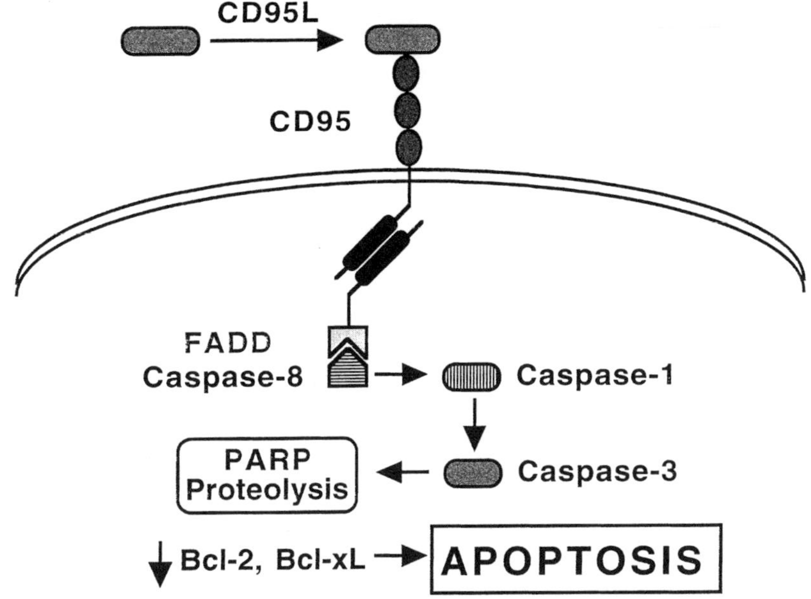

Fig. 2. Diagrammatic representation of the CD95/CD95L pathway in neuroblastomas. Upregulation of the ligand or receptor (or both) in response to chemotherapy leads to activation of the signaling pathway, with activation of caspases, proteolytic cleavage of PARP, downregulation of Bcl-2 and Bcl-X_L and, ultimately, apoptosis.

apoptosis is mediated by upregulation of CD95L and interaction with CD95 in an autocrine or paracrine manner (*see* Apoptosis, CD95, and Response of Neuroblastomas to Chemotherapy; *12••*). The downstream effectors for the apoptotic response are the caspases (*see* Expression and Function of Caspase Family Genes in Neuroblastomas).

PCD AND Bcl-2
IN THE DEVELOPING NERVOUS SYSTEM

PCD plays a crucial role in the embryonic development of the nervous system (*13*). During neurogenesis, PCD occurs in defective and redundant cells, to create a permissive environment for axonal growth (*14*). Bcl-2, a protein that functions to inhibit apoptosis, is expressed in a developmentally regulated manner in neural tissue (*15–17*). Studies using cultured sympathetic neurons and embryonic sensory neurons have demonstrated Bcl-2 can protect cultured neural cells from apoptosis induced by withdrawal of NGF (*15,18•*). Bax appears to be required for this process to occur (*19•*). Agents that prevent PCD can rescue PC-12 pheochromocytoma cells from neurotrophic factor deprivation *(20)*. Bcl-X$_L$, an apoptosis-inhibiting member of the Bcl-2 family, can also rescue primary sympathetic neurons from apoptosis induced by NGF withdrawal *(21)*. Thus, regulation of the entire Bcl-2/Bax/Bcl-X$_L$ family appears to be important in determining whether neural cells survive or die.

EXPRESSION AND FUNCTION OF Bcl-2 FAMILY MEMBERS IN NEUROBLASTOMAS

Several studies have found that Bcl-2 is widely expressed in neuro-blastoma tumors *(22•,23–27)* and tumor-derived cell lines *(28•,29•)*. Moreover, the level of Bcl-2 expression in tumors is inversely related to the apoptotic index, as measured by morphologic assessments and terminal deoxynucleotidal transferase-mediated dUTP-biotin nick end labeling (TUNEL) assays of DNA fragmentation *(23,27,29•,30,31)*. There have been conflicting reports about the relationship between Bcl-2 expression and prognostic features associated with this disease. One group reported an association between Bcl-2 expression, *MYCN* amplification, and unfavorable histology *(25)*. Other investigations, how-ever, have failed to confirm these correlations *(24,26)*. In neuroblastoma cell lines, both Bcl-2 and Bcl-X$_L$ (but not Bcl-X$_S$) are expressed *(28•)*. Bcl-2 is primarily expressed in lines of chromaffin lineage; Bcl-X$_L$ is expressed in both chromaffin and nonchromaffin lineage neuroblastoma cell lines *(28•,29•)*. Generally, neuroblastoma cell lines that express high levels of one protein express lower levels of the other *(28•)*.

EXPRESSION AND FUNCTION OF CASPASE FAMILY GENES IN NEUROBLASTOMAS

Caspases are cysteine proteases that play an important effector role in mediating PCD that has been initiated by a variety of other mecha-nisms. At least 10 caspases have been identified (caspase-1 to caspase-10), although each goes by one or several other names. The prototypic mammalian caspase is interleukin-1-converting enzyme (ICE, or cas-pase-1), a homolog of the cysteine protease ced-3, which is a required protein for normal PCD during development of the nematode. Two groups have looked at the expression and function of selected caspases in neuroblastomas *(32•,33)*.

One group examined the expression of caspase-3 (also known as CPP32) in a neuroblastoma cell line, after exposure to staurosporin (Table 1), an agent known to induce PCD *(33)*. They demonstrated proteolytic processing of caspase-3, intracellular relocation, and pro-cessing of the protein substrate poly-ADP-ribose polymerase (PARP)

Table 1
Examples of Apoptosis-Inducing Agents and Their Mechanism of Action

Agent	*Mechanism of action*
Differentiating agents: retinoids	
9-*cis*-retinoic acid	Mechanism unknown
13-*cis* retinoic acid	
All *trans* retinoic acid	
Fenretinide	
DNA strand-breaking agents	
Cisplatin	CD95/CD95-L
Doxorubicin	
Etoposide (VP-16)	
Cyclophosphamide (4HC)	
Protein kinase inhibitors	
Staurosporine	Caspase activation
Other	
Peroxynitrate	PARP activation

within 3–6 h after staurosporin exposure. The second group measured caspase-1, -2, and -3 in 52 primary neuroblastomas by Northern, Western, and immunohistochemistry, and/or compared these findings to other clinical and biological features *(32•)*. They found high levels of expression and nuclear localization of caspase-1 and caspase-3 in favorable neuroblastomas with high TrkA expression and no *MYCN* amplification. Furthermore, caspases-1 and -3 were translocated to the nucleus in apoptotic cells, suggesting that these proteins were playing an important role in the PCD occurring in neuroblastomas, and potentially with spontaneous regression occasionally seen in these tumors.

RETINOIDS AND APOPTOSIS IN NEUROBLASTOMA

An array of pharmacological agents with variable mechanisms of action can induce apoptosis in neuroblastoma cells in vitro (Table 1). Of this group, the most thorough understanding has emerged from studies on the retinoids. All-*trans* retinoic acid (ATRA) can induce neuroblastoma cells to differentiate or undergo apoptosis, depending on the cell phenotype *(34•)*. Three phenotypic variants of neuroblastoma cells in culture have been described. N-type cells constitute the neuroblastic phenotype. These cells have neuritic processes and display biochemical characteristics of neuronal cells. S-type cells are substrate-adherent, epithelial in phenotype, and display biochemical features of immature Schwann, glial, or melanocytic cells. The third phenotypic variant is I-type cells, with intermediate morphology and biochemical features of both N- and S-type cells. S-type cells do not express Bcl-2, and are particularly sensitive to induction of apoptosis by ATRA. This is in contrast to N-type cells, which express Bcl-2, and are less sensitive to RA-induced apoptosis *(35,36)*. Neuroblastoma cells that differentiate in response to RA are resistant to chemotherapy-induced apoptosis. This resistance is associated with upregulation of Bcl-2 *(36)*.

The effects of the retinoids are agent-specific and dose-dependent. 13-*cis* RA is capable of slowing growth and inducing differentiation, at least in vitro. However, its effects on inducing PCD in neuroblastomas are less clear *(34•,37)*. 9-*cis* RA is more effective than ATRA at inducing apoptosis in neuroblastoma cells *(38)*. Successful induction of apoptosis by 9-*cis* RA requires that it be cleared from the cells by a washout technique in culture, which may have implications for in vivo dosing with this agent *(38)*. Finally, there is a novel retinoid called *N*-(4-hydroxyphenyl)retinamide (fenretinide), which appears to be particularly potent at inducing apoptosis *(39,40)*. The mechanism through which the retinoids induce apoptosis has not been clearly defined. There have been no reports demonstrating activation of the caspase pathway in response to retinoids in neuroblastoma cells.

APOPTOSIS, CD95, AND RESPONSE OF NEUROBLASTOMAS TO CHEMOTHERAPY

A variety of DNA-damaging agents have been shown to induce apoptosis in neuroblastoma cells in vitro *(28•,36,41••)*. Several studies have indicated that neuroblastoma cells undergo apoptosis in response to DNA-damaging agents after arrest in G2/M of the cell cycle *(42,43)*. Neuroblastoma cells treated with these agents show classic morphologic features of apoptosis and evidence of DNA damage. The CD95 system has recently been shown to mediate chemotherapy-induced apoptosis in neuroblastoma *(12••)*. CD95, a member of the TNFR superfamily, triggers cell death in a variety of cell types. In neuroblastoma cells, doxorubicin, cisplatin, and etoposide induce expression of CD95, as well as the ligand, CD95L (Fig. 2; *12••*). Chemotherapy-induced apoptosis mediated by CD95 in neuroblastoma involves activation of ced-3-like proteases, resulting in cleavage of PARP, an enzyme involved in DNA repair *(12••)*. These results suggest that autocrine or paracrine activation of CD95/CD95L leads to a death response in neuroblastoma cells after treatment with these specific chemotherapeutic drugs.

Both Bcl-2 and Bcl-X$_L$ can modulate chemotherapy-induced apoptosis in neuroblastoma cells. Single-gene transfection studies have shown that deregulated expression of Bcl-2 or Bcl-X$_L$ in a dose-dependent manner can confer resistance to apoptosis induced by cisplatin or 4-hydroxycyclophosphamide, and delay apoptosis induced by etoposide *(28•,41••,44)*.

Not all cytotoxic agents induce apoptosis by engaging CD95. Betulinic acid (BA), a pentacyclic triterpene, induces apoptosis in a number of cell types, including neuroblastoma, and this appears to be independent of the CD95 and p53 pathways *(45)*. BA-induced apoptosis is associated with increased caspase activity, enzymatic processing of the upstream caspase component, FLICE (caspase-8), and PARP cleavage. BA-induced apoptosis, however, is independent of CD95/CD95L activation. BA-induced apoptosis is associated with induction of the apoptosis-facilitating proteins, Bax and Bcl-X$_S$. Both Bcl-2 and Bcl-X$_L$ can inhibit apoptosis induced by BA in a manner similar to its protection from DNA-damaging agents *(12••)*.

Caspase activation appears to be the final common pathway for induction of apoptosis by most agents. Staurosporin, an alkaloid inhibitor of protein kinases, is also able to induce apoptosis in neuroblastoma cells *(46)*. Staurosporin-induced apoptosis is associated with proteolytic processing of CPP32 (caspase-3), PARP activation, and PARP cleavage. In some systems, however, PARP activation is not always associated with proteolytic processing of PARP. Peroxynitrate (PN), which forms through a reaction of superoxide with nitrous oxide, has recently been shown to induce apoptosis in the NSC34 neuroblastoma spinal cord cell line *(47)*. In this system, PN-induced apoptosis was associated with morphologic features of apoptosis and increased PARP immunoreactivity, but there was no evidence of PARP cleavage. It has been suggested that PN induces DNA damage that stimulates PARP, which depletes intracellular energy stores, resulting in cell death *(47)*. This possibility will require further investigation; however, these results suggest that additional pathways may function to mediate specific drug-induced apoptosis in neuroblastoma.

Therapeutic strategies based on in vitro studies of neuroblastoma cells have led to several current clinical trials in patients with neuroblastoma. Most notable are the protocols utilizing retinoic acid (RA) in the treatment of advanced stage disease, which are based on in vitro assays indicating RA can inhibit neuroblastoma proliferation. A phase 1 trial utilizing 13 *cis*-RA following bone marrow transplantation for

Stage 4 neuroblastoma has suggested that the drug is well tolerated, with minimal toxicity *(48)*, and its therapeutic efficacy is currently under investigation.

An improved understanding of the molecular pathways controlling the survival/death response of neuroblastoma cells should provide insight into mechanisms that could ultimately be targeted for novel therapeutic intervention. In support of this possibility are studies indicating that neuroblastoma cells, regardless of their Bcl-2 or Bcl-X_L expression, are sensitive to induction of apoptosis by adenoviral-mediated expression of the death-facilitating protein Bcl-X_S *(44)*. The possibility of therapeutically engaging the death pathway directly is intriguing, but will require the development of targeted gene therapy approaches. Finally, therapeutic interventions aimed at either blocking the Trk ligand–receptor interaction, or selectively inhibiting the Trk family of receptors, may provide a therapeutic approach that is more effective and less toxic than currently used chemotherapeutic agents.

ACKNOWLEDGMENTS

This work was supported in part by grants NS-34514 (G.M.B.) and CA-69276 (V.P.C.) from the National Institutes of Health, and from the Audrey E. Evans Endowed Chair (G.M.B.).

REFERENCES

• of special interest
•• of outstanding interest

1. Brodeur GM, Castleberry RP. Neuroblastoma. In: *Principles and Practice of Pediatric Oncology.* Pizzo PA, Poplack DG, eds. J. B. Lippincott; Philadelphia. 1993; pp. 739–767.
2. Brodeur GM, Maris JM, Yamashiro DJ, Hogarty MD, White PS. Biology and genetics of human neuroblastomas. *J Pediatr Hematol Oncol* 1997; 19: 93–101.
3. Brodeur GM, Nakagawara A, Yamashiro DJ, Ikegaki N, Liu X-G, Lee CP, Evans AE. Expression of TrkA, TrkB, and TrkC in human neuroblastomas. *J Neuro-Onc* 1996; 12: 37–41.
••4. Nakagawara A, Arima-Nakagawara M, Scavarda NJ, Azar CG, Cantor AB, Brodeur GM. Association between high levels of expression of the TRK gene and favorable outcome in human neuroblastoma. *New Engl J Med* 1993; 328: 847–854.
   First paper to demonstrate a relationship between high TrkA expression and favorable outcome in neuroblastomas. It also demonstrated that neurotrophin deprivation in high TrkA-expressing tumors could lead to PCD.
••5. Nakagawara A, Azar CG, Scavarda NJ, Brodeur GM. Expression and function of TRK-B and BDNF in human neuroblastomas. *Mol Cell Biol* 1994; 14: 759–767.

First report of TrkB expression in neuroblastomas and the first to suggest that the co-expression of TrkB and BDNF may represent an autocrine survival pathway for aggressive neuroblastomas.

6. Yamashiro DJ, Nakagawara A, Ikegaki N, Liu X-G, Brodeur GM. Expression of TrkC in favorable human neuroblastomas. *Oncogene* 1996; 12: 37–41.
7. Ryden M, Sehgal R, Dominici C, Schilling FH, Ibanez CF, Kogner P. Expression of mRNA for the neurotrophin receptor trkC in neuroblastomas with favourable tumour stage and good prognosis. *Br J Cancer* 1996; 74: 773–779.
8. Rabizadeh S, Oh J, Zhong L-T, Yang J, Bitler CM, Butcher LL, Bredesen DE. Induction of apoptosis by the low-affinity NGF receptor. *Science* 1993; 261: 345–348.
9. Barrett GL, Bartlett PF. The p75 nerve growth factor receptor mediates survival or death depending on the stage of sensory neuron development. *Proc Natl Acad Sci USA* 1994; 91: 6501–6505.
•10. Carter BD, Kaltschmidt C, Kaltschmidt B, Offenhauser N, Bohm-Matthaei B, Baeuerle PA, Barde Y-A. Selective activation of NF-κB by nerve growth factor through the neurotrophin receptor p75. *Science* 1996; 272: 542–545.
Demonstrated an important role for NF-κB signaling by NGF through the p75 receptor.
•11. Bunone G, Mariotti A, Compagni A, Morandi E, Della Valle G. Induction of apoptosis by p75 neurotrophin receptor in human neuroblastoma cells. *Oncogene* 1997; 14: 1463–1470.
Demonstrated that expression of p75 may be important for the induction of apoptosis in neuroblastomas.
12. Fulda S, Sieverts H, Friesen C, Herr I, Debatin KM. CD95 (APO-1/Fas) system mediates drug-induced apoptosis in neuroblastoma cells. *Cancer Res* 1997; 57: 3823–3829.
This important paper demonstrated the role of the CD95 pathway in mediating drug-induced apoptosis. Generally, it had been presumed that chemotherapy killed cells based on their mechanism of action, but many may die earlier as a result of activating the death program.
13. Ellis R, Yuan Y, Horvitz H. Mechanisms and functions of cell death. *Ann Rev Cell Biol* 1991; 7: 663–698.
14. Oppenheim R. Cell death during development of the nervous system. *Ann Rev Neurosci* 1991; 14: 453–501.
15. Garcia I, Martinou I, Tsujimoto Y, Martinou J-C. Prevention of programmed cell death of sympathetic neurons by the bcl-2 proto-oncogene. *Science* 1992; 258: 302–305.
16. Greenlund LJS, Korsmeyer SJ, Johnson EMJ. Role of BCL-2 in the survival and function of developing and mature sympathetic neurons. *Neuron* 1995; 15: 649–661.
17. Abe-Dohmae S, Harada NKY, Tanaka R. Bcl-2 gene is highly expressed during neurogenesis in the central nervous system. *Biochem Biophys Res Comm* 1993; 191: 915–921.
•18. Allsopp TE, Wyatt S, Paterson HF, Davies AM. Proto-oncogene bcl-2 can selectively rescue neurotrophic factor-dependent neurons from apoptosis. *Cell* 1993; 73: 295–307.
Bcl-2 was shown to be important in preventing apoptosis caused by neurotrophin deprivation in neuronal cells.

•19. Deckwerth TL, Elliot JL, Knudson CM, Johnson EMJ, Snider WD, Korsmeyer SJ. BAX is required for neuronal death after trophic factor deprivation and during development. *Neuron* 1996; 17: 401–411.
Bax was shown to be an important mediator of apoptosis caused by neurotrophic factor deprivation in neuronal cells.

20. Batistatou A, Greene L. 1-Aurintricarboxylic acid rescues PC-12 cells and sympathetic neurons from cell death by nerve growth factor deprivation: correlation with suppression of endonuclease activity. *J Cell Biol* 1991; 115: 461–471.

21. Gonzalez-Garcia M, Garcia I, Ding L, O'Shea S, Boise LH, Thompson CB, Nunez G. Bcl-x is expressed in embryonic and postnatal neural tissues and functions to prevent neuronal cell death. *Proc Natl Acad Sci USA* 1995; 92: 4304–4308.

•22. Hoehner JC, Hedborg F, Wiklund HJ, Olsen L, Pahlman S. Cellular death in neuroblastoma: In situ correlation of apoptosis and bcl-2 expression. *Int J Cancer* 1995; 62: 19–24.
Showed an inverse correlation between Bcl-2 expression and apoptosis in neuroblastoma tumor samples, and a gradient related to the tumor blood supply.

23. Hoehner JC, Gestblom C, Olsen L, Pahlman S. Spatial association of apoptosis-related gene expression and cellular death in clinical neuroblastoma. *Br J Cancer* 1997; 75: 1185–1194.

24. Ramani P, Lu QL. Expression of bcl-2 gene product in neuroblastoma. *J Pathol* 1994; 172: 273–278.

25. Castle VP, Heidelberger KP, Bromberg J, Ou X, Dole M, Nunez G. Expression of the apoptosis-suppressing protein bcl-2 in neuroblastoma is associated with unfavorable histology and N-myc amplification. *Am J Pathol* 1993; 143: 1545–1550.

26. Ikegaki N, Katsumata M, Tsujimoto Y, Nakagawara A, Brodeur GM. Relationship between bcl-2 and myc gene expression in human neuroblastoma. *Cancer Let* 1995; 91: 161–168.

27. Ikeda H, Hirato J, Akami M, Matsuyama S, Suzuki N, Takahashi A, Kuroiwa M. Bcl-2 oncoprotein expression and apoptosis in neuroblastoma. *J Pediatr Surg* 1995; 30: 805–808.

•28. Dole MG, Jasty R, Cooper MJ, Thompson CB, Nunez G, Castle VP. Bcl-xL is expressed in neuroblastoma cells and modulates chemotherapy-induced apoptosis. *Cancer Res* 1995; 55: 2576–2582.
Demonstrated that expression of the apoptosis suppressing protein Bcl-X$_L$ helps to protect cells from chemotherapy-induced apoptosis.

•29. Reed JC, Meister L, Tanaka S, Cuddy M, Yum S, Geyer C, Pleasure D. Differential expression of bcl2 protooncogene in neuroblastoma and other human tumor cell lines of neural origin. *Cancer Res* 1991; 51: 6529–6538.
First report of differential Bcl-2 expression in neuroblastomas and related tumors.

30. Hanada M, Krajewski S, Tanaka S, Cazals-Hatem D, Spengler BA, Ross RA, Biedler JL, Reed JC. Regulation of bcl-2 oncoprotein levels with differentiation of human neuroblastoma cells. *Cancer Res* 1993; 53: 4978–4986.

31. Oue T, Fukuzawa M, Kusafuka T, Kohmoto Y, Imura K, Nagahara S, Okada A. In situ detection of DNA fragmentation and expression of bcl-2 in human neuroblastoma: relation to apoptosis and spontaneous regression. *J Pediatr Surg* 1996; 31: 251–257.

•32. Nakagawara A, Nakamura Y, Ikeda H, Hiwasa T, Kuida K, Su MS, et al. High levels of expression and nuclear localization of interleukin-1 beta converting enzyme (ICE) and CPP-32 in human neuroblastomas. *Cancer Res* 1997; 57: 4578–4584.
   First report characterizing patterns of expression of caspase family members in neuroblastomas, and relating this to evidence of apoptosis in these tumors.

 33. Posmantur R, McGinnis K, Nadimpalli R, Gilbertsen RB, Wang KK. Characterization of CPP32-like protease activity following apoptotic challenge in SH-SY5Y neuroblastoma cells. *J Neurochem* 1997; 68: 2328–2337.

•34. Melino G, Thiele CJ, Knight RA, Piacentini M. Retinoids and the control of growth/death decisions in human neuroblastoma cell lines. *J Neuro-Oncol* 1997; 31: 65–83.
   Describes the importance of retinoids in regulating neuroblastoma cell survival.

 35. Melino G, Annicchiarico-Petruzzelli M, Piredda L, Candi E, Gentile V, Davies PJ, Piacentini M. Tissue transglutaminase and apoptosis: sense and antisense transfection studies with human neuroblastoma cells. *Mol Cell Biol* 1994; 14: 6584–6596.

 36. Lasorella A, Iavarone A, Israel MA. Differentiation of neuroblastoma enhances bcl-2 expression and induces alterations of apoptosis and drug resistance. *Cancer Res* 1995; 55: 4711–4716.

 37. Melino G, Draoui M, Bellincampi L, Bernassola F, Bernardini S, Piacentini M, Reichert U, Cohen P. Retinoic acid receptors alpha and gamma mediate the induction of "tissue" transglutaminase activity and apoptosis in neuroblastoma cells. *Exp Cell Res* 1997; 235: 55–61.

 38. Lovat PE, Irving H, Annicchiarico-Petruzzelli M, Bernassola F, Malcolm AJ, Pearson AD, Melino G, Redfern C. Apoptosis of N-type neuroblastoma cells after differentiation with 9-cis-retinoic acid and subsequent washout. *J Natl Cancer Inst* 1997; 89: 446–452.

 39. Di Vinci A, Geido E, Infusini E, Giaretti W. Neuroblastoma cell apoptosis induced by the synthetic retinoid N-(4-hydroxyphenyl)retinamide. *Int J Cancer* 1994; 59: 422–426.

 40. Ponzoni M, Bocca P, Chiesa V, Decensi A, Pistoia V, Raffaghello L, Rozzo C, Montaldo PG. Differential effects of N-(4-hydroxyphenyl)retinamide and retinoic acid on neuroblastoma cells: apoptosis versus differentiation. *Cancer Res* 1995; 55: 853–861.

••41. Dole MG, Nunez G, Merchant AK, Maybaum J, Rode C, Bloch C, Castle VP. Bcl-2 inhibits chemotherapy-induced apoptosis in neuroblastoma. *Cancer Res* 1994; 54: 3253–3259.
   This important paper demonstrated the role that Bcl-2 plays in preventing apoptosis in response to chemotherapy. This showed that Bcl-2 not only protected cells from physiological mechanisms of PCD, but that it could also protect cells from chemotherapy-induced apoptosis.

 42. Piacentini M, Fesus L, Melino G. Multiple cell cycle access to the apoptosis death programme in human neuroblastoma cells. *Fed Eur Biochem Soc* 1993; 320: 150–154.

 43. Cece R, Barajon I, Tredici G. Cisplatin induces apoptosis in SH-SY5Y human neuroblastoma cell line. *Anticancer Res* 1995; 15: 777–782.

 44. Dole MG, Clarke MF, Holman P, Benedict M, Lu J, Jasty R, et al. Bcl-xS enhances adenoviral vector-induced apoptosis in neuroblastoma cells. *Cancer Res* 1996; 56: 5734–5740.

45. Fulda S, Friesen C, Los M, Scaffidi C, Mier W, Benedict M, et al. Betulinic acid triggers CD95 (APO-1/Fas)-and p53-independent apoptosis via activation of caspases in neural tumors. *Cancer Res* 1997; 57: 4956–4964.
46. Boix J, Llecha N, Vj Y, Comella J. Characterization of the cell death process induced by staurosporine in human neuroblastoma cell lines. *Neuropharmacology* 1997; 36: 811–821.
47. Cookson M, Ince P, Shaw P. Peroxynitrite and hydrogen peroxide induced cell death in the NSC34 neuroblastoma spinal cord cell line: role of poly (ADP-ribose) polymerase. *J Neurochem* 1998; 70: 501–508.
48. Villablanca JG, Kahn AA, Avramis VI, Seeger RC, Matthay KK, Reynolds CP. Phase I trial of 13-cis-retinoic acid in children with neuroblastoma following bone marrow transplantation. *J Clin Oncol* 1995; 13: 894–901.

21 Response of Testicular Tumors to Chemotherapy

Relative Role of Drug–DNA Interactions and Apoptosis

Christine M. Chresta, PhD

CONTENTS

ABSTRACT

Testicular germ cell tumors (GCT) are one of the few solid tumors that can be cured by chemotherapy, even when metastatic. The success of therapy is probably related to the unusual susceptibility of this tumor type to undergo apoptosis. The DNA damage threshold for drug-induced apoptosis is very low and the kinetics of programmed cell death are unusually rapid, being more similar to those of hematological malignancies than those of other solid tumors. This chapter

From: *Apoptosis and Cancer Chemotherapy*
Edited by: J. A. Hickman and C. Dive © Humana Press Inc., Totowa, NJ

Table 1
Anticancer Agents that Have Produced Complete Responses
in Testicular Germ Cell Tumors

Drug	Mechanism of action	Ref. and date
Actinomycin D ±	Transcription	(1), 1960
Methotrexate or	Antimetabolite	
Chlorambucil	Alkylation/crosslink	
Bleomycin	Free radical SSB and DSB	(35), 1980
Vinblastine	Microtubule (destabilizer)	
Cisplatin ±	Platination/crosslink	(36), 1974
Bleomycin		(37), 1977
Vinblastine		
Etoposide ±	Topoisomerase II poison	(38), 1980
Cisplatin		(39), 1987
Bleomycin		
Taxol	Microtubule (stabilizer)	(40), 1996

Summary of some of the clinical trials that have resulted in complete responses in patients with testicular GCTs. It demonstrates that this disease is responsive to agents of many different mechanisms of action so that Taxol, cisplatin, and etoposide have all shown activity against this disease as single agents.

outlines the genetic alterations that occur during conversion of the normal germ cell into a pluripotential tumor, and discusses how they could contribute to drug sensitivity. It also analyzes the relative roles that drug–DNA interactions and apoptosis play in chemosensitivity. Recent studies employing isogenic cell line pairs of GCTs differing in expression of p53, and the Bcl-2 family members show that these proteins play an important role in both drug-induced apoptosis and long-term loss of reproductive potential in this disease.

INTRODUCTION

Testicular cancer has been a model for a curable neoplasm since 1960 *(1)*. Unlike other common solid tumors, the majority of male germ cell tumors (GCTs) can be cured by chemotherapy, even when metastatic. Early clinical studies provide important clues to the basis of drug sensitivity. The reader is encouraged to refer to the recent review by Einhorn *(2•)*, whose seminal work on cisplatin resulted in a major improvement in treatment of this disease. Clinical studies highlight that, although cisplatin is probably the most effective drug against testicular tumors, the sensitivity of GCTs is not limited to platinating and crosslinking agents (Table 1). Testicular tumors have also been cured by DNA strand-break-inducing agents and microtubule poisons. This collateral sensitivity to diverse agents suggests that sensitivity is not likely to be related to a particular drug-repair pathway or target protein. Rather, these studies suggest perturbation of the cell is sensed and coupled to activation of cell death at a lower threshold than seen in more resistant tumor types.

Several laboratories are attempting to determine if GCTs have unique genetic changes/defects that contribute to tumor formation and the chemosensitivity. Obviously, such genetic alterations could provide a vital target for drug hunters to improve therapy in more chemoresistant solid tumors. This chapter focuses on the genetic and phenotypic alterations in GCTs that could contribute to the unusually chemosensitive phenotype, concentrating, first, on in vivo data, and second, on data from model systems in vitro.

CHARACTERISTIC GENETIC ALTERATIONS IN GCT: ORIGIN OF GCTS AND KARYOTYPE ANALYSIS

Testicular GCTs and the premalignant lesion, carcinoma *in situ* (CIS), arise from primordial germ cells, which are the direct precursors of sperm. Malignant transformation of germ cells can result in development of seminomas, which resemble primordial germ cells, or of non-seminomatous GCTs, which exhibit embryonal-like differentiation patterns.

All GCTs, including CIS, have an unusual karyotype: They are polyploid, ranging in chromosome number from hyperdiploid to tetraploid *(3,4)*. These features of genetic instability and polyploidy indicate that there may be defects in cell-cycle checkpoints, resulting in a lack of fidelity during DNA replication and cell division. The two most characteristic chromosomal abnormalities in GCTs are trisomy of the long arm of chromosome 1 and one or more copies of an isochromosome i(12p). This isochromosome results in multiple copies of genes on the short arm of chromosome 12, and, in some cases, reduced copies of the long arm 12q *(3)*. Chaganti et al. *(5,6)* have identified several candidate oncogenes on i(12p), including c-*ki-ras*-2 and cyclin D2. It is not yet known if the increased copy number of these genes contributes to tumorigenesis, although it is possible that overexpression of cyclin D2 could perturb cell cycle control through complexing with the cyclin-dependent kinases 4 and 6, which inactivate Rb. This would allow cells to proliferate with damaged DNA, and could result in genetic instability in GCTs. In theory, it could also enhance drug sensitivity, if cells damaged by chemotherapy fail to arrest at checkpoints. The author will return to this point later in relation to in vitro studies.

Other unique features of testicular tumors are the genetic alterations that do not occur during tumorigenesis. In contrast to the majority of solid tumors, mutations are rarely found in the *p53* or *pRb* genes, although the expression of both these proteins is altered *(7)*. Rb protein expression is very low, but the expression of wt p53 protein is unusually high *(6,8–12)*. Recent studies of the small percentage of patients with relapsed disease have uncovered the presence of an increased frequency of p53 mutations in this subset of patients, suggesting wt p53 probably contributes to susceptibility of GCTs to therapy *(13••)*. This is supported by a study of cell lines derived from these tumors with mutant p53, which were found to be resistant to drug-induced apoptosis *(13••)*.

Unlike ectopically overexpressed wild-type p53, the p53 of GCTs is controlled. High levels of p53 in GCTs does not result in corresponding overexpression of p53-dependent genes *(14,15•)*. In fact, the p53-dependent, cyclin-dependent kinase inhibitor, p21, which in several situations has been demonstrated to inhibit p53-dependent apoptosis *(16,17•)*, is expressed at very low levels in GCTs *(6,14)*. This inverse relationship between p53 and p21 expression could be related to the exquisite apoptosis susceptibility of this tumor type.

DRUG SENSITIVITY OF GCT: RELATIVE ROLES OF DRUG DNA INTERACTIONS AND SUSCEPTIBILITY TO APOPTOSIS

Over the past decade, each of the steps involved in drug action have been investigated in model systems of GCTs. In some studies, GCTs were compared to drug-resistant solid tumors (bladder and colon carcinoma), and, in others, GCT cell lines ranging in drug sensitivity were analyzed.

The drug cisplatin provides an interesting case study. Initial platination of DNA is, paradoxically, lower in GCTs than in more resistant bladder and colon tumors *(18–20)*. This suggests that enhanced drug uptake or reduced detoxification are not the basis of the unusual chemosensitivity of GCTs *(18,21)*. However, the majority of GCT cell lines display a reduced capacity to remove cisplatin-induced intra- and interstrand crosslinks *(18,20,22)*. The persistence of the adducts apparently results from a defect in the incision step of nucleotide excision repair (NER) *(23••)*. However, two findings suggest a defect in NER is not the sole reason for drug sensitivity in GCTs. First, some GCT cell lines (denoted by a star in Fig. 1A) exhibit normal repair, yet are still hypersensitive to cisplatin *(18–20,22)*. Second, although defects in NER could enhance cell killing by cisplatin and alkylating agents,

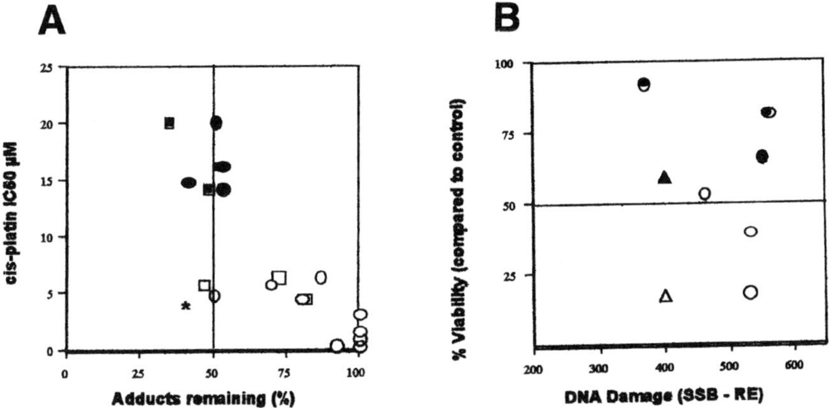

Fig. 1. Lack of correlation between DNA damage and death. **(A)** Comparison of loss of viability (IC50) and percent cisplatin DNA adducts remaining in TGCT and TCC cell lines following 24 *(circle)* or 48 h *(square)* incubation in drug-free medium. DNA damage was measured either by atomic absorption spectroscopy or antibodies to cisplatin DNA adducts. Cytotoxicity was assessed by colony formation. Each point represents an individual cell line. The GCTs are represented by *open* symbols and the bladder tumors by *closed* symbols. Data is summarized from refs. *19, 20,* and *22.* *There are two GCT cell lines that show similar repair capacity to TCCs, but still show enhanced sensitivity *(19,20,22)*. **(B)** Comparison of DNA strand breaks produced by the indicated concentrations of etoposide *(circles)* and bleomycin *(triangles),* with the loss of viability produced by the same concentrations of these agents. DNA damage measured by alkaline elution is expressed in rad equivalents, and viability assessed by colony formation is expressed as a percentage relative to the untreated control. Each point represents an individual cell line. The GCTs are represented by *open* symbols and the bladder tumors by *closed* symbols. Data for etoposide are from ref. *14•.*

it would be unlikely to explain the collateral sensitivity to γ-irradiation, bleomycin, and etoposide. As can be seen in Fig. 1B, concentrations of etoposide or bleomycin that result in similar levels of DNA damage (400–500 rad equivalents) in GCTs, and in transitional carcinoma of the bladder (TCC), consistently resulted in much greater loss of viability in GCTs (Fig. 1B). Furthermore, in contrast to the studies with cisplatin, there is no evidence for a defect in repair of DNA strand breaks (C. M. Chresta, unpublished data).

Taken together, the clinical and in vitro evidence of collateral sensitivity to multiple agents suggest that it is the response to DNA damage (the downstream events) that differs in GCT, and results in such extreme chemosensitivity. This has recently been examined, and is discussed in detail below.

CELLULAR RESPONSE TO DNA DAMAGE

Following drug treatment, the cancer cell has two possible fates: to arrest in cell cycle, repair damage, then continue to divide; or to die with or without attempt to repair. The fate is to some extent determined by the level of damage suffered by the cell. However, as can be seen in Fig. 1B, cell death is activated at lower levels of DNA damage in GCTs than in resistant cell types. Why is the threshold of cell death lower? One possibility is that GCTs do not arrest at cell-cycle checkpoints. It has been found that, after etoposide treatment, GCTs fail to arrest at the G1/S border, and Rb remains hyperphosphorylated (inactivated) *(14)*. Evasion of this checkpoint may occur for two reasons. First, GCTs do not significantly upregulate p21 protein following DNA damage *(14)*. Since the stoichiometry of p21 to the cyclin-dependent kinase (CDK) complex is important for inhibition of the CDK, low levels of p21 could explain the lack of G1 checkpoint in this cell type. Alternatively, overexpression of D2-type cyclins, resulting from the presence of isochromosome 12p *(see above),* could result in activation of CDKs and override the G1 checkpoint *(6)*. Low levels of p21 are related to less-differentiated types of GCT, and appear to be inversely correlated to the expression of cyclin-D2 *(6)*. Low levels of p21 could also contribute to chemosensitivity, as has been reported in colon tumors *(17•,24)*.

APOPTOSIS AND GENES
THAT REGULATE CELL DEATH

GCTs undergo rapid apoptosis in response to chemotherapy *(14,25)*. In contrast to the majority of solid tumors, which undergo a stable cell-cycle arrest, and then commence apoptosis 24–48 h after drug treatment, GCTs activate apoptosis as early as 4 h after drug treatment. Huddart et al. *(25)* noted that cisplatin (15 μM) results in activation of apoptosis 6–9 h after drug treatment, and that, by 24 h, >90% of cells were dead. Apoptosis and loss of reproductive potential occurs in GCTs at levels of DNA damage that are actually sublethal to most tumor types *(14)*. However, the important questions are, "Why are GCTs exquisitely sensitive to apoptosis?" and "Is apoptosis susceptibility important for long-term survival?" As discussed in Chapter 1, susceptibility to apoptosis does not always correlate with loss of long-term viability.

p53

One possible basis for the unusual apoptosis susceptibility of GCTs is the presence of wild-type p53. Extremely high levels of wild-type p53 protein are maintained in testicular tumors. However, the p53 is not inactive: When the cells are challenged with a DNA-damaging agent, p53 becomes transcriptionally active *(13–15•)*.

Several studies have attempted to determine if p53 is important for apoptosis susceptibility in GCTs. Some have compared cell lines derived from different patients; others employ cell lines derived from murine testicular tumors. The murine studies, which compared wt p53-expressing teratocarcinomas to those from p53-deletion mutant mice, were all in agreement that p53 is required for drug-induced apoptosis *(15•,26)*. Lutzker and Levine *(15•)* initially demonstrated that p53 was required for etoposide-induced apoptosis, and Zamble et al. *(26)* subsequently demonstrated cisplatin-induced apoptosis was also p53-dependent. The results from studies of human testicular tumor cell lines are more controversial. A study by Houldsworth et al. *(13••)* supported a role for p53: Cell lines that express mutant p53 (derived from tumors of relapsed patients) were resistant to cisplatin-induced apoptosis. However, Burger et al. *(27,28)* have shown that a GCT cell line (NCCIT), with mutant p53, is sensitive to cisplatin-induced apoptosis, but another (2102EP), with wtp53, is apoptosis-resistant. This controversy probably results from variability in the genetic background of the tumors. The importance of p53 will not be truly dissected until cells differing only in p53 status are derived and analyzed in xenograft models. It is impossible to rule out the importance of a gene that controls apoptosis by merely comparing nonisogenic lines: These cells probably differ in expression of several proteins involved in the numerous steps on the pathway to cell death. The most significant finding is that p53 mutations occur more frequently (20% of cases) in the tumors of the small subset of patients, who do not respond to therapy. This supports a potentially important role for p53 in chemosensitivity in the clinic *(13••)*.

Another area of controversy (*see* Chapters 1–3) is whether p53 status and apoptosis susceptibility affect long-term survival. The only study on this subject in GCTs thus far used murine teratocarcinoma cell lines treated with cisplatin *(26)*. Despite immense differences in susceptibility to apoptosis, the p53 wild-type and negative cells did not show differences in a clonogenic survival assay. However, there are two points to note before deciding that p53 plays no role in chemosensitivity in this disease. First, clonogenic assays do not always reflect the in vivo response of tumors to therapy: In some situations, apoptosis assays have been a better predictor of in vivo response *(17•,29)*. Second, results with cisplatin may differ from those with other agents, because p53 is involved in repair of cisplatin damage. In several situations, p53 has been shown to protect cells from cisplatin cytotoxicity, because p53-

dependent genes enhance nucleotide excision repair. These opposing effects of p53 on drug sensitivity may confound assays of long-term survival *(30–32)*.

BCL-2 FAMILY MEMBERS

A second factor that could account for the low threshold of drug-induced apoptosis is whether regulators of apoptosis, such as the Bcl-2 family members, are differentially expressed or activated in testicular tumors. GCTs in culture have a high Bax:Bcl-2 ratio, compared to resistant tumors, and it has been suggested this could contribute to chemosensitivity *(14)*. In support of this hypothesis, studies of clinical samples have shown underexpression of Bax in drug-resistant GCTs *(13••)*. In addition, studies of ovarian germ cells have shown Bax to play an essential role in the response to chemotherapy *(33•)*. However, as yet, an actual requirement for Bax in the drug response of testicular GCTs has not been tested.

Low Bcl-2 expression, or its complete absence, has been noted in both cell line models and clinical samples of GCTs *(11,12,14,27,28,34)*. To determine directly if low levels of Bcl-2 affected chemosensitivity, Bcl-2 was expressed ectopically in the GCT cell line 833K. Paradoxically, overexpression of Bcl-2 in GCTs does not confer drug-resistance, but, instead, increased drug sensitivity. Drug-induced apoptosis and loss of cloning potential are both augmented by overexpression of Bcl-2. This suggests that in the germ cell setting Bcl-2 may be regulated differently, and is unable to confer a survival advantage. In contrast, Bcl-X_L does confer drug resistance in the GCTs *(34)*. These findings indicate that comparison of Bcl-2 expression in nonisogenic cell lines by immunoblotting may not predict the activity of Bcl-2, and may explain some of the controversial findings in this field *(27,28)*.

CONCLUSIONS AND OUTLOOK

In vivo and in vitro studies have shown GCTs to be collaterally sensitive to activation of apoptosis by many agents. The recent findings of p53 mutations and low Bax expression in drug-resistant patients supports the idea of a role for these proteins in drug sensitivity. Defects in nucleotide excision repair may also enhance the response of GCTs to cisplatin. It will be of interest to determine whether the molecular defect responsible for inefficient NER is present in the tumors of patients that relapse. Potentially, this defect could be used as a marker to design therapy with agents unaffected by NER, such as taxol and etoposide. The proapoptotic activity of Bcl-2 in the GCT cell line 833K was surprising *(34),* and it suggests the possibility that survival signaling pathways may be differentially affecting the activity of Bcl-2 family members in this tumor type. Obviously, how the activity of these proteins is controlled is of immense importance for treatment of resistant tumors. To convert overexpressed antiapoptotic Bcl-2 to inactive or proapoptotic Bcl-2 would be extremely useful in anticancer therapy.

REFERENCES

• of special interest
•• of outstanding interest

1. Li MC, Whitmore WF, Golbey R, Grabstald H. Effects of combined drug therapy on metastatic cancer of the testis. *J Am Med Assoc* 1960; 174: 145–153.
•2. Einhorn LH. Testicular cancer: an oncological success story. *Clin Cancer Res* 1997; 3: 2630–2632.
   Up-to-date review on the treatment of testicular germ cell tumors.
3. Chaganti RSK, Houldsworth J. The cytogenetic theory of pathogenesis of human adult male germ cell tumors. *APMIS* 1998; 106: 80–84.
4. Andrews PW. Human teratocarcinomas. *Biochim Biophys Acta* 1988; 948: 17–36.
5. Dmitrovsky E, Murty V, Moy D, Miller W, Nanus D, Albino A, et al. Isochromosome 12p in non-seminoma cell lines: karyotype amplification of c-ki-ras2 without point-mutational activation. *Oncogene* 1990; 5: 543–548.
6. Houldsworth J, Reuter V, Bosl GJ, Chaganti RSK. Aberrant expression of cyclin D2 is an early event in human male germ cell tumorigenesis. *Cell Growth Differ* 1997; 8: 293–299.
7. Lutzker SG. p53 tumour suppressor gene and germ cell neoplasia. *APMIS* 1998; 106: 85–89.
8. Strohmeyer TG, Reissman P, Cordon-Cardo C, Hartmann M, Ackermann R, Slamon DJ. Correlation between retinoblastoma gene expression and differentiation in human testicular tumors. *Proc Natl Acad Sci USA* 1991; 88: 6662–6666.
9. Bartek J, Bartkova J, Vojtesek B, Staskova Z, Lukas J, Rejthar A, et al. Aberrant expression of the p53 oncoprotein is a common feature of a wide spectrum of human malignancies. *Oncogene* 1991; 6: 1699–1703.
10. Bartkova J, Bartek J, Lukas J, Vojtesek B, Staskova Z, Rejthar A, et al. p53 protein alterations in human testicular cancer including pre-invasive intratubular germ-cell neoplasia. *Int J Cancer* 1991; 49: 196–202.
11. Heidenreich A, Schenkman NS, Sesterhenn IA, Mostofi KF, Moul JW, Srivastava S, Engelmann UH. Immunohistochemical and mutational analysis of the p53 tumour suppressor gene and the bcl-2 oncogene in primary testicular germ cell tumours. *APMIS* 1998; 106: 99,100.
12. Soini Y, Paakko P. Extent of apoptosis in relation to p53 and bcl-2 expression in germ cell tumors. *Hum Pathol* 1996; 27: 1221–1226.
••13. Houldsworth J, Xiao H, Murty VVVS, Chen W, Ray B, Reuter VE, Bosl GJ, Chaganti RSK. Human male germ cell tumor resistance to cisplatin is linked to TP53 gene mutation. *Oncogene* 1998; 16: 2345–2349.
   Important paper that analyzed the tumors of relapsed patients and demonstrated that p53 mutations do occur in this subset of patients.

14. Chresta CM, Masters JRW, Hickman JA. Hypersensitivity of human testicular tumours to etoposide-induced apoptosis is associated with functional p53 and a high Bax:Bcl-2 ratio. *Cancer Res* 1996; 56: 1834–1841.

•15. Lutzker SG, Levine AJ. A functionally inactive p53 protein in teratocarcinoma cells is activated by either DNA damage or cellular differentiation. *Nature Med* 1996; 2: 804–810.
   The first to show that loss of p53 decreased etoposide-induced apoptosis in murine GCTs.

16. Polyak K, Waldman T, He TC, Kinzler KW, Vogelstein B. Genetic determinants of p53-induced apoptosis and growth arrest. *Genes Dev* 1996; 10: 1945–1952.

•17. Waldman T, Zhang Y, Dillehay L, Yu J, Kinzler K, Vogelstein B, Williams J. Cell-cycle arrest versus cell death in cancer therapy. *Nature Med* 1997; 3: 1034–1036.
   Highlighted the importance of apoptosis in the response of solid tumor xenografts to therapy. It also showed that the p53-dependent protein p21 could protect tumors from therapy.

18. Bedford P, Fichtinger-Schepman AMJ, Shellard SA, Walker MC, Masters JRW, Hill BT. Differential repair of platinum-DNA adducts in human bladder and testicular tumor continuous cell lines. *Cancer Res* 1988; 48: 3019–3024.

19. Sark MWJ, Timmer-Bosscha H, Meijer C, Uges DRA, Sluiter WJ, Peters WHM, Mulder NH, deVries EGE. Cellular basis for differential sensitivity to cis-platin in human germ cell tumour and colon carcinoma cell lines. *Br J Cancer* 1995; 71: 684–690.

20. Koberle B, Grimaldi K, Sunters A, Hartley JA, Kelland LR, Masters JRW. DNA repair capacity and cis-platin sensitivity of human testis tumour cells. *Int J Cancer* 1997; 70: 551–555.

21. Masters JRW, Thomas R, Hall AG, Hogarth L, Matheson EC, Cattan EC, Lochrer H. Sensitivity of testis tumour cells to chemotherapeutic drugs: role of detoxifying pathways. *Eur J Cancer* 1996; 32A: 1248–1253.

22. Hill BT, Scanlon KJ, Hansson J, Harstrick A, Pera M, Fichtinger-Schepman AMJ, Shellard SA. Deficient repair of cisplatin-DNA adducts identified in human testicular teratoma cell lines established from tumours from untreated patients. *Eur J Cancer* 1994; 30: 832–837.

23. Koberle B, Masters JRW, Hartley JJ, Wood RD. Reduced capacity for initial stages of nucleotide excision repair in two testis tumor cell lines. *Proc Am Assoc Cancer Res* 1998; 39: 535.
   Important paper analyzing the molecular basis of the defect in NER that could enhance cis-platin sensitivity of GCTs.

24. Wouters BG, Giacci AJ, Denko NC, Brown JM. Loss of p21waf-1/cip-1 sensitizes tumors to radiation by an apoptosis-independent mechanism. *Cancer Res* 1997; 57: 4703–4706.

25. Huddart RA, Titley J, Robertson D, Williams GT, Horwich A, Cooper CS. Programmed cell death in response to chemotherapeutic agents in human germ cell tumour lines. *Eur J Cancer* 1995; 31: 739–746.

26. Zamble DB, Jacks T, Lippard, SJ. p53-dependent and -independent responses to cisplatin in mouse testicular teratocarcinoma cells. *Proc Natl Acad Sci USA* 1998; 95: 6163–6168.

27. Burger H, Nooter K, Boersma AWM, Kortland CJ, Stoter G. Lack of correlation between cisplatin-induced apoptosis, p53 status and expression of Bcl-2 family proteins in testicular germ cell tumour cell lines. *Int J Cancer* 1997; 73: 592–599.

28. Burger H, Nooter K, Boersma AWM, Kortland CJ, van-den-Berg AP, Stoter G. Expression of p53, p21, Bcl-2, Bax, Bcl-X, and Bak in radiation-induced apoptosis in testicular germ cell tumor lines. *Int J Radiat Oncol* 1998; 41: 414–424.

29. Lamb JR, Friend SH. Which guesstimate is the best guesstimate? Predicting chemotherapeutic outcomes. *Nature Med* 1997; 3: 962–963.

30. Fan S, Smith ML, Rivet DJ, Duba D, Zhan Q, Kohn KW, Fornace AJ, and O'Connor PM. Disruption of p53 function sensitizes breast cancer MCF-7 cells to cis-platin and pentoxifylline. *Cancer Res* 1995; 55: 1649–1654.

31. Fan SF, Chang JK, Smith ML, Duba D, Fornace AJ, and O'Connor PM. Cells lacking CIP1/WAF1 genes exhibit preferential sensitivity to cisplatin and nitrogen mustard. *Oncogene* 1997; 14: 2127–2136.

32. Smith ML, Kontny HU, Zhan Q, Sreenath A, O'Connor PM, Fornace AJ. Antisense GADD45 expression results in decreased DNA repair and sensitizes cells to U.V.-irradiation or cisplatin. *Oncogene* 1996; 13: 2255–2263.

•33. Perez GI, Knudson CM, Leykin L, Korsmeyer SJ, Tilley JL. Apoptosis-associated signalling pathways are required for chemotherapy-mediated germ cell destruction. *Nature Med* 1997; 3: 1228–1232.
In vivo evidence supporting a role for apoptosis in chemosensitivity of germ cells.

34. Arriola EL, Rodriguez-Lopez A, Hickman JA, Chresta CM. Overexpression of Bcl-2 sensitizes germ cell tumours to drug-induced apoptosis. *Oncogene* 1998; in press.

35. Samuels ML, Johnson DE. Adjuvent therapy of testis cancer: the role of vinblastine and bleomycin. *J Urol* 1980; 124: 369–371.

36. Higby DJ, Wallace HJ, Albert D, Holland JF. Diamminodichloroplatinum in the chemotherapy of testicular tumors. *J Urol* 1974; 112: 100–104.

37. Einhorn LH, Donohue JP. Combination chemotherapy with *cis*-diamminedichloroplatinum, vinblastine, and bleomycin in disseminated testicular cancer. *Ann Int Med* 1977; 87: 293–298.

38. Fitzharris BM, Kaye SB, Saverymuttu S, Newlands ES, Barrett A, Peckham MJ, McElwain TJ. VP16-213 as a single agent in advanced testicular tumors. *Eur J Cancer* 1980; 16: 1193–1197.

39. Williams SD, Birch R, Irwin L, Greco A, Loehrer PJ, Einhorn LH. Disseminated germ cell tumors: chemotherapy with cisplatin plus bleomycin plus either vinblastine or etoposide. *N Engl J Med* 1987; 316: 1435–1440.

40. Bokemeyer C, Hartmann JT, Kuczyk MA, Truss MC, Beyer J, Jonas U, Kanz L. The role of paclitaxel in chemosensitive urological malignancies: current strategies in bladder cancer and testicular germ-cell tumors. *World J Urol* 1996; 14: 354–359.

CONCLUDING REMARKS

Has the Information Explosion on Apoptosis Affected Clinical Practice?

Much has been learned and written on cell death via apoptosis in the last several years. Has this knowledge been "translated" into clinical practice? Clearly there is an awareness among oncologists that some tumors, particularly lympoid tumors, as well as certain other chemosensitive tumors, e.g., testicular tumors or "Wilms" tumor, choose to die when confronted with DNA damage caused by X-rays or chemotherapeutic agents, explaining the high cure rates in these diseases. As described in this book, apoptosis involves activation of caspases mediated by different signals among them interaction to Fas ligand with the Fas receptor.

In one subset of lymphomas, indolent or "low-grade" lymphoma (nodular or diffuse small cell lymphoma), elevated levels of Bcl-2, an antiapoptotic molecule, may be a major cause of lack of curability of this disease, or perhaps these cells find a "survival niche" as described in one of the chapters in this book. This molecule and its family members, in work reviewed in several of the chapters here, may also explain in part why certain other tumors, e.g., colon cancers, do not readily apoptose after chemotherapy treatment, but can survive this insult and repair critical DNA damage. The role of Bcl-2 in preventing apoptosis has led to a major effort to downregulate Bcl-2 in tumors, mainly by introducing antisense molecules into tumor cells, with some encouraging results in low-grade lymphomas. It remains to be seen if this simple strategy will overcome the inherent resistance of solid tumors that express Bcl-2 to chemotherapy.

Major culprits in lack of solid tumor sensitivity to chemotherapy also include the presence of p53 mutants, nonfunctional pRb, or both abnormalities. In most studies these abnormalities are associated with poor prognosis, and perhaps decreased sensitivity to chemotherapeutic agents. Paclitaxel may be an exception to this statement. Restoring wild-type p53 or pRb to tumor cells with these defects slows tumor growth and results in apoptosis, and surviving tumor cells appear to be more sensitive to drugs. These strategies are now being pursued with enthusiasm in the clinic, but patients usually die from disseminated disease, and the problem of getting these genes into all cancer cells has not yet been solved. Our own studies show that cells lacking pRb and with mutant p53 upregulate DNA synthesis target enzymes (dihydrofolate reductase, thymidylate synthase) as a consequence of increased levels of E2F, not sequestered by hypophosphorylated pRb *(1)*. Paradoxically, transfected cells overexpressing E2F-1 are more sensitive to the Topo II inhibitor doxorubicin *(2)*. The picture remains complex and difficult to translate easily to the clinic.

What about other tumors that harbor p16 and p19 deficiencies or ras mutations, or overexpress cyclin D1? Inhibitors of ras farnesylation are currently in early clinical trial, and the outcomes of these trials are eagerly awaited. What about the effect of mismatch repair deficiency on radiation and chemosensitivity? All of these questions are being addressed and could have an important impact on how we treat cancer patients. One outcome of these studies would be to be able to stratify patients and select therapy based on their genetic makeup. A perhaps more important and ultimate goal would be to find the Achilles heel for each tumor, and trick it into committing suicide via apoptosis, or perhaps even better, to deliver an apoptosis-producing molecule selectively targeted to tumor cells.

Clearly, next decade will be an exciting one, and novel and more selective treatments for cancer patients will be generated. Much more has to be learned about apoptosis and nonapoptotic cell death, but the foundation for translation of these advances to the clinic is nearing completion.

__Joseph R. Bertino, MD__

REFERENCES

1. Li WW, Fan J, Hochhouser D, Banerjee, D, Zielinski Z, Almasan A, Yin Y, Kelly R, Wahl GM, Bertino JR. Lack of functional retinoblastoma protein mediates increased resistance to antimetabolites in human sarcoma cell lines. *Proc. Natl. Acad. Sci. USA* 1995; 92: 10,436–10,440.
2. Banerjee D, Schneiders B, Fu JZ, Adhikari D, Zhao SC, Bertino JR. Role of E2F-1 in chemosensitivity. *Cancer Res.* 1998; 58: 4292–4296.